CAMBRIDGE TRACTS IN MATHEMATICS

General Editors
H. BASS, H. HALBERSTAM, J.F.C. KINGMAN
J.E.ROSEBLADE & C.T.C. WALL

---

82. *Polycyclic groups*

T0275617

DANIEL SEGAL

Reader in Mathematics
University of Manchester Institute of Science & Technology

Polycyclic groups

CAMBRIDGE UNIVERSITY PRESS

CAMBRIDGE

LONDON NEW YORK NEW ROCHELLE

MELBOURNE SYDNEY

CAMBRIDGE UNIVERSITY PRESS
Cambridge, New York, Melbourne, Madrid, Cape Town, Singapore, São Paulo

Cambridge University Press
The Edinburgh Building, Cambridge CB2 2RU, UK

Published in the United States of America by Cambridge University Press, New York

www.cambridge.org
Information on this title: www.cambridge.org/9780521241465

First published 1983
This digitally printed first paperback version 2005

A catalogue record for this publication is available from the British Library

Library of Congress Catalogue Card Number: 82–9476

ISBN-13 978-0-521-24146-5 hardback
ISBN-10 0-521-24146-4 hardback

ISBN-13 978-0-521-02394-8 paperback
ISBN-10 0-521-02394-7 paperback

This book is dedicated to my parents

Contents

Preface

Nothing is simpler than a cyclic group. So if we build a group, starting from the identity, by a finite number of iterated extensions with cyclic groups, we would expect its structure to be pretty transparent. Such a group is called *polycyclic*.

In one sense, of course, polycyclic groups do have a transparent structure. But in the last few years, some remarkably intricate mathematics has been brought to bear on the study of these groups. Of course, the question of whether the end justifies the means is ultimately a matter of personal taste; to me, the picture which has begun to emerge is an attractive one. Working in this subject has given me a lot of pleasure, and if a little of that gets across to the reader of this book then the effort of writing it will have been well worth while.

This is not an encyclopaedic work on polycyclic groups. A number of thoroughly deserving topics have been omitted altogether, or merely touched on in the text (some of these, with references, are mentioned in the appendix). My guiding aim has simply been to present a connected account of some interesting mathematics, and throughout I have laid more stress on the ideas than on the results. In consequence, some of the results are given in lesser generality than they might be, and some of the proofs are leisurely where they could have been slick. Having been thus exposed to the basic techniques, the newcomer to this subject should be in a position to invent his/her own improvements.

More specifically, the purpose of the book is twofold. The earlier chapters are intended to provide a convenient and self-contained reference for the body of 'classical' results on polycyclic groups; Chapters 1–5 form an introductory course, suitable for the beginning research student (perhaps leaving out sections D and E of Chapter 4). The second half of the book is an introduction to more advanced topics, including the 'isomorphism problem' (Chapters 6–8) and the recent finiteness theorem of Grunewald–Pickel–Segal regarding groups with isomorphic finite quotients (Chapters 9 and 10). The final chapter, by way of light relief, offers various examples of polycyclic groups to illustrate some of the themes discussed before.

The results of the later chapters depend heavily on the theory of linear groups, on algebraic number theory, and on the theory of algebraic groups. I have endeavoured to present the necessary material in a reasonably general form, in order to emphasize that our results are but minor applications of some powerful and important mathematics. Some algebraic number theory is also used in Chapters 2 and 4; rather than include proofs of the relevant elementary results, I refer the reader to standard textbooks for material which every algebraist should know. However, the important theorem of Schmidt–Chevalley on the congruence subgroup property of unit groups in rings of algebraic integers is proved in full, modulo 'standard' results, in section E of Chapter 4.

Few of the results in this book are really new; some of the arguments may be. References to the literature are given in the 'Notes' sections at the end of each chapter; but the absence of a reference for a specific result does not imply any claim of originality (it is quite likely due to the author's ignorance). However, I believe that most of the theory of Chapter 7 and the last part of Chapter 8 are fairly original; the argument of Chapter 10 is also rather an improvement on the published version.

The 'Notes' sections also give fairly copious suggestions for further reading. These are particularly important in the case of Chapter 6, which deals with certain aspects of torsion-free finitely generated nilpotent groups (called '$\mathfrak{T}$-groups' throughout the text); there is an extensive and elegant theory of these groups which needs a book to itself – such books exist, and should be read as a necessary complement to this one by anyone wishing to learn about polycyclic groups.

Exercises are scattered liberally throughout the book. These are often an essential part of the text; with the generous hints they are almost all supposed to be very easy (the reader who wants to think harder should in the first instance ignore the hints!)

It may be helpful if I suggest here some 'subsequences' of the book which tell a reasonably connected story.

Core course Chapters 1, 2 and 5

Second course Chapter 3, Chapter 4 (sections A, B and C), Chapter 11.

Advanced course Chapter 4, Chapter 6, Chapter 7 (section A), Chapter 8 (sections A, B and C).

Special topics Chapter 7, Chapter 8 (section D), Chapters 9 and 10.

Acknowledgements

Anyone who works on polycyclic groups owes a debt to the 'founding fathers': Kurt Hirsch, who started the whole thing off in the 1930's, Reinhold Baer, A.I. Mal'cev and Philip Hall. Most of what I know about the subject I learnt from Bert Wehrfritz, Otto Kegel, Karl Gruenberg, Jim Roseblade and Fred Pickel; I thank them all for teaching me. I must also acknowledge a special debt to my friend and long-time collaborator Fritz Grunewald: what I have learnt from him is not easily summarised, but it has affected my whole outlook on mathematics.

Bert Wehrfritz read the whole manuscript with great care, and I am extremely grateful to him for his many helpful comments, resulting in a number of corrections and improvements. A word of appreciation is due to the audiences in Queen Mary College, Bielefeld and Manchester who sat through my lectures on parts of this work and helped me to sort it all out; particular thanks are due to Dave Warhurst, whose notes of a course I gave in Manchester formed the skeleton for the later parts of the book, and to Mrs Margaret Hillock for help with the typing.

Finally, I wish to thank the kindly editor, Dr Roseblade, for accepting this book for the Cambridge Tracts.

Notation

$A \leq B$	A is a subgroup of B
$A < B$	A is a proper subgroup of B
$A \lhd B$	A is a normal subgroup of B (groups); A is an ideal of B (rings)
$A \leq_f B \, (A \lhd_f B)$	A is subgroup (normal subgroup) of finite index in B
$\langle X \rangle$	group generated by the set X
$G^n = \langle g^n \vert g \in G \rangle$	if G is a group and n is a positive integer
$A \times B$	direct product of A and B
$\underset{i \in I}{\mathrm{Dr}} A_i$	restricted direct product of the A_i
$\prod_{i \in I} A_i$	Cartesian product of the A_i
$\oplus_{i \in I} A_i$	direct sum of the A_i
$A] B$	semi-direct product of (normal subgroup) A by B
$x^y = y^{-1} x y$	if x and y belong to the same group
$[x, y] = x^{-1} x^y$	
$[x_1, \ldots, x_n] = [[x_1, \ldots, x_{n-1}], x_n]$ for $n > 2$	
$[X, Y] = \langle [x, y] \vert x \in X, y \in Y \rangle$	
$[X_1, \ldots, X_n] = [[X_1, \ldots, X_{n-1}], X_n]$ for $n > 2$	
G'	derived group of G
$\gamma_i(G)$	ith term of the lower central series of G
$G^{(n)}$	nth term of the derived series of G
$\zeta_i(G)$	ith term of the upper central series of G
$\mathrm{Fitt}(G)$	Fitting subgroup of G
$\mathrm{Aut}\, G$	Automorphism group of G
$M_n(R)$	ring of $n \times n$ matrices over ring R
$GL_n(R)$	group of invertible matrices in $M_n(R)$
$D_n(R)$	group of diagonal matrices in $GL_n(R)$
$\mathrm{Tr}_n(R)$	group of upper-triangular matrices in $GL_n(R)$
$\mathrm{Tr}_1(n, R)$	group of matrices in $\mathrm{Tr}_n(R)$ with all diagonal entries 1
R^*	group of units of ring R
R^+	additive group of ring R
$\mathrm{End}_R(E)$	endomorphism ring of R-module E
$\mathrm{Aut}_R(E)$	automorphism group of R-module E
$C_G(X) = \{ g \in G \vert x^g = x \ \forall x \in X \}$, the centralizer of X in G	

$N_G(X) = \{g \in G \mid X^g = X\}$, the normalizer of X in G

$\mathbb{N}$ the natural numbers (excluding zero)

$\mathbb{Z}$ the integers

$\mathbb{Q}$ the rational numbers

$\mathbb{R}$ the real numbers

$\mathbb{C}$ the complex numbers

$\mathbb{Z}_p$ the p-adic integers

$\mathbb{Q}_p$ the p-adic numbers

C_∞ the infinite cyclic group

C_n the cyclic group of order n

S_n the symmetric group of degree n (sometimes identified with the group of all $n \times n$ permutation matrices in $GL_n(\mathbb{Z})$)

1
The elements

Here we shall introduce our principal objects of study, the polycyclic groups, and derive some of their simpler properties. The main purpose of the chapter is to illustrate the use of a variety of quite elementary techniques which play a humble but necessary role in this subject.

A. The maximal condition and solubility

Definition   A group G has *max* if one of the following holds:

(a) every family of subgroups of G has a maximal member;
(b) every strictly ascending chain of subgroups of G is finite;
(c) every subgroup of G is finitely generated.

Exercise 1   Prove that (a), (b) and (c) are equivalent. (This is really universal algebra. Use Zorn's Lemma for '$(b) \Rightarrow (a)$'.)

Proposition 1   Suppose $N \triangleleft G$. Then G has *max* if and only if both N and G/N have *max*.

Proof 'Only if' is very easy. For the converse, consider an ascending chain $(H_i)_{i \in \mathbb{N}}$ of subgroups of G; that is, $H_1 \leq H_2 \leq \ldots \leq G$. If N and G/N have *max*, than there exists $n \in \mathbb{N}$ such that $H_i \cap N = H_n \cap N$ and $H_i N = H_n N$ for all $i \geq n$. We deduce that $H_i = H_n$ for all $i \geq n$ and the conclusion is that G has *max*. (The deduction goes like this: suppose $i \geq n$; then

$$H_i = H_i \cap (H_i N) = H_i \cap (H_n N)$$
$$= H_n(H_i \cap N) \text{ by the modular law, since } H_i \geq H_n$$
$$= H_n(H_n \cap N) = H_n.)$$

Examples   of groups having *max*:

(a) finite groups;
(b) C_∞, the infinite cyclic group.

Until recently, the only known groups with *max* were those built up from
(*a*) and (*b*) using Proposition 1. In 1978, E. Rips and A. Yu. Ol'shanskii
independently announced the construction of infinite simple groups having
both *max* and *min* (every descending chain of subgroups is finite). These
groups are both complex and mysterious, and we shall say no more about
this development. Instead we concentrate on the groups mentioned above,
the *polycyclic-by-finite* groups. The reason for this name will shortly
become clear.

Suppose $\mathscr{P}$ is a property of groups. A group G is called poly-$\mathscr{P}$ if there
exists a finite chain of subgroups

$$1 = G_0 \lhd G_1 \lhd \dots \lhd G_{n-1} \lhd G_n = G \tag{1}$$

such that each of the factor groups G_i/G_{i-1} has the property $\mathscr{P}$.

Let $\mathscr{Q}$ be another property of groups. A group G is a $\mathscr{P}$-by-$\mathscr{Q}$ group if G
has a normal subgroup N such that N has $\mathscr{P}$ and G/N has $\mathscr{Q}$.

Proposition 2 The following properties of a group are
equivalent:

(*a*) poly-(cyclic or finite);
(*b*) polycyclic-by-finite;
(*c*) (poly-C_∞)-by-finite.

Here 'polycyclic' means 'poly-cyclic', and 'poly-C_∞' is short for 'poly-
(infinite cyclic or trivial)'. Obviously (*c*) $\Rightarrow$ (*b*) and (*b*) $\Rightarrow$ (*a*). To show that
(*a*) $\Rightarrow$ (*c*) we make some preliminary observations.

Lemma 1 Let H be a finitely generated group and B a finite
group. Then H has only finitely many normal subgroups N with $H/N \cong B$.

Proof There are only finitely many distinct homomorphisms of
H onto B, since there are at most $|B|$ possible images for each of the finitely
many generators of H. The result now follows from the fact that every
$N \lhd H$ with $H/N \cong B$ arises as the kernel of such a homomorphism.

Lemma 2 Suppose H is a finitely generated group and
$K \lhd_f H \lhd G$ for some group G. Then K has a subgroup K^0 of finite index
with $K^0 \lhd G$.

Proof If $g \in G$ then $K^g \lhd H$ and $H/K^g \cong H/K$. By Lemma 1 there
are only finitely many distinct groups among the K^g as g runs through G; call

them $K_1, \ldots, K_n$. Then $K^0 = K_1 \cap \ldots \cap K_n$ is normal in G and has index at most $|H:K|^n < \infty$ in H.

Lemma 3 Suppose a group A has a finite normal subgroup B such that $A/B \cong C_\infty$. Then A has an infinite cyclic normal subgroup of finite index.

Proof We have $A = B\langle x \rangle$ for some $x \in A$. Since B is finite, so is Aut B. The automorphism $b \mapsto b^x (b \in B)$ therefore has finite order, e say. Then x^e lies in the centre of A, so $C = \langle x^e \rangle \vartriangleleft A$. Also

$$|A:C| = |A:BC| \cdot |BC:C| \le e|B| < \infty.$$

Proof of Proposition 2 Suppose the group G has a finite series (1) with each factor G_i/G_{i-1} either finite or cyclic; we are to show that G has a normal poly-C_∞ subgroup K of finite index. If $n = 1$, there is nothing to prove. If $n > 1$ we argue by induction. Thus we assume inductively that G_{n-1} has a normal poly-C_∞ subgroup L of finite index. By Lemma 2, with L for K and G_{n-1} for H, there exists $L^0 \vartriangleleft_f G_{n-1}$ with $L^0 \le L$ and $L^0 \vartriangleleft G$. We now distinguish two cases.

Case 1 G/G_{n-1} finite. Then $L^0 \vartriangleleft_f G$, and we take $K = L^0$; (why is L^0 poly-C_∞? See Exercise 2 below!).

Case 2 $G/G_{n-1} \cong C_\infty$. Now apply Lemma 3, with G/L^0 for A and G_{n-1}/L^0 for B. This shows that there exists $K/L^0 \vartriangleleft_f G/L^0$ with $K/L^0 \cong C_\infty$. Since L^0 is poly-C_∞, K is also poly-C_∞ and the proof is finished.

Exercise 2 Suppose $\mathscr{P}$ is a property of groups such that every subgroup of a group with $\mathscr{P}$ also has $\mathscr{P}$. Show that every subgroup of a poly-$\mathscr{P}$ group is again poly-$\mathscr{P}$.

Before moving on to soluble groups in general, a fact about abelian groups:

Lemma 4 An abelian group is polycyclic if and only if it is finitely generated.

Proof 'Only if' is clear. Suppose conversely that $A = \langle a_1, \ldots, a_n \rangle$ is abelian. Then

$$1 \le \langle a_1 \rangle \le \langle a_1, a_2 \rangle \le \ldots \le \langle a_1, \ldots, a_{n-1} \rangle \le \langle a_1, \ldots, a_n \rangle = A$$

is a series with cyclic factors, showing A to be polycyclic.

A group is called *soluble* if it is poly-abelian.

Definition   For a group G, the *derived group* G' of G is the subgroup generated by all *commutators*

$$[x, y] = x^{-1}y^{-1}xy$$

with $x, y \in G$;

$$G' = \langle [x, y] \mid x, y \in G \rangle = [G, G].$$

For $n > 1$, define

$$G^{(n)} = (G^{(n-1)})',$$

where $G^{(0)} = G$, $G^{(1)} = G'$.

Evidently, for any group G the *derived series*

$$G \geq G' \geq G^{(2)} \geq \ldots \geq G^{(n)} \geq G^{(n+1)} \geq \ldots$$

is a descending series of characteristic subgroups of G. The derived group G' is characterised by the properties

G/G' is abelian, and

if $N \triangleleft G$ and G/N is abelian, then $N \geq G'$.

Proposition 3   Let G be a group.

(i) If G is soluble and $H \leq G$, then H is soluble.
(ii) If $N \triangleleft G$, then G is soluble if and only if both N and G/N are.
(iii) G is soluble if and only if $G^{(n)} = 1$ for some positive integer n.

Proof   Exercise.

The least n for which $G^{(n)} = 1$ is called the *derived length* of the soluble group G.

Proposition 4   A soluble group has *max* if and only if it is polycyclic.

Proof   'If' we already know, and 'only if' follows from Lemma 4.

In contrast to Lemma 4, not every finitely generated soluble group has *max*. Let us construct a counterexample. Take

$$A = \operatorname*{Dr}_{i=-\infty}^{\infty} \langle a_i \rangle$$

to be the restricted direct product of infinitely many infinite cyclic groups, and let x be the automorphism of A defined by

$$a_i^x = a_{i+1} \quad (-\infty < i < \infty).$$

Then x has infinite order and generates a group $\langle x \rangle \cong C_\infty$ of automorphisms of A. Now form the semi-direct product

$$G = A] \langle x \rangle.$$

Then G is soluble, and G is generated by two elements, namely a_0 and x. But the series

$$1 < \langle a_0 \rangle < \langle a_0, a_1 \rangle < \ldots < \langle a_0, a_1, \ldots, a_n \rangle < \ldots$$

shows that G does *not* have *max*. This group illustrates two constructions of wide usefulness: written additively, A becomes an $\langle x \rangle$-module, and as such it is isomorphic to the *group ring* $\mathbb{Z}\langle x \rangle$; and G itself is a so-called *wreath product*,

$$G = \langle a_0 \rangle \wr \langle x \rangle \cong C_\infty \wr C_\infty.$$

B. Nilpotency

We move on now to discuss *nilpotency*, a concept which will be of central importance throughout.

Definition The *centre* of a group G is

$$\zeta_1(G) = \{ g \in G \mid gx = xg \text{ for all } x \in G \}.$$

Evidently $\zeta_1(G)$ is a characteristic abelian subgroup of G. An element of G is called *central* if it lies in $\zeta_1(G)$. Now set $\zeta_0(G) = 1$ and define recursively

$$\zeta_i(G)/\zeta_{i-1}(G) = \zeta_1(G/\zeta_{i-1}(G)) \text{ for } i \geq 1.$$

Thus we obtain the *upper central series*

$$1 = \zeta_0(G) \leq \zeta_1(G) \leq \ldots \leq \zeta_n(G) \leq \zeta_{n+1}(G) \leq \ldots,$$

an ascending series of characteristic subgroups of G. It follows from the definition that for $n \geq 1$ and $x \in G$,

$$x \in \zeta_n(G) \Leftrightarrow [x, g] \in \zeta_{n-1}(G) \ \forall g \in G.$$

More generally, a series of subgroups

$$1 = H_0 \leq H_1 \leq \ldots \leq H_k = G \qquad (2)$$

is called a *central series* of G if for every $n \in \{1, \ldots, k\}$,

$$x \in H_n \Rightarrow [x, g] \in H_{n-1} \ \forall g \in G.$$

Proposition 5 The following are equivalent:

(a) the series (2) is a central series of G;

(b) $H_n \triangleleft G$ for $1 \leq n \leq k$, and $H_n/H_{n-1} \leq \zeta_1(G/H_{n-1})$ for $1 \leq n \leq k$;

(c) $H_n \triangleleft G$ for $1 \leq n \leq k$ and for each n, the action of G by conjugation on the factor H_n/H_{n-1} is trivial.

The proof is immediate. We call a group G *nilpotent* if G has a (finite) central series.

Dual to the upper central series, one defines the *lower central series* of a group G,

$$G = \gamma_1(G) \geq \gamma_2(G) \geq \ldots \geq \gamma_n(G) \geq \gamma_{n+1}(G) \geq \ldots \tag{3}$$

by setting $\gamma_1(G) = G$ and for $n \geq 1$,

$$\gamma_{n+1}(G) = [\gamma_n(G), G] = \langle [x, g] \mid x \in \gamma_n(G), g \in G \rangle.$$

Then (3) is a descending central series of characteristic subgroups of G.

Proposition 6   Suppose (2) is a central series of G. Then $H_n \leq \zeta_n(G)$ and $H_{k-n} \geq \gamma_{n+1}(G)$ for $0 \leq n \leq k$.

Proof   Exercise.

An immediate consequence of Propositions 5 and 6 is now

Proposition 7   The following are equivalent for a group G:

(a) G is nilpotent;
(b) $\zeta_c(G) = G$ for some $c \in \mathbb{N}$;
(c) $\gamma_{c+1}(G) = 1$ for some $c \in \mathbb{N}$.

If G is nilpotent, its *nilpotency class* is the length of a shortest central series of G (the length of the series (2) being the number k, i.e. the number of 'gaps'). It is now an easy exercise to improve Proposition 7 to:

$$G \text{ is nilpotent of class } \leq c \Leftrightarrow \zeta_c(G) = G \Leftrightarrow \gamma_{c+1}(G) = 1.$$

Proposition 8   Subgroups and quotient groups of a nilpotent group are nilpotent. A direct product of finitely many nilpotent groups is nilpotent.

Exercise 3   Verify the above proposition. Construct counter-examples to disprove the following statements: (i) '$N \lhd G$, N nilpotent and G/N nilpotent $\Rightarrow G$ nilpotent'; (ii) 'a direct product of infinitely many nilpotent groups is nilpotent'.

Exercise 4   Let G be a group and $N \neq 1$ a normal subgroup of G.
(i) Show that if G is nilpotent then $N \cap \zeta_1(G) \neq 1$. (ii) Show that if G is soluble then G has an abelian normal subgroup A with $1 \neq A \leq N$. (iii) If G is

nilpotent and N is maximal among all abelian normal subgroups of G, then $C_G(N) = N$. (iv) If G is nilpotent then every subgroup of G is *subnormal*, i.e. if $H \leq G$ then there is a series $H = H_0 \lhd H_1 \lhd ... \lhd H_k = G$.

(*Hint*: (i) Consider the least j for which $N \cap \zeta_j(G) \neq 1$. (ii) Consider the derived series of N. (iii) If $C_G(N) > N$, consider the subgroup $N\langle x \rangle$ of G, where $xN \in \zeta_1(G/N) \cap C_G(N)/N$. (iv) Try $H_i = H\zeta_i(G)$.)

Nilpotent groups form a class intermediate between abelian groups and soluble groups. Let us consider some examples.

(1) *Finite p-groups* (The letter p denotes a prime number, and a finite p-group is a group whose order is a power of p.) If $G \neq 1$ is a finite p-group then $\zeta_1(G) \neq 1$, and $G/\zeta_1(G)$ is again a p-group, of smaller order than G. Therefore $\zeta_c(G) = G$ for some c and so G is nilpotent. For this, and the following result, see any book on finite group theory:

Theorem A finite group is nilpotent if and only if it is a direct product of p-groups (for various primes p).

(2) Let k be a field and n a positive integer. Denote by

$$\text{Tr}_1(n, k)$$

the group consisting of all upper-triangular $n \times n$ matrices over k with all diagonal entries equal to 1; this is the *upper unitriangular group* of degree n over k. The group $\text{Tr}_1(n, k)$ is nilpotent of class $n - 1$, as we shall see later.

The last example is generalised in

Proposition 9 Let E be a ring (with 1) and I an ideal of E with $I^n = 0$. Put

$$G = 1 + I = \{1 + x | x \in I\} \subseteq E.$$

(i) G is a subgroup of the group of units of E.

(ii) Put $G_i = 1 + I^i$ for $i = 1, ..., n$; then

$$1 = G_n \leq G_{n-1} \leq ... \leq G_1 = G$$

is a central series of G, so G is nilpotent of class at most $n - 1$.

(iii) For each $i < n$, the factor group G_i/G_{i+1} is isomorphic to the additive group of I^i/I^{i+1}.

(As usual, I^i denotes the ideal consisting of all finite sums $\sum x_1 x_2 ... x_i$ with each $x_j \in I$.)

Proof (i) For $x \in i$ we have

$$(1 + x)(1 - x + x^2 - ... \pm x^{n-1}) = 1$$

so each element of G is a unit in E; and

$$(1+x)(1+y) = 1 + (x+y+xy)$$

so G is closed under multiplication.

(ii) Note that each G_i is a subgroup of G, by the argument of part (i). If $g \in G_i$ and $h \in G$ then $g = 1 + x$, $h = 1 + y$ with $x \in I^i$ and $y \in I$. Now

$$gh - hg = xy - yx \in I^{i+1}$$

so $$[g,h] - 1 = g^{-1}h^{-1}(gh - hg) \in I^{i+1}$$

and so $[g,h] \in 1 + I^{i+1} = G_{i+1}$.

(iii) Define a map $\theta : G_i \to I^i/I^{i+1}$ by

$$g\theta = (g-1) + I^{i+1}.$$

Then for g and $h \in G_i$,

$$\begin{aligned}(gh)\theta &= (gh - 1) + I^{i+1} \\ &= (g-1) + (h-1) + (g-1)(h-1) + I^{i+1} \\ &= (g-1) + (h-1) + I^{i+1}\end{aligned}$$

since $g - 1 \in I^i$, $h - 1 \in I^i \subseteq I$. Thus θ is a homomorphism. It is evident that θ is surjective and that $\ker \theta = G_{i+1}$.

To see how this generalises the case of $\mathrm{Tr}_1(n,k)$, consider a commutative ring R (with 1), a free R-module $V = R \oplus \ldots \oplus R$ (n summands), and take

Put
$$E = \mathrm{End}_R(V) = M_n(R).$$

$$V_i = 0 \oplus \ldots \oplus 0 \oplus R \oplus \ldots \oplus R \leq V$$
$$\leftarrow n-i \rightarrow \quad \leftarrow i \rightarrow$$

and take

$$I = \{\alpha \in E \mid V_i\alpha \subseteq V_{i-1} \quad \text{for } i = 1,\ldots,n\}.$$

Thus I consists of all matrices over R of the form

$$\begin{bmatrix} 0 & * & \cdots & & * \\ & 0 & & & \vdots \\ \vdots & & \ddots & & \\ & & & & * \\ 0 & \cdots & & 0 & 0 \end{bmatrix}$$

and we write $I = \mathrm{Tr}_0(n,R)$. Clearly $I^n = 0$, so the proposition assures us that the group

$$G = 1 + I = \mathrm{Tr}_1(n,R)$$

is nilpotent.

Exercise 5   (i) Show that I^i consists exactly of all matrices of the

form

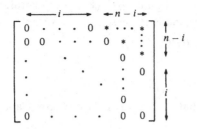

(the non-trivial part is that every such matrix belongs to I^i; this is where we need the hypothesis that $1 \in R$). (ii) Deduce that $G_i/G_{i+1} \cong R^+ \oplus \ldots \oplus R^+$ with $n-i$ summands, where R^+ denotes the additive group of R and $G_i = 1 + I^i$. (iii) Show that $G_i = \zeta_{n-i}(G)$ for $i = 1, \ldots, n$, hence that G is nilpotent of class exactly $n-1$. (*Hint*: argue by induction on $n-i$. Observe that for $g, h \in G$, $g^{-1}h \in G_m \Leftrightarrow g - h \in I^m$; apply this with $h = \alpha^{-1} g \alpha$, where α is a matrix of the form

$$\alpha_{pq} = \delta_{pq} + \delta_{pk}\delta_{(k+1)q}.)$$

(iv) If R is a field, then G is torsion-free if char $R = 0$, and G has exponent dividing p^{n-1} if char $R = p \neq 0$.

In the situation we have been considering, the R-module V is also a G-module; its submodules V_i are G-submodules and the induced action of G on each factor V_i/V_{i-1} is *trivial*. It is both possible and useful to generalize this setup to the 'non-linear' case: so let V be a group and G a group acting by automorphisms on V, and suppose

$$1 = V_0 \leq V_1 \leq \ldots \leq V_n = V \tag{4}$$

is a series of G-invariant normal subgroups of V such that the induced action G on each factor V_i/V_{i-1} is trivial. Then (4) is called a G-*central* series of V, G is said to *stabilize* the series (4), and we say that G acts *nilpotently* on V. The special case where $V = G$ and the action is by conjugation is already familiar – G acts nilpotently on itself if and only if G is nilpotent.

Proposition 10 If a group G acts faithfully on a group V and stabilizes a series of normal subgroups in V of length n, then G is nilpotent of class at most $n - 1$.

Proof Say G stabilizes the series (4) above. We argue by induction on n. If $n = 1$ then $G = 1$ is nilpotent of class 0 (i.e. $\gamma_1(G) = 1$). Suppose $n > 1$ and put

$$C = C_G(V/V_1) \cap C_G(V_{n-1}).$$

Now $G/C_G(V/V_1)$ acts faithfully on V/V_1 and $G/C_G(V_{n-1})$ acts faithfully on V_{n-1}; applying the inductive hypothesis to these groups we obtain that

$$\gamma_{n-1}(G/C_G(V/V_1)) = \gamma_{n-1}(G/C_G(V_{n-1})) = 1,$$

whence

$$\gamma_{n-1}(G) \le C_G(V/V_1) \cap C_G(V_{n-1}) = C.$$

Thus it will suffice now to show that $[C, G] = 1$. Take $c \in C$, $g \in G$, and $v \in V$. Then

$$v^g = wv$$

for some $w \in V_{n-1}$; and $v^{-1}v^c \in V_1$, so $(v^{-1}v^c)^g = v^{-1}v^c$. Therefore

$$v^c = v(v^{-1}v^c)^g = w^{-1}v^{cg}$$

so

$$v^{cg} = wv^c = (wv)^c = v^{gc}$$

(note that $w = w^c$ since $w \in V_{n-1}$). As v was arbitrary and the action of G is faithful, we deduce that $cg = gc$. Thus $[c, g] = 1$ as required.

Let us list some corollaries.

Corollary 1   For every group H, and all i and j,
$$[\gamma_i(H), \gamma_j(H)] \le \gamma_{i+j}(H).$$

Proof Take

$$V = \gamma_i(H)/\gamma_{i+j}(H), G = H/C_H(V)$$

and let G act on V via conjugation. Then G acts faithfully on V and stabilizes the series of length j whose lth term is $\gamma_{i+j-l}(H)/\gamma_{i+j}(H)$. The proposition shows that G is nilpotent of class at most $j - 1$, thus $\gamma_j(H) \le C_H(V)$. But this means nothing other than $[\gamma_i(H), \gamma_j(H)] \le \gamma_{i+j}(H)$.

Corollary 2   If H is a group with $\gamma_n(H) = 1$ then $H^{(d)} = 1$ where $d = 1 + [\log_2 n]$.

For repeated applications of Corollary 1 give $H^{(i)} \le \gamma_{2^i}(H), i = 1, 2, \ldots, d$.

Corollary 3   For every group H and every i,
$$[\zeta_i(H), \gamma_i(H)] = 1.$$

Proof Apply Proposition 10 to $G = H/C_H(\zeta_i(H))$ acting by conjugation on $V = \zeta_i(H)$.

Next, we examine more closely the situation where a group stabilizes a series of length 2. First, a

Definition   Let A be a group and B an A-module. A *derivation* of A into B is a map $\delta : A \rightarrow B$ such that

$$(xy)\delta = (x\delta)^y + y\delta \quad \forall x, y \in A.$$

The set of all derivations of A into B is denoted

Der (A, B).

Remark   (i) Pointwise addition makes Der(A, B) into an abelian group, a subgroup of the Cartesian power $B^{|A|}$.

(ii) If A acts trivially on B, then derivations are just homomorphisms, so in this case

$$\text{Der}\,(A, B) = \text{Hom}\,(A, B) = \text{Hom}\,(A/A', B);$$

(as B is commutative, we can identify any homomorphism of A into B with the induced homomorphism from A/A' into B).

Proposition 11   Suppose the group G acts faithfully on a group V and stabilizes a series of length 2:

$$1 \leq W \vartriangleleft V.$$

Consider $Z = \zeta_1(W)$ as a V/W-module, with V acting by conjugation. Then G is isomorphic to a subgroup of Der$(V/W, Z)$.

Proof   For $g \in G$ define $\delta_g : V \rightarrow W$ by

$$v\delta_g = v^{-1}v^g, \quad v \in V.$$

The reader can then verify that

(a) $Wv_1 = Wv_2 \Rightarrow v_1\delta_g = v_2\delta_g$, so δ_g induces a map $\Delta_g : V/W \rightarrow W$;
(b) $v\delta_g \in \zeta_1(W) \ \forall v \in V$, so Δ_g maps V/W into $Z = \zeta_1(W)$;
(c) $v\delta_{gh} = (v\delta_g) \cdot (v\delta_h) \ \forall v \in V, g \in G, h \in G$;
(d) $\Delta_g : V/W \rightarrow Z$ is a derivation;
(e) $\Delta_g = 1$ (the constant map $Wv \mapsto 1 \ \forall v \in V) \Leftrightarrow g = 1$.

(c), (d) and (e) show that the map $g \mapsto \Delta_g$ is an injective homomorphism from G into Der$(V/W, Z)$, and the proposition follows.

This also has a number of corollaries; let H be any group and write $Z_i = \zeta_i(H)$.

Corollary 4   For each i, Z_{i+1}/Z_i is isomorphic to a subgroup of Hom$(H/H', Z_i/Z_{i-1})$.

Proof Apply Proposition 11 with $V = H/Z_{i-1}$, $W = Z_i/Z_{i-1}$, and $G = Z_{i+1}/Z_i$ acting by conjugation on V.

Corollary 5 If Z_1 is torsion-free (respectively has exponent dividing m), then the same holds for each factor Z_{i+1}/Z_i.

This follows from Corollary 4 by induction on i and an obvious property of the functor Hom. An immediate consequence is

Corollary 6 If H is nilpotent of class c and Z is torsion-free (respectively has exponent dividing m), then H is torsion-free (respectively has exponent dividing m^c).

These corollaries show that the upper central factors of a group inherit certain kinds of property from the centre of the group. Something analogous can be said for the lower central factors:

Proposition 12   Let G be a group and write $G_i = \gamma_i(G)$. For each $i > 1$, there is an epimorphism

$$\psi_i : (G_{i-1}/G_i) \otimes (G_1/G_2) \rightarrow G_i/G_{i+1},$$

induced by the map

$$(G_i g, G_2 h) \mapsto G_{i+1}[g, h] \tag{5}$$

($\otimes$ denotes the tensor product of abelian groups).

Before proving this let us state an 'operator' version of it. Suppose Γ is a group acting by automorphism on G; then Γ acts on each of the factor groups G_j/G_{j+1}, and we can make $(G_{i-1}/G_i) \otimes (G_1/G_2)$ into a Γ-module by extending the action

$$(a \otimes b)^\gamma = a^\gamma \otimes b^\gamma.$$

With this convention, we have

Proposition 12 (supplement) If Γ acts by automorphisms on G, then the map ψ_i is an epimorphism of Γ-modules.

Proof of Proposition 12   The map (5) from $(G_{i-1}/G_i) \times (G_1/G_2)$ $\rightarrow G_i/G_{i+1}$ is well defined, because if $x \in G_i$ and $y \in G_2$ then, for $g \in G_{i-1}$ and $h \in G$,

$$[xg, yh] = [x, h]^g [g, h] [x, y]^{gh} [g, y]^h$$

$$\in G_{i+1}[g, h]$$

by Corollary 1. The map is bilinear because

$$[g_1 g_2, h] = [g_1, h]^{g_2} [g_2, h],$$
$$[g, h_1 h_2] = [g, h_2][g, h_1]^{h_2}$$

and G_i/G_{i+1} is a central factor of G. The map does therefore induce a homomorphism ψ_i as claimed, and it is clear from the definition that it respects the action of Γ as claimed in the 'supplement'. Since ψ_i is a homomorphism, its image is a subgroup of G_i/G_{i+1}; but Im ψ_i contains the set $\{G_{i+1}[g,h]\,|\,g\in G_i, h\in G\}$ which generates G_i/G_{i+1}, so ψ_i is surjective as required.

Corollary 7   If G/G' is finitely generated then so is $\gamma_i(G)/\gamma_{i+1}(G)$ for each $i \geq 1$.

For the tensor product of two finitely generated abelian groups is again finitely generated, so the result follows by induction on i.

Exercise 6   Let G and Γ be as in the supplement to Proposition 12. Suppose G/G' is finitely generated as a Γ-module. Does it follow that each $\gamma_i(G)/\gamma_{i+1}(G)$ is finitely generated as a Γ-module? Careful!

Corollary 8   If G is a nilpotent group and G/G' is finitely generated then G is finitely generated, indeed polycyclic.

This follows from Corollary 7 and Lemma 4. Regarding the generation of a nilpotent group one can actually say a bit more:

Exercise 7   Suppose G is nilpotent and $G = G'H$ for some subgroup H. Show that then $H = G$.
(*Hint*: Put $Z_i = \zeta_i(G)$. Show that if $H < G$ then $Z_{i-1}H < Z_iH = G$ for some i. Then show that $G'H \leq Z_{i-1}H$, a contradiction.)

Corollary 9   Suppose G/G' has finite exponent dividing m. Then so has each of the factor groups $\gamma_i(G)/\gamma_{i+1}(G)$. If also G is nilpotent of class c, then G has finite exponent dividing m^c. If in addition G is finitely generated, then G is finite.

Proof   The first claim is immediate from Proposition 12, and the second follows at once. For the final claim, note that a finitely generated abelian group of finite exponent is necessarily finite.

Corollary 10   Let H be a nilpotent group and denote by
$$\tau(H)$$
the set of all element of finite order in H. Then $\tau(H)$ is a subgroup of H, it is characteristic in H, and $H/\tau(H)$ is torsion-free. If H is finitely generated then $\tau(H)$ is finite.

Proof Take $x, y \in \tau(H)$ and put $G = \langle x, y \rangle$. Then G/G' has finite exponent (namely the l.c.m. of the orders of $G'x, G'y$ in G/G'), so by Corollary 9 the group G is finite. Therefore $G \subseteq \tau(H)$, so xy and x^{-1} lie in $\tau(H)$. Thus $\tau(H)$ is a subgroup of H. That $\tau(H)$ is characteristic and that $H/\tau(H)$ is torsion-free is then obvious. The final claim follows from Corollaries 8 and 9.

By way of an extended exercise, here is a generalisation of a theorem proved by P. Hall in the context of finite p-groups:

Proposition 13   Let G be a nilpotent group and Γ a group of automorphisms of G. Suppose that G/G' has finite exponent dividing a power of some positive integer m, and that Γ acts trivially on the factor group $G/G'G^m$. Then Γ has finite exponent dividing a power of m.

Corollary 11   If G is a finite p-group, then the centraliser in Aut G of $G/G'G^p$ is a p-group.

Outline proof of Proposition 13   Put $C = C_\Gamma(G/G')$ and $A = G/G'$. Write A additively and suppose $m^k A = 0$.

(a) Γ/C acts faithfully on A and stabilizes the series
$$A \geq mA \geq \ldots \geq m^{k-1}A \geq m^k A = 0;$$
so Γ/C has exponent dividing m^{k-1}.

(b) C stabilizes the lower central series of G, in which each factor has exponent dividing a power of m.

(c) C has exponent dividing a power of m (an inductive argument using Proposition 11).

A frequently useful application of Proposition 12 (supplement) is given by

Proposition 14   Let G be a group and Γ a group acting by automorphisms on G. If Γ acts nilpotently on G/G', then Γ acts nilpotently on each factor $\gamma_i(G)/\gamma_{i+1}(G)$.

This will follow once we have established

Lemma 5   If A and B are Γ-modules on which Γ acts nilpotently, then Γ acts nilpotently on $A \otimes B$ (made into a Γ-module via $(a \otimes b)^\gamma = a^\gamma \otimes b^\gamma$).

Proof Let $0 = A_0 < A_1 < \ldots < A_n = A$ and

$0 = B_0 < B_1 < \ldots < B_m = B$ be Γ-central series in A and B respectively. Put

$$C_q = \sum_{i+j=q} (A_i \times B_j)^\dagger,$$

where $(A_i \times B_j)^\dagger$ denotes the natural image of $A_i \times B_j$ in $A \otimes B$. If $a \in A_i$ and $b \in B_j$ then for $\gamma \in \Gamma$ we have

$$(a \otimes b)^\gamma - (a \otimes b) = a^\gamma \otimes b^\gamma - a \otimes b$$

$$= a^\gamma \otimes (b^\gamma - b) + (a^\gamma - a) \otimes b$$

$$\in (A_i \times B_{j-1})^\dagger + (A_{i-1} \times B_j)^\dagger \le C_{q-1},$$

where $q = i + j$. Since C_q is generated additively by elements of the form $a \otimes b$, it follows that C_q is a Γ-submodule of $A \otimes B$ and that Γ acts trivially on C_q / C_{q-1}. Thus Γ stabilizes the series $(C_q)_{0 \le q \le m+n}$ in $A \otimes B$.

Corollary 12 Suppose a group H has a nilpotent normal subgroup N such that H/N' is nilpotent. Then H is nilpotent.

Proof Apply Proposition 14 with $G = N$ and $\Gamma = H$ acting by conjugation.

To conclude this section on nilpotency, we mention *Fitting's theorem*:

Proposition 15 Let $N_1, \ldots, N_k$ be finitely many nilpotent normal subgroups of a group G. Then the subgroup $\langle N_1, \ldots, N_k \rangle = N_1 N_2 \ldots N_k \le G$ is nilpotent.

Proof An obvious induction reduces to the case $k = 2$. Write $N = N_1$, $M = N_2$, and put $Z_i = \zeta_i(M)$ for $i = 0, 1, \ldots, c$ where $\zeta_c(M) = M$. Then M acts trivially and N acts nilpotently (by conjugation) on each of the factors $(Z_i \cap N)/(Z_{i-1} \cap N)$; therefore MN acts nilpotently on each of these factors and hence also on $Z_c \cap N = M \cap N$. So it will suffice to check that $MN/(M \cap N)$ is nilpotent. But

$$MN/(M \cap N) \cong M/(M \cap N) \times N/(M \cap N),$$

a direct product of two nilpotent groups, so the result follows.

Definition The *Fitting subgroup* of a group G is
Fitt$(G) = \langle N | N \lhd G$ and N nilpotent$\rangle$.

Corollary 13 If G is polycyclic-by-finite then Fitt(G) is nilpotent.

For in this case Fitt(G) is generated by finitely many nilpotent normal subgroups of G; alternatively, let F be any maximal nilpotent normal subgroup and observe that F must contain every other nilpotent normal subgroup of G by the proposition, consequently $F = \text{Fitt}(G)$.

C. Some theorems about polycyclic groups

We now prove some 'classic' theorems, to show how far one can get with only the most elementary arguments. To prove a result about polycyclic groups, one often argues by induction. There are various ways to do this, all essentially equivalent:

(a) induction on the length of a series with cyclic or finite factors (e.g. the proof of Proposition 2);

(b) induction on the derived length;

(c) induction on the *Hirsch number*: this is the number of infinite cyclic factors in a series with cyclic or finite factors, and is an invariant of polycyclic-by-finite groups;

(d) 'Noetherian induction': suppose we want to prove that all polycyclic groups with property $\mathscr{P}$ also have property $\mathscr{Q}$, where $\mathscr{P}$ is a property inherited by quotient groups. Assume that a counterexample G exists. Then $G = G/1$ has $\mathscr{P}$ but not $\mathscr{Q}$; since G has *max*, there exists among the normal subgroups of G one, N say, which is maximal subject to G/N not having $\mathscr{Q}$. The group $\bar{G} = G/N$ now has $\mathscr{P}$, and every proper quotient group of $\bar{G}$ has $\mathscr{Q}$. The 'induction step' consists in showing that a group $\bar{G}$ with these properties must itself have $\mathscr{Q}$; having shown this, we get a contradiction to the defining property of N. Thus the counterexample G cannot exist after all.

The Hirsch number of a polycyclic-by-finite group G is denoted $h(G)$.

Exercise 8 Show that $h(G)$ is an invariant of G, i.e. it is independent of the series used to define it. Suppose $H \leq G$ and $N \lhd G$; show that

$$h(H) \leq h(G);$$
$$h(H) = h(G) \Leftrightarrow |G:H| < \infty ;$$
$$h(G) = h(N) + h(G/N).$$

Lemma 6 An infinite polycyclic-by-finite group has an infinite free abelian normal subgroup.

Proof Let G be the group. By Proposition 2, G has a poly-C_∞ normal subgroup H of finite index. H is soluble, of derived length d say.

Then $H^{(d-1)}$ is abelian, torsion-free and finitely generated, and $\neq 1$, so it is infinite and free abelian. Also $H^{(d-1)}$ is normal in G because it is characteristic in H.

Exercise 9 Show that in Lemma 6, 'normal' can be strengthened to 'characteristic'.

Definition Let $\mathscr{P}$ be a property of groups. A group G is *residually* $\mathscr{P}$ if
$$\bigcap\{N\,|\,N\triangleleft G \text{ and } G/N \text{ has } \mathscr{P}\} = 1;$$
in other words, if every non-identity element of G lives in some $\mathscr{P}$-image of G.

Exercise 10 (i) Suppose that subgroups of groups with $\mathscr{P}$ necessarily also have $\mathscr{P}$. Show that subgroups of residually-$\mathscr{P}$ groups are residually-$\mathscr{P}$. (ii) Show by examples that one can't replace 'subgroups' by 'quotient groups' in (i). (iii) Take $\mathscr{P}$ to be either 'finite' or 'finite p-group'. Suppose G is a polycyclic-by-finite group which has a residually-$\mathscr{P}$ normal subgroup K such that G/K has $\mathscr{P}$. Show that G is residually $\mathscr{P}$. Generalize this!

The question of just which quotient groups of a residually-$\mathscr{P}$ group are themselves residually-$\mathscr{P}$ is quite a subtle and interesting one, and quite a lot can be said about it. We do not pursue this here (but see Exercise 13, below).

Theorem 1 Polycyclic-by-finite groups are residually finite.

Proof Let us argue by induction on the Hirsch number (as Hirsch did). So consider a polycyclic-by-finite group G, with Hirsch number h. If $h = 0$ then G is finite, so residually finite. Suppose $h > 0$, and assume the result known for groups of smaller Hirsch number. The group G is infinite, hence by Lemma 6 G has an infinite free abelian normal subgroup A say. If m is any positive integer, the group $A^m = \langle a^m\,|\,a\in A\rangle$ is normal in G, being characteristic in A, and has finite index in A. Therefore
$$h(G/A^m) = h(G) - h(A^m) = h(G) - h(A) < h,$$
hence by inductive hypothesis
$$\bigcap\{N/A^m\,|\,N/A^m\triangleleft G/A^m \text{ and } G/N \text{ finite}\} = A^m/A^m;$$
i.e.
$$\bigcap\{N\,|\,A^m \leq N \triangleleft_f G\} = A^m.$$
Taking the intersection over all positive integers m gives
$$\bigcap\{N\,|\,N\triangleleft_f G\} \leq \bigcap_m A^m = 1;$$

the final equality is obvious if we write A additively, for $A = \mathbb{Z} \oplus \ldots \oplus \mathbb{Z}$ gives $A^m = m\mathbb{Z} \oplus \ldots \oplus m\mathbb{Z}$. This completes the proof.

Theorem 1 has a useful generalisation:

Exercise 11   Let G be a polycyclic-by-finite group and H a subgroup of G. Then
$$H = \bigcap \{L | H \leq L \leq_f G\} = \bigcap \{NH | N \vartriangleleft_f G\}.$$
(*Hint*: The second equality is clear if we note that $H \leq L \leq_f G \Rightarrow NH \leq L$ where $N = \bigcap_{g \in G} L^g \vartriangleleft_f G$. Now taking A as before, an induction argument reduces the problem to showing that $H = \bigcap_m A^m H = B$, say. But $A^m H \cap A = A^m(H \cap A)$, and these groups intersect in $H \cap A$ since $A/(H \cap A)$ is residually finite; so $B \cap A = H \cap A$. The result then follows because $H \leq B \leq AH$.)

Once we know that a group is residually finite, or as the Russians say 'approximated by finite groups', it makes sense to try and derive information about the group from consideration of its finite quotients. This is entirely analogous to the activity of number theorists when they show that certain equations are soluble in integers, or rationals, provided they are 'soluble mod p' for all primes p. The analogy is in some instances more than merely linguistic, and will in fact be one of the *leitmotivs* of later chapters. As a simple foretaste, here is another of Hirsch's classic results:

Theorem 2   Let G be polycyclic-by-finite. If every finite quotient of G is nilpotent, then G is nilpotent.

Proof   The reader should flesh out the following sketch. We argue by induction on the Hirsch number. We may suppose that G has a free abelian normal subgroup $A \cong \mathbb{Z} \oplus \ldots \oplus \mathbb{Z}$ (n summands), and that G/A^p is nilpotent for every prime p. Then G acting by conjugation stabilizes some finite chain of subgroups in A/A^p; but such a chain can have length no more than n, so $[A, \underbrace{G, \ldots, G}_{n}] \leq A^p$. This holds for each prime p, therefore

$$[A, \underbrace{G, \ldots, G}_{n}] \leq \bigcap_p A^p = 1.$$

But $\gamma_m(G) < A$ for some m, hence $\gamma_{m+n}(G) = 1$.

A nice application is the generalization of *Frattini's theorem*. Recall that the *Frattini subgroup* of a group G is

$$\Phi(G) = \bigcap \{M | M \text{ is a maximal proper subgroup of } G\}.$$

Theorem 3   The Frattini subgroup of a polycyclic-by-finite group is nilpotent.

Proof   Let G be polycyclic-by-finite and $F = \Phi(G)$. By Theorem 2, it will suffice to show that every finite image of F is nilpotent. In view of Lemma 2, we may restrict attention to a normal subgroup N of G with $N \lhd_f F$, and have to show that F/N is nilpotent. Since G/N is residually finite and F/N is finite, there exists $K \lhd_f G$ with $K \cap F = N$. Then $KF/K \le \Phi(G/K)$ (think about the maximal subgroups of G/K); and Frattini's theorem (which is Theorem 3 for finite groups) shows that $\Phi(G/K)$ is nilpotent. It follows that

$$F/N = F/(K \cap F) \cong KF/K$$

is nilpotent, as desired.

In conclusion, here is a more subtle version of Theorem 1:

Theorem 4   (i) Let G be a finitely generated nilpotent group. If G is torsion-free then G is residually a finite p-group for every prime p. If $\tau(G)$ is a p-group for some prime p, then G is residually a finite p-group.

(ii) Let G be a polycyclic-by-finite group. Then for every prime p, G has a residually-finite-p normal subgroup of finite index.

Corollary 14   Polycyclic groups are (residually nilpotent)-by-finite.

Exercise 12   Construct an example of a torsion-free polycyclic group which is not residually nilpotent.
(*Hint*: take $A = \mathbb{Z} \oplus \mathbb{Z}$ and consider the semi-direct product $A] \langle x \rangle$, where x is a suitable matrix in $GL_2(\mathbb{Z})$.)

Proof of Theorem 4   (i) Let G be finitely generated, nilpotent and torsion-free. Choose a prime p. Put $Z = \zeta_1(G)$; then G/Z is torsion-free (by Corollary 5) and has smaller nilpotency class than G, so arguing by induction on the class we may suppose that G/Z is residually finite-p (if $G = Z$ this is trivially true). So to prove that G has this property, it will suffice to show that for each $z \in Z, z \ne 1$, there exists $N \lhd G$ with G/N a finite p-group and $z \notin N$. Since G is torsion-free, Z is free abelian, and therefore $z \notin Z^{p^m}$ for a sufficiently large power p^m of p. Since $Z^{p^m} \lhd G$, we may choose a normal subgroup N of G containing Z^{p^m}, not containing z, and maximal subject to these conditions. Let $\bar{}: G \to G/N$ denote the natural map. Then every non-identity normal subgroup of $\bar{G}$ contains the element $\bar{z}$, and since $\bar{G}$ is residually finite, $\bar{G}$ must be finite. Thus $\bar{G}$ is a finite nilpotent group, and as such it is the direct product of a p-subgroup P and a p'-subgroup Q, say (a p'-group is one whose order is coprime to p). Then Q is a normal subgroup of $\bar{G}$, so, if $Q \neq 1$ then $\bar{z} \in Q$; but $\bar{z}^{p^m} = 1$ since $z^{p^m} \in Z^{p^m} \leq N$, and $\bar{z} \neq 1$, so $\bar{z}$ cannot belong to the p'-group Q. Thus $Q = 1$ and $G/N = \bar{G} = P$ is a finite p-group as desired.

Now let G be finitely generated and nilpotent, and suppose that $T = \tau(G)$ is a p-group. By the first part, G/T is residually finite-p; so it will suffice to find for each $t \in T, t \neq 1$, a normal subgroup N of G with G/N a finite p-group and $t \notin N$. Choosing $N \lhd G$ maximal subject to $t \notin N$, we can argue exactly as we did above to show that G/N is a finite p-group.

(ii) Let G be polycyclic-by-finite, and choose a prime p. We have to find $H \lhd_f G$ with H residually a finite p-group. If G is finite we can take $H = 1$, so suppose G infinite and argue by induction on the Hirsch number. Now G has a normal poly-C_∞ subgroup G_1 of finite index, and if $H_1 \lhd_f G_1$ is residually finite-p then H_1 has a subgroup H of finite index with $H \lhd G$, and H is residually finite-p. So replacing G by G_1, we may assume that G is poly-C_∞. Thus G has a normal subgroup K with G/K infinite cyclic, say $G = K\langle x \rangle$ for some $x \in G$. As $h(K) = h(G) - 1$, we may suppose by inductive hypothesis that K has a normal subgroup K_1 of finite index with K_1 residually finite-p, and as usual we can take $K_1 \lhd G$. For each $i > 1$, put

$$K_i = \gamma_i(K)\, K^{p^{i-1}}.$$

Thus K_1/K_i is a finite p-group for each i; and $\bigcap_{i=1}^{\infty} K_i = 1$, because every $N \lhd K_1$ with K_1/N a finite p-group satisfies $N \geq K_i$ for some i. Now let $e > 0$ be the order of the automorphism induced by x on the factor K_1/K_2, and put $y = \langle x^e \rangle$, so

$$C_{\langle x \rangle}(K_1/K_2) = \langle x^e \rangle = \langle y \rangle.$$

Applying Proposition 13, with K_1/K_i for G and $\langle y \rangle / C_{\langle y \rangle}(K_1/K_i)$ for Γ, we find that

$$y^{p^{m(i)}} \in C_{\langle y \rangle}(K_1/K_i)$$

for some $m(i)$, depending on i. The group $K_1 \langle y^{p^{m(i)}} \rangle / K_i$ is then just the direct product $(K_1/K_i) \times \langle y^{p^{m(i)}} \rangle$, which is easily seen to be residually finite-p. As this group is a normal subgroup of index $p^{m(i)}$ in $K_1 \langle y \rangle / K_i$, it follows that $K_1 \langle y \rangle / K_i$ is residually finite-p. But $\bigcap_i K_i = 1$ so $K_1 \langle y \rangle$ is itself residually finite-p. Evidently $K_1 \langle y \rangle$ has finite index in G, so it contains a subgroup H with $H \lhd_f G$, and the proof is finished.

Remark Theorems 1 and 4 will receive much more intuitive proofs later on: both will be virtually immediate consequences of L. Auslander's embedding theorem, which first appears in Chapter 5.

Exercise 13 (i) If H is residually a finite p-group, then every finite subgroup of H is a p-group. (ii) If H is residually finite-p, $N \lhd H$ and $C = C_H(N)$, then H/C is residually finite-p. (iii) Let G be residually finite-p and let $x, y \in G$. Show that if x^h commutes with y for some $h \in \mathbb{N}$, then x^{p^e} commutes with y where p^e exactly divides h. (iv) Show that if G is a torsion-free nilpotent group, $x, y \in G$, and some non-trivial power of x commutes with y, then x commutes with y.

(*Hint*: For (ii), put $\mathscr{S} = \{K \lhd H \,|\, H/K \text{ is a finite } p\text{-group}\}$. Observe that

$$[N, \bigcap_{K \in \mathscr{S}} CK] \le \bigcap_{K \in \mathscr{S}} [N, CK] = \bigcap_{K \in \mathscr{S}} [N, K] \le \bigcap_{K \in \mathscr{S}} K = 1;$$

deduce that $C = \bigcap_{K \in \mathscr{S}} CK$. For (iii), put $N = C_G(x^h)$, $C = C_G(N)$ and $H = N_G(N)$. Then apply parts (i) and (ii) to the finite group $C\langle x \rangle /C \le H/C$. For (iv), apply Theorem 4(i) to the group $\langle x, y \rangle$.)

D. Notes

Much of the material in this chapter is 'folklore'. For supplementary reading, I would recommend Hall (1969), Chapters 1, 2 and 3; Robinson (1972), Chapters 1, 2, 3 and 9.

The first systematic study of polycyclic groups was made by K.A. Hirsch in the series of papers Hirsch (1938)–(1954). In particular, Theorem 1 appeared in (1952), Theorem 2 in (1946) and Theorem 3 in (1954).

For an instructive and readable account of Frattini subgroups in infinite soluble groups, see Roseblade (1973).

Exercise 11 is due to Mal'cev (1958). Theorem 4(i) is from Gruenberg (1957), and Theorem 4(ii) was discovered independently by Wehrfritz (1970) and Šmel'kin (1968).

Much of section B, on nilpotency, originates with Mal'cev and Hall. Proposition 10 is due to Kalužnin. Proposition 12 is due to Robinson (1968); this paper also gives various applications different from the ones we have mentioned.

The reader may have noticed that although we took the trouble to introduce derivations (for Proposition 11), all the subsequent applications concerned situations where the derivations were just homomorphisms. We shall meet derivations again, playing an essential role, in several of the later chapters.

2
Mal'cev's theorems

In this chapter we obtain some general results of a qualitative nature about the structure of polycyclic groups, at a somewhat deeper level than those of Chapter 1, by beginning to exploit the 'linear' aspect of these groups. At its simplest, this comes down to the observation that if A/B is a free abelian factor, of rank n say, in a group G (with $B < A$ both normal subgroups of G), then the action of G by conjugation on this factor affords a representation of G in $\text{Aut}(A/B) = GL_n(\mathbb{Z})$. In later chapters we probe more deeply into the precise nature of the link between polycyclic groups and linear groups over $\mathbb{Z}$.

A. Rationally irreducible modules

In the study of a finite soluble group G, a fruitful line of attack is to pick a minimal normal subgroup $N \neq 1$ of G and investigate the action of G on N. If we try the same thing when G is infinite and polycyclic, we find that usually there is no such N; if N does exist, it is always finite and so makes an insignificant contribution to the structure of G (for example we have $h(G/N) = h(G)$). What we do, instead, is to consider a free abelian normal subgroup $A \neq 1$ of G of minimal rank. A then has the structure of a *rationally irreducible* G/A-module, and about such things much can be said. In fact we shall see in section C that $G/C_G(A)$ is abelian-by-finite; and Proposition 1 which follows shortly completely describes the structure of A for the case where $G/C_G(A)$ is abelian.

This is a convenient place to introduce the *group ring* of a group G, over a commutative ring R (with 1). This is the ring RG whose elements are finite formal sums

$$r_1 g_1 + \cdots + r_n g_n = \sum_{g \in G} r(g)g \tag{1}$$

with $r(g) \in R$ for all $g \in G$ and $r(g) = 0$ for all but finitely many $g \in G$. Addition is defined componentwise, and multiplication by specifying that, formally, $gr = rg$ for $r \in R$ and $g \in G$: thus

$$\left(\sum r(g)g\right)\left(\sum s(g)g\right) = \sum_{g \in G} \left(\sum_{xy=g} r(x)s(y)\right)g.$$

22

Identifying $g \in G$ with the element $1_R \cdot g \in RG$, and $r \in R$ with $r \cdot 1_G \in RG$, we have G as a subgroup and R as a subring of RG; this makes the expression on the left of (1) into an ordinary sum of products of elements in the ring RG.

It is easy to see that RG is characterised by the following 'universal property': given any ring E, a ring homomorphism $\rho : R \to E$, and a group homomorphism $\gamma : G \to E^*$, the group of units of E, such that $g\gamma$ commutes with $r\rho$ for all $g \in G$ and $r \in R$, then there exists a unique ring homomorphism $\theta : RG \to E$ such that $\theta|_R = \rho$ and $\theta|_G = \gamma$. We apply this to the situation where G acts by R-module automorphisms on an R-module M; the corresponding homomorphisms $\rho : R \to E = \text{End}_{\mathbb{Z}} M$ and $\gamma : G \to E^* = \text{Aut}_{\mathbb{Z}} M$ extend simultaneously to $\theta : RG \to \text{End}_{\mathbb{Z}} M$. Thus M has a natural structure as RG-module, so natural indeed that one usually considers 'RG-module' and 'R-module with G acting by R-automorphisms' as synonymous expressions.

We turn now to the main topic of this section. Suppose G is a group and $A \cong \mathbb{Z}^n$ a G-module, for some finite n. The $\mathbb{Q}$-vector space $\mathbb{Q} \otimes A \cong \mathbb{Q}^n$ can be identified with the *module of fractions* consisting of 'fractions' $m^{-1}a$, $0 \neq m \in \mathbb{Z}$ and $a \in A$, and it has a natural G-module structure given by

$$(m^{-1}a)g = m^{-1}(ag), \qquad g \in G.$$

Choosing a $\mathbb{Z}$-basis for A represents the action of G by matrices in $GL_n(\mathbb{Z})$, and the action of G on $\mathbb{Q} \otimes A$ is given by considering the corresponding matrices as belonging to $GL_n(\mathbb{Q})$. Thus $\mathbb{Q} \otimes A$ is a $\mathbb{Q}G$-module.

Definition Let G be a group and A a G-module. A is *rationally irreducible* if (i) $A \cong \mathbb{Z}^n$ for some finite n, and (ii) $\mathbb{Q} \otimes A$ is a simple $\mathbb{Q}G$-module. We abbreviate 'rationally irreducible' to 'r.i.'.

Exercise 1 Show that a G-module $A \cong \mathbb{Z}^n$ is r.i. if and only if every non-zero G-submodule of A has finite index in A. Deduce that a non-identity finitely generated free abelian normal subgroup A of minimal rank in a group H is r.i. as an H/A-module, with H/A acting by conjugation.

Proposition 1 Let G be an abelian group and A a r.i. G-module, on which G acts faithfully. Then there exist an algebraic number field k, with ring of integers $\mathfrak{o}$, and embeddings $\bar{} : A \to \mathfrak{o}^+$, $\bar{} : G \to \mathfrak{o}^*$ such that

$$(ag)^- = \bar{a}\bar{g} \tag{2}$$

for all $a \in A$ and $g \in G$. Both $\bar{A}$ and the subring $\mathbb{Z}[\bar{G}]$ generated by $\bar{G}$ have finite index in $\mathfrak{o}$.

Recall that an *algebraic number field* is a finite extension field of $\mathbb{Q}$ and an *algebraic integer* is any zero of a monic polynomial with coefficients in $\mathbb{Z}$. The algebraic integers lying in a given algebraic number field k form a subring $\mathfrak{o}$ of the field, called its *ring of integers*. If $[k:\mathbb{Q}] = n$, then

$$\mathfrak{o}^+ \cong \mathbb{Z}^n, \qquad \mathbb{Q} \cdot \mathfrak{o} = k$$

($\mathfrak{o}^+$ denotes the additive group of the ring $\mathfrak{o}$). It follows that for any finitely generated additive subgroup M of k there exists a non-zero integer m such that $mM \subseteq \mathfrak{o}$. For all of this, see any textbook of algebraic number theory or commutative algebra.

Proof of Proposition 1   The $\mathbb{Q}G$-module $\mathbb{Q} \otimes A$ is simple. Choose a non-zero element $c \in A$ and define

$$\lambda : \mathbb{Q}G \to \mathbb{Q} \otimes A$$

by

$$r\lambda = cr, \qquad r \in \mathbb{Q}G.$$

Let K be the annihilator in $\mathbb{Q}G$ of the element c. Since $\mathbb{Q} \otimes A$ is simple, the map λ is surjective and ker $\lambda = K$ is a maximal ideal of $\mathbb{Q}G$; in fact K is equal to the annihilator of $\mathbb{Q} \otimes A$ in $\mathbb{Q}G$. There is an induced isomorphism $\bar{\lambda} : \mathbb{Q}G/K \to \mathbb{Q} \otimes A$. Since $\mathbb{Q}G/K$ is simple as $\mathbb{Q}G$-module, it is a simple commutative ring, therefore a field; write $k = \mathbb{Q}G/K$. Suppose now that $A \cong \mathbb{Z}^n$. Then $k \cong \mathbb{Q} \otimes A \cong \mathbb{Q}^n$, so $[k:\mathbb{Q}] = n$ and k is an algebraic number field. Now let

$$\theta : \mathbb{Q} \otimes A \to k$$

be the inverse of $\bar{\lambda}$, and denote by

$$\psi : \mathbb{Q}G \to k$$

the natural map $r \mapsto r + K (r \in \mathbb{Q}G)$. Then θ is an additive isomorphism and $\psi|_G$ is a multiplicative homomorphism. $A\theta$ is thus a finitely generated additive subgroup of k, so $m \cdot A\theta \subseteq \mathfrak{o}$ for some non-zero integer m. Define

$$^{-} : A \to \mathfrak{o}^+$$

by

$$\bar{a} = m \cdot a\theta, \qquad a \in A.$$

Since $\bar{A} \cong A \cong \mathbb{Z}^n \cong \mathfrak{o}^+$ we have at once that $|\mathfrak{o}^+ : \bar{A}|$ is finite. For the embedding $^{-} : G \to \mathfrak{o}^*$ we just take the map $\psi|_G$; we have to check that this does send G into $\mathfrak{o}^*$ and that it is injective. Now if $g \in G$,

$$g\psi = 1 \Rightarrow g - 1 \in K \Rightarrow A(g-1) = 0 \Rightarrow g \in C_G(A) = 1,$$

so $\psi|_G$ is injective. The action of g on A can be represented by some matrix $\mu \in GL_n(\mathbb{Z})$; if f is the characteristic polynomial of μ then $f(\mu) = 0$, by the

Cayley–Hamilton theorem, so the element $f(g) \in \mathbb{Q}G$ belongs to K. Therefore $f(g\psi) = 0$, and as f is a monic polynomial over $\mathbb{Z}$, $g\psi \in \mathfrak{o}$. Likewise, $(g\psi)^{-1} = g^{-1}\psi \in \mathfrak{o}$, consequently $g \in \mathfrak{o}^*$ as desired. By construction, every element of $k = \mathbb{Q}G/K$ has some non-zero integer multiple lying in the subring $(\mathbb{Z}G + K)/K = (\mathbb{Z}G)\psi = \mathbb{Z}[\bar{G}]$; since $\mathfrak{o}^+ \cong \mathbb{Z}^n$ it follows that $\mathfrak{o}^+/\mathbb{Z}[\bar{G}]$ is finite.

Finally, we must check that (2) holds. It is immediate from the definition that $k = \mathbb{Q}G/K$ is a G-module, the action of $g \in G$ being multiplication by $\bar{g}$, and that $\bar{\lambda} : k \to \mathbb{Q} \otimes A$ is a G-module isomorphism. θ is therefore also a G-module isomorphism, so if $a \in A$ and $g \in G$ then

$$(ag)^- = m \cdot (ag)\theta = (ma)\theta \cdot \bar{g} = \bar{a}\bar{g}.$$

Full use of Proposition 1 will be made in later chapters, where the subtler 'arithmetical' aspects of polycyclic groups come under scrutiny. We content ourselves now with a single application, using the qualitative part of Dirichlet's famous *Units Theorem: if $\mathfrak{o}$ is the ring of integers of an algebraic number field, then its group of units $\mathfrak{o}^*$ is finitely generated.* This is much the easier part of the Units Theorem (the harder part gives the exact rank of $\mathfrak{o}^*$); for a proof see any textbook of algebraic number theory.

Proposition 2 If an abelian group G acts faithfully on a r.i. G-module, then G is finitely generated.

This follows at once from Proposition 1 and the Units Theorem, since, in the previous notation, $G \cong \bar{G} \leq \mathfrak{o}^*$.

B. Soluble automorphism groups

We can now give two striking theorems of Mal'cev.

Theorem 1 Every soluble group of automorphisms of a polycyclic-by-finite group is polycyclic.

Theorem 2 A soluble group is polycyclic if each of its abelian 2-step subnormal subgroups is finitely generated.

Definition A subgroup A is *2-step subnormal* in a group G if there is a normal subgroup $B \lhd G$ with $A \lhd B$.

Before proceeding with the proofs, let us single out the most significant special case of Theorem 1: since $\text{Aut}\,\mathbb{Z}^n = GL_n(\mathbb{Z})$ and $\mathbb{Z}^n$ is polycyclic, we have

Corollary 1   Every soluble subgroup of $GL_n(\mathbb{Z})$ is polycyclic.

The converse of this result, completing the abstract characterisation of soluble $\mathbb{Z}$-linear groups, will be proved in Chapter 5.

For the proof of Theorem 1, we consider a polycyclic-by-finite group G and a soluble subgroup Γ of $\mathrm{Aut}\,G$, and have to show that Γ is polycyclic. *Step* 1   Suppose G and Γ are both abelian. Then G is a finitely generated abelian group, so $T = \tau(G)$ is finite, of exponent m say. It follows that $G^m \cap T = 1$; so if G is infinite, it certainly possesses an infinite Γ-invariant free abelian subgroup. We argue by induction on $h(G)$ to show that Γ is finitely generated. If $h(G) = 0$ then G is finite, so Γ is finite and there is nothing to prove. Suppose $h(G) > 0$; then G is infinite, so we can choose an infinite Γ-invariant subgroup A of G of minimal rank. By Exercise 1, A is r.i. as Γ-module, so Proposition 2 shows that $\Gamma/C_\Gamma(A)$ is finitely generated. Also $h(G/A) = h(G) - h(A) < h(G)$, so by inductive hypothesis we may suppose that $\Gamma/C_\Gamma(G/A)$ is finitely generated. Put $\Delta = C_\Gamma(A) \cap C_\Gamma(G/A)$. Then Γ/Δ is isomorphic to a subgroup of $\Gamma/C_\Gamma(A) \times \Gamma/C_\Gamma(G/A)$, hence is finitely generated; and it remains to show that Δ is finitely generated. For this, apply Proposition 11 of Chapter 1, which shows that Δ is isomorphic to a subgroup of $\mathrm{Hom}(G/A, A)$. The latter is finitely generated because if G/A can be generated by r elements, then $\mathrm{Hom}(G/A, A)$ embeds into $A \times \ldots \times A$ with r factors.
Step 2   Suppose next that G is any polycyclic-by-finite group, but that Γ is still abelian. We again argue by induction on $h(G)$ and can assume that G is infinite. By Exercise 9 of Chapter 1, G has a characteristic infinite free abelian subgroup A say. By the first step, above, $\Gamma/C_\Gamma(A)$ is finitely generated, and by inductive hypothesis $\Gamma/C_\Gamma(G/A)$ is finitely generated. That Γ is itself finitely generated now follows as above (only $\mathrm{Hom}(G/A, A)$ must be replaced by $\mathrm{Der}(G/A, A)$ in the present context).

Before going further we record the following useful observation:

Proposition 3   Let G be a soluble group and A an abelian normal subgroup of G. Then G has a normal subgroup N which is nilpotent of class at most 2 and satisfies

$$A \leq \zeta_1(N) = C_G(N).$$

Proof   By Zorn's Lemma, A is contained in a maximal abelian normal subgroup of G, which we may as well take in place of A; so let us assume that A is maximal among abelian normal subgroups of G. If $C_G(A) = A$, take $N = A$. If not, then

$$1 \neq C_G(A)/A \lhd G/A,$$

so since G is soluble there exists an abelian normal subgroup $N/A \lhd G/A$

with $1 \neq N/A \leq C_G(A)/A$. Choose N maximal subject to these conditions. Then $N \lhd G$, and $N' \leq A$ so $[N',N] \leq [A,N] \leq [A,C_G(A)] = 1$, thus N is nilpotent of class at most 2. It remains to show that $A \leq C \leq N$ where $C = C_G(N)$; in fact we show that $A = C$. Now $N \cap C = \zeta_1(N)$ is abelian, normal in G, and contains A, so by the maximality of A we have $N \cap C = A$. Suppose $C \neq A$. Then C/A contains an abelian normal subgroup X/A of G/A, with $X > A$. Since N/A and X/A are abelian and $[N,X] \leq [N,C] = 1$, the group NX/A is abelian as well as normal in G/A. The choice of N ensures that then $NX = N$, i.e. $X \leq N$. But then $X = X \cap C \leq N \cap C = A$, contradicting $X > A$. Thus C must be equal to A as required.

We can now conclude the proof of Theorem 1; we prove Theorem 2 simultaneously. Let P_d denote the statement 'every soluble group of automorphisms of derived length at most d of a polycyclic-by-finite group is polycyclic', and let Q_d stand for 'a soluble group of derived length at most d is polycyclic if all its abelian 2-step subnormal subgroups are finitely generated'. The proposition P_1 has been proved in 'Step 2', above; and Q_1 is self-evident. Now suppose $d > 1$ and assume inductively that P_{d-1} and Q_{d-1} have been established. To deduce Q_d, let H be a group with $H^{(d)} = 1$ and all its abelian 2-step subnormal subgroups finitely generated. By Proposition 3, H has a normal subgroup N which is nilpotent of class at most 2 and satisfies

$$H^{(d-1)} \leq C_H(N) = \zeta_1(N).$$

Let Q be a maximal abelian normal subgroup of N. Then Q is finitely generated, since $Q \lhd N \lhd H$, and $C_N(Q) = Q$ by Exercise 4 of Chapter 1. Thus N/Q acts faithfully by conjugation on Q. Also N/Q is abelian, since $N' \leq \zeta_1(N) \leq C_N(Q) = Q$, consequently by P_1 the group N/Q is finitely generated. Therefore N is polycyclic. Now $H/C_H(N)$ acts faithfully on N, and $H^{(d-1)} \leq C_H(N)$, so P_{d-1} shows that $H/C_H(N)$ is polycyclic. Since $C_H(N) \leq N$ it follows that H is polycyclic. Thus Q_d holds.

Obviously P_d follows at once from P_1 and Q_d, and so Theorems 1 and 2 both follow by induction on derived length.

Exercise 2 (i) Show that if G is a soluble group then

$$C_G(\mathrm{Fitt}(G)) \leq \mathrm{Fitt}(G)$$

(use Proposition 3). (ii) Let G be a soluble group in which every abelian 2-step subnormal subgroup is finite. Show that G is finite.

C. Soluble linear groups

We now step back somewhat from the narrow topic of polycylic groups, and consider quite generally what can be said of a soluble group of matrices over a field. At the end, by specializing the result obtained, we shall find some important applications to polycyclic groups.

Throughout, k will denote an arbitrary field. A subgroup $G \le GL_n(k)$ is called *irreducible* if the kG-module k^n is simple, otherwise G is called *reducible*. If k^n is a direct sum of simple kG-submodules, G is said to be *completely reducible*. Now suppose G is reducible, so $V = k^n$ has a proper non-zero kG-submodule W. Let $(w_1, \ldots, w_r)$ be a basis for W and extend it to a basis $(u_1, \ldots, u_s, w_1, \ldots, w_r)$ for V (so $r + s = n$). The action of an element $g \in G$ on W is represented w.r.t. the basis (w_i) by a matrix $g\sigma \in GL_r(k)$, and the action of g on V/W is represented w.r.t. the basis $(u_i + W)$ by a matrix $g\pi \in GL_s(k)$. Thus we have homomorphism

$$\sigma: G \to GL_r(k), \quad \pi: G \to GL_s(k).$$

If $x \in GL_n(k)$ is the relevant basis-change matrix, then the matrix of $g \in G$ w.r.t. the new basis $(u_1, \ldots, u_s, w_1, \ldots, w_r)$ of V is given by

$$x^{-1}gx = \begin{bmatrix} g\pi & * \\ 0 & g\sigma \end{bmatrix} \tag{2}$$

(where * represents some $s \times r$ matrix). If W has a *complement* in V, so $V = U \oplus W$ for some kG-module U, and if we choose $(u_1, \ldots, u_s)$ to be a basis for U, then the submatrix * in (2) will be the zero matrix.

Now take W as above of smallest possible dimension, and repeat with V/W and $G\pi$ in place of V and G. This gives

Lemma 1 Let G be any subgroup of $GL_n(k)$. Then there exist homomorphisms $\sigma_i : G \to GL_{r_i}(k)(i = 1, \ldots, t; r_1 + \ldots + r_t = n)$ such that each $G\sigma_i$ is irreducible, and a matrix $x \in GL_n(k)$ such that

$$g^x = \begin{bmatrix} g\sigma_t & * & \cdot & \cdot & \cdot & * \\ & g\sigma_{t-1} & & & & \cdot \\ & & \cdot & & & \cdot \\ & & & \cdot & & * \\ 0 & & & & & g\sigma_1 \end{bmatrix} \quad \text{for all } g \in G.$$

If G is completely reducible, each g^x is actually a block-diagonal matrix $\mathrm{diag}(g\sigma_t, g\sigma_{t-1}, \ldots, g\sigma_1)$.

To go further we must consider irreducible groups in more detail. The easiest soluble groups are obviously the abelian ones; so let us take an irreducible abelian subgroup A of $GL_n(k)$. The kA-module $V = k^n$ is then simple, so as in Proposition 1 above we can identify V with a certain finite extension field of k, namely the field kA/K where K is the annihilator of V.

Now suppose, for simplicity, that the field k was algebraically closed. Then k has no finite extensions at all; so the scenario under discussion can only occur when $n = \dim_k V$ is equal to 1. Thus we deduce

Lemma 2 Assume the field k algebraically closed.

(i) If $GL_n(k)$ has any irreducible abelian subgroup, then $n = 1$.

(ii) If A is a completely reducible abelian subgroup of $GL_n(k)$, then there exists $x \in GL_n(k)$ such that A^x consists of diagonal matrices.

Part (ii) follows from Lemma 1, for on taking $G = A$ we see from part (i) that the degree r_i of the group $A\sigma_i$ is equal to 1 for each i.

In fact, Lemmas 1 and 2 show that if k is algebraically closed, then every abelian subgroup of $GL_n(k)$ can be put into *upper triangular* form by a suitable change of basis. Can we do the same for any soluble group? Evidently, from Lemma 1, it would suffice to do this for an irreducible soluble group. We know that a soluble group is well endowed with abelian normal subgroups, so it will be sensible to investigate what an abelian normal subgroup of an irreducible group looks like.

More generally, let G be a group, N a normal subgroup of G, and V a simple kG-module, finite-dimensional over k. Then V has simple kN-submodules, namely the non-zero kN-submodules of minimal k-dimension. Let U be one of them. Then

$$UkG = \sum_{g \in G} Ug = V \tag{3}$$

since V is simple for kG. Each Ug is a simple kN-submodule of V; for

$$Ug \cdot kN = U(kN)^{g^{-1}}g = U \cdot kN \cdot g = Ug$$

since $N \lhd G$, and if $0 < W <_{kN} Ug$ then $0 < Wg^{-1} <_{kN} U$ which is impossible as U is simple for kN. There is a simple general principle which now applies:

Lemma 3 Let V be a module and $(U_i)_{i \in I}$ a family of simple submodules with $V = \sum_{i \in I} U_i$. Then there is a subset J of I such that $V = \bigoplus_{j \in J} U_j$.

Proof Let J be a subset of I maximal with respect to the property that the sum $\sum_{j \in J} U_j$ is a direct sum. That such a set J exists is obvious in the finite-dimensional case which is of interest to us; in the general case, it is an easy exercise using Zorn's Lemma. To show that $\sum_{j \in J} U_j = V$ is also an easy exercise, which I leave to the reader.

Going back to (3) above, we deduce that

$$V = \bigoplus_{i=1}^{t} U g_i \tag{4}$$

for some $g_1, \ldots, g_t \in G$; of course $t = \dim V / \dim U$ is finite here. Thus N is *completely reducible*. Now the action of G permutes the summands in (3) among themselves; it will not in general permute the direct summands in (4), but on grouping these together according to their isomorphism type as kN-modules we shall see that it does something almost as good.

Definition   Let N be a group and V a kN-module. Let W be a simple kN-submodule of V. The *homogeneous component* of V belonging to W is

$$V_W = \sum \left\{ Y \underset{kN}{\leq} V \,\middle|\, Y \underset{kN}{\cong} W \right\}.$$

Lemma 3 shows that V_W is a direct sum of isomorphic simple kN-submodules. Moreover, we have

Lemma 4   Let $V = U_1 \oplus \ldots \oplus U_t$ be a direct sum of simple kN-submodules U_i. Then every simple kN-submodule of V is isomorphic to one of the U_i; the homogeneous components of V are given by

$$V_{U_i} = \oplus \{ U_j \mid U_j \cong U_i \}; \tag{5}$$

and V is the direct sum of its homogeneous components.

Proof   If $W \leq V$ is simple, then at least one of the projection maps of W into the direct summands $U_1, \ldots, U_t$ must be an isomorphism, so $W \cong U_i$ for some i. Denote the module on the r.h.s. of (5) by V_i. Obviously $V_i \leq V_{U_i}$. For the reverse inclusion, suppose $Y \leq V$ and $Y \cong U_i$. For each j, the projection map of Y into U_j must be either zero or an isomorphism; since $Y \cong U_i$, the only non-zero projections of Y are those into U_j with $U_j \cong U_i$. Thus $Y \leq V_i$, and so $V_{U_i} = V_i$ as required. The final claim of the lemma is then immediate.

Using Lemma 4 we can re-label the elements $g_1, \ldots, g_t$ occurring in (4) so that

$$\{ g_1, \ldots, g_t \} = \{ g_{11}, \ldots, g_{1r_1}, \ldots, g_{m_1}, \ldots, g_{mr_m} \}$$

where the

$$V_i = \bigoplus_{j=1}^{r_i} U g_{ij}, \quad i = 1, \ldots, m \tag{7}$$

are exactly the homogeneous components of the kN-module V. This establishes most of what is known as *Clifford's theorem*:

Proposition 4 Let $G \leq GL_n(k)$ be irreducible and $N \lhd G$. Then N is completely reducible, and the action of G permutes the homogeneous components of k^n as kN-module transitively among themselves.

Proof We must check that for $g \in G$ and V_i as in (7), $V_i g = V_{i(g)}$ where $i \mapsto i(g)$ is a permutation of the set $\{1, \ldots, m\}$. Now a kN-module isomorphism $\theta : Ug_{i1} \to Ug_{ij}$ gives a kN-module isomorphism $\theta^g : Ug_{i1}g \to Ug_{ij}g$, which sends an element $v = ug_{i1}g$ to $(vg^{-1})\theta \cdot g = (ug_{i1}\theta)g$. Therefore $V_i g$ is a sum of submodules isomorphic to $Ug_{i1}g$, and hence, by Lemma 4, $V_i g \subseteq V_{i(g)}$ for some $i(g) \in \{1, \ldots, m\}$. Similarly $V_{i(g)}g^{-1} \subseteq V_i$, so $V_i g = V_{i(g)}$. It is then easy to see that $i \mapsto i(g)$ is a permutation of $\{1, \ldots, m\}$. It is also clear that $i(gh) = i(g)(h)$ for g and $h \in G$, so we do get a permutation action of G on the set of homogeneous components V_i; this action is evidently transitive.

Corollary 2 Let G and N be as above. If every subgroup of index at most n in G is irreducible, then k^n is a direct sum of isomorphic simple kN-submodules.

For the stabilizer H, say, in G of the homogeneous component V_1 has index $m \leq n$, so H is irreducible. But V_1 is a non-zero kH-submodule of k^n, so $V_1 = k^n$.

Let us apply this corollary to the original problem. If N is abelian and k is algebraically closed, then the simple kN-submodules of k^n are all 1-dimensional over k, by Lemma 2; thus in this case k^n has a basis of eigenvectors for N. To say that the simple kN-submodules are all isomorphic then means that each element of N has all its eigenvalues equal. Thus N consists entirely of *scalar* matrices. As a final trick, let us pin things down further by restricting attention to *matrices of determinant* 1: there are not very many scalar matrices with this property! Recall that

$$SL_n(k) = \{g \in GL_n(k) \mid \det g = 1\}.$$

Proposition 5 Assume that k is algebraically closed. Let G be a soluble subgroup of $SL_n(k)$ such that every subgroup of index at most n in G is irreducible. Then G is finite, of order at most

$$v(n) = n^{2n-2} \cdot (n^{2n-2} \cdot n^{2n-2}!)! \tag{8}$$

Proof By Proposition 3, G has a normal subgroup N, nilpotent of class at most 2, such that $C_G(N) = \zeta_1(N) = Z$ say. Since Z is abelian and

normal in G, Z consists of scalar matrices. But if $z = \lambda \cdot 1_n \in Z$ with $\lambda \in k$, then
$$1 = \det z = \lambda^n,$$
so $z^n = 1$. Thus Z has exponent dividing n, and so N has exponent dividing n^2 (Chapter 1, Corollary 6). Next, let A be a maximal abelian normal subgroup of N. Then A is completely reducible, by two applications of Proposition 4, so by Lemma 2 there exists a matrix $x \in GL_n(k)$ such that A^x is diagonal. If $a = \mathrm{diag}(\lambda_1, \ldots, \lambda_n) \in A$, then

$$1_n = (a^{n^2})^x = \mathrm{diag}(\lambda_1^{n^2}, \ldots, \lambda_n^{n^2}),$$

so each eigenvalue λ_i is an n^2th root of unity in k. But k contains at most n^2 distinct n^2th roots of unity, so for the element a there are at most
$$(n^2)^{n-1} = n^{2n-2}$$
possibilities (note that $\lambda_1, \ldots, \lambda_{n-1}$ determine λ_n since $\det a = 1$). Thus $|A| \le n^{2n-2}$. The rest of the argument now follows a familiar pattern:

$$|N/A| = |N/C_N(A)| \le |\mathrm{Aut}\ A| \le n^{2n-2}!,$$

so

$$|N| \le n^{2n-2} \cdot n^{2n-2}!$$

Then

$$|G/Z| = |G/C_G(N)| \le |\mathrm{Aut}\ N| \le (n^{2n-2} \cdot n^{2n-2}!)!.$$

But $Z \le A$, so $|Z| \le n^{2n-2}$ and it follows that $|G| \le v(n)$. (Compare Exercise 2, above).

The hard work is now over. For if $G \le GL_n(k)$ has a reducible subgroup H of index $\le n$, we have

$$H^x = \begin{bmatrix} H\pi & * \\ 0 & H\sigma \end{bmatrix} \tag{9}$$

for some $x \in GL_n(k)$ and $H\pi \le GL_s(k)$, $H\sigma \le GL_r(k)$ with $r < n$ and $s < n$; then some sort of inductive argument will deal with the groups $H\pi$ and $H\sigma$. The result of such an analysis is Theorem 3, below; let us see first what happens on removing the restriction to $SL_n(k)$ in Proposition 5:

Corollary 3   Assume that k is algebraically closed. Let G be a soluble subgroup of $GL_n(k)$ such that every subgroup of index at most n in G is irreducible. Then

$$|G : G \cap k^* 1_n| \le v(n).$$

Here $v(n)$ is given in (8), above, and $k^* 1_n$ denotes the group of all scalar matrices in $GL_n(k)$.

Proof Put $G_1 = G \cdot k^* 1_n \cap SL_n(k)$. Then $G \leq G_1 \cdot k^* 1_n$; for if $g \in G$ then det g has an nth root x, say, in k^*, and then $g \cdot x^{-1} 1_n \in G_1$. Therefore

$$|G:G \cap k^* 1_n| \leq |G_1|,$$

and the corollary will follow provided G_1 satisfies the hypotheses of Proposition 5. It is evident that G_1 is soluble. Suppose H is a subgroup of index at most n in G_1. Putting $H_1 = H \cdot k^* 1_n \cap G$ we have

$$|G:H_1| = |G \cap G_1 \cdot k^* 1_n : G \cap H \cdot k^* 1_n| \leq |G_1 \cdot k^* 1_n : H \cdot k^* 1_n|$$

$$\leq |G_1 : H| \leq n,$$

so by hypothesis H_1 is irreducible. Since every kH-submodule of k^n is automatically a kH_1-submodule, we see that H is irreducible. Thus Proposition 5 may be applied to G_1, as required.

We are now ready to analyse the structure of a soluble linear group. A subgroup $G \leq GL_n(k)$ is called *triangularizable over* k if there is a matrix $x \in GL_n(k)$ such that

$$G^x \leq \mathrm{Tr}\,(n, k);$$

$\mathrm{Tr}\,(n, k)$ denotes the group of all *upper-triangular* matrices in $GL_n(k)$,

$$\mathrm{Tr}\,(n, k) = \{g = (g_{ij}) \in GL_n(k) | g_{ij} = 0 \quad \text{if} \quad i > j\}.$$

The group $G \leq GL_n(k)$ is just *triangularizable* if G is triangularizable over the algebraic closure of k. Note that this is equivalent to being triangularizable over some finite extension field of k, since the n^2 entries of a suitable matrix x will generate such a finite extension.

Theorem 3 There is a function $\mu : \mathbb{N} \to \mathbb{N}$ such that for any field k, every soluble subgroup of $GL_n(k)$ has a triangularizable subgroup of index at most $\mu(n)$.

Proof Let us carry out the procedure suggested at (9) above and see what $\mu(n)$ will have to be. So let G be a soluble subgroup of $GL_n(k)$. We can equally well consider G as a subgroup of $GL_n(\bar{k})$ where $\bar{k}$ is the algebraic closure of k, so without loss of generality let us assume that k is algebraically closed. Suppose first that every subgroup of index at most n in G is irreducible. Then Corollary 3 shows that $|G:G \cap k^* I_n| \leq v(n)$. Scalar matrices are already triangular, so in this case any function μ such that

$$\mu(n) \geq v(n) \tag{10}$$

fulfils the requirement. Suppose on the other hand that G has a reducible subgroup H with $|G:H| \leq n$. Let

$$\pi : H \to GL_s(k), \quad \sigma : H \to GL_r(k)$$

be as in (9) above, with $r + s = n$ and $r < n, s < n$. Assuming that $\mu(r)$ and $\mu(s)$ are known, there will be triangularizable subgroups $K_1 \leq H\pi$ and $K_2 \leq H\sigma$ with

$$|H\pi:K_1| \leq \mu(s), \quad |H\sigma:K_2| \leq \mu(r).$$

The subgroup

$$K = K_1\pi^{-1} \cap K_2\sigma^{-1} \leq H$$

is then easily seen to be triangularizable (look at (9) again!). Its index in H is at most $\mu(r)\,\mu(s)$, so we have

$$|G:K| \leq |G:H|\ |H:K| \leq n\mu(r)\,\mu(s).$$

Thus a satisfactory value for $\mu(n)$ in this case is given by

$$\mu(n) = n \cdot \max_{0 < r < n} \mu(r)\mu(n - r). \tag{11}$$

To start the induction off we can obviously take $\mu(1) = 1$. Putting (10) and (11) together we then make a recursive definition: for $n > 1$,

$$\mu(n) = \max \left\{ v(n), n \cdot \max_{0 < r < n} \mu(r)\mu(n - r) \right\}.$$

Theorem 3 shows that all soluble linear groups are 'essentially' triangular. For any commutative ring R, denote by

$$\mathrm{Tr}\,(n, R)$$

the group of all invertible $n \times n$ matrices over R with all below-diagonal entries equal to zero.

Proposition 6 $\mathrm{Tr}_1\,(n, R) \lhd \mathrm{Tr}\,(n, R)$ and

$$\mathrm{Tr}\,(n, R)/\mathrm{Tr}_1\,(n, R) \cong \underbrace{R^* \times \ldots \times R^*}_{n}.$$

Proof Let $\psi : \mathrm{Tr}(n, R) \to D_n(R)$ be the map sending a matrix to its diagonal. It is easy to see that ψ is a surjective homomorphism with $\ker \psi = \mathrm{Tr}_1\,(n, R)$, and the proposition follows.

We saw in Chapter 1 that $\mathrm{Tr}_1(n, R)$ is nilpotent, of class at most $n - 1$, hence soluble of derived length at most $1 + [\log_2 (n - 1)]$. Therefore $\mathrm{Tr}(n, R)$ is soluble of derived length at most $2 + [\log_2(n - 1)]$. If R^+ is finitely generated, then we saw that $\mathrm{Tr}_1\,(n, R)$ is polycyclic; so if R^* is also finitely generated we deduce that $\mathrm{Tr}(n, R)$ is polycyclic. This gives an alternative approach to the results of section B, outlined in Exercises 3 and 4 below. First, some other consequences of Theorem 3.

Corollary 4 Every soluble linear group is (nilpotent-by-abelian)-by-finite.

Corollary 5 A soluble linear group of degree n has derived length bounded by a function of n.

Corollary 4 is immediate from Theorem 3 and Proposition 6. For Corollary 5, suppose $G \le GL_n(k)$ is soluble and H is a triangularizable subgroup of index at most $\mu(n)$ in G. Then there exists $H^0 \lhd G$ with $H^0 \le H$ and $|G:H^0| \le \mu(n)!$, and $G^{(r)} \le H^0$ where $r = 2 + [\log_2 |G:H^0|]$. The discussion above shows that $H^{0(s)} = 1$ with $s = 2 + [\log(n-1)]$; putting all this together we deduce Corollary 5.

Proposition 7 If G is a triangularizable subgroup of $GL_n(k)$ then G' acts nilpotently on the module k^n.

Proof Let $\bar{k}$ denote the algebraic closure of k. Then $G^x \le \mathrm{Tr}(n, \bar{k})$ for some $x \in GL_n(\bar{k})$, so $(G')^x \le \mathrm{Tr}_1(n, \bar{k})$. Therefore G' acts nilpotently on $\bar{k}^n$; for if x^{-1} maps the standard basis of $\bar{k}_n$ to a basis $(u_1, \ldots, u_n)$ say, then G' fixes each u_i modulo $\sum_{j=i+1}^n \bar{k} u_j$. But G' leaves the k-subspace k^n of $\bar{k}^n$ invariant, so G' also acts nilpotently on k^n.

Corollary 6 Let G be a soluble group and A a rationally irreducible G-module. Then $G/C_G(A)$ is abelian-by-finite.

Proof We assume without loss of generality that $C_G(A) = 1$. The action of G on A then embeds G into $GL_n(\mathbb{Z})$, where n is the rank of A. By Theorem 3, G has a triangularizable subgroup H of finite index, which may take normal in G. By Proposition 7, H' acts nilpotently on $V = \mathbb{Q} \otimes A$, so $C_V(H') > 0$. But $C_V(H')$ is a $\mathbb{Q}G$-submodule of V since $H' \lhd G$; since V is simple for $\mathbb{Q}G$ it follows that $C_V(H') = V$. Thus $H' \le C_H(V) = 1$ and H is abelian.

This result leads to what is perhaps the most important of Mal'cev's theorems on polycyclic groups.

Theorem 4 Every polycyclic group is (nilpotent-by-abelian)-by-finite.

Now we have alluded to the fact that every polycyclic group is isomorphic to a linear group, so Theorem 4 looks like no more than a special case of Corollary 4: but its importance lies in the fact that one has to use Theorem 4 in order to prove the linearity of polycyclic groups, as we shall see in Chapter 5.

Proof of Theorem 4   Argue by induction on the Hirsch number of a polycyclic group G. If $h(G) = 0$, G is finite and there is nothing to prove. Suppose $h(G) > 0$. Then G has an infinite free abelian normal subgroup A, and we choose A of minimal rank so that A becomes a r.i. G-module. Then by Corollary 6, G has a normal subgroup H_1 of finite index such that

$$[A, H_1'] = 1.$$

By inductive hypothesis, G/A has a normal subgroup H_2/A of finite index such that

$$\gamma_m(H_2') \leq A$$

for some m. Then $H = H_1 \cap H_2$ is normal in G, $|G:H|$ is finite, and

$$\gamma_{m+1}(H') \leq [\gamma_m(H_2'), H_1'] \leq [A, H_1'] = 1.$$

The theorem follows.

Exercise 3   (i) Suppose G is a triangularizable subgroup of $GL_n(\mathbb{Z})$. Show that there exist an algebraic number field k, with ring of integers $\mathfrak{o}$, and a matrix $x \in GL_n(k)$ such that $G^x \leq \operatorname{Tr}(n, \mathfrak{o})$. (*Hint:* Suppose $G^y \leq \operatorname{Tr}(n, k)$ with $y \in GL_n(k)$. Then $my \in M_n(\mathfrak{o})$, $my^{-1} \in M_n(\mathfrak{o})$ for some $m \in \mathbb{Z}$, $m \neq 0$. Show that $x = y \cdot \operatorname{diag}(1, m^2, m^4, \ldots m^{2n-2})$ will do the job.) (ii) Deduce that every soluble subgroup of $GL_n(\mathbb{Z})$ is polycyclic, using the Units Theorem, Theorem 3, and Proposition 6.

Exercise 4   Deduce Theorems 1 and 2 directly from Corollary 1, using Proposition 3.

For a final exercise, let us show that a soluble linear group is not in general triangularizable:

Exercise 5   (i) Show that if char $k = 0$, every triangularizable finite subgroup of $GL_n(k)$ is abelian; if char $k = p$, every triangularizable finite p'-subgroup of $GL_n(k)$ is abelian. (*Hint:* consider the possible torsion elements in $\operatorname{Tr}_1(n, k)$.) (ii) Deduce that the function $\mu(n)$ in Theorem 3 must tend to infinity with n, by constructing for each sufficiently large n a finite soluble subgroup of $GL_n(k)$ having no abelian subgroup of index $< f(n)$, where $f(n) \to \infty$ as $n \to \infty$. (*Hint:* Take primes p and q with $pq \leq n$. The wreath product $G = C_p \wr C_q$ is the semi-direct product of the group ring

$GF(p)C_q$ with the group C_q. Show that if $H \leq G$ with $|G:H| < q$ and $|G:H| < p^{q-1}$, then H is not abelian. G is generated by an element x of order p and an element y of order q, and can be embedded in the symmetric group S_{pq} by

$$x \mapsto (12 \ldots p)$$
$$y \mapsto (1 \ p+1 \ldots (q-1)p+1)(2 \ p+2 \ldots (q-1)p+2) \ldots (p \ 2p \ldots qp).$$

Finally, S_{pq} has a natural representation in $GL_{pq}(k)$, acting by permutations on the standard basis of k^{pq}.)

D. Notes

For the elementary facts about algebraic number fields and their rings of integers, see for example Stewart & Tall (1979); Janusz (1973), Chapter 1; Cohn (1978), Chapters 2, 7 and 11; Lang (1965), Chapter 9 (see Exercise 7 of that chapter for the qualitative form of the Units Theorem).

As supplementary reading on linear groups, I would recommend Wehrfritz (1973*b*), Chapters 1–4.

Theorems 1, 2, 3 and 4 are due to Mal'cev (1951). In that paper Mal'cev discusses wider classes of soluble groups as well as polycyclic ones. The proof of Theorem 3 which I have given here, in a heuristic spirit, is more or less Mal'cev's original one; a similar account is given in Hall (1969), Chapter 9. Of course, the function μ given in Theorem 3, and the bound derived from it for the derived length of a soluble linear group, are very crude. For the best possible result regarding the latter, see Chapter 3 of Wehrfritz (*op. cit.*), and references therein. Presumably the best possible value for μ is also known, but I have not located it.

A quite different proof of Theorem 3, but without the bound on the index, is given in Chapter 5 of Wehrfritz (*op. cit.*), under the name of the Lie–Kolchin Theorem. This is a beautiful argument using the Zariski topology in a linear group (see Chapter 8, below); Wehrfritz has pointed out to me that one can make essentially the same argument work, with no mention of topology, if in that proof one defines the 'identity component' of a subgroup G of $GL_n(k)$ to be

$$G^0 = \bigcap \{H | H \leq_f G \text{ and } H = G \cap E \text{ for some } k\text{-subalgebra } E \text{ of } M_n(k)\}.$$

Corollary 5 was proved long before Mal'cev's theorem, in Zassenhaus (1938).

For a weaker form of Theorem 2, which does not depend on the Units Theorem, see Exercise 3 in Chapter 3 (which is due to D.J.S. Robinson).

3
Extensions

Given two groups N and G, the task of *extension theory* is to determine all *extensions* of N by G: that is, groups E having N as a normal subgroup and $E/N \cong G$. This is a highly developed subject, but its general recipes are difficult to apply in a really illuminating way when dealing with infinite groups. An exception is when the theory tells us that the extension necessarily *splits*, i.e. that E has a subgroup H with $N \cap H = 1$ and $NH = E$ (so that $H \cong E/N \cong G$); such an H is called a *complement* for N in E. After some necessary preparations in section A, section B gives a splitting result of this kind for the case where N is abelian, G is nilpotent, and both are finitely generated.

Suppose E is a polycyclic group and $N = \text{Fitt}(E)$. We know from Chapter 2 that then $G = E/N$ is abelian-by-finite; taking nilpotent groups and abelian-by-finite groups as fairly well understood, it would be very satisfactory if there was a straightforward way to describe all extensions of such an N by such a G. Unfortunately (or perhaps fortunately, for the interest of the subject), the answer given by general extension theory is too complicated to be of much practical use; in particular we shall see in the final chapter that such an extension can be very far from having to split. There is however an alternative way to get an overview of the structure of E, provided by the main result proved in section C: the group E has a nilpotent subgroup C such that NC has finite index in E. We might call C a 'nilpotent almost-supplement' for $N = \text{Fitt}(E)$ in E. As well as providing an attractive application for the methods of section B, this result is important for two reasons. Firstly, it eliminates the major 'extension problem' from the theory of polycyclic groups, by showing that every such group is a *finite* extension of a group built rather simply out of two nilpotent groups; and secondly, it is the key to the construction of 'semi-simple splittings', a rather sophisticated procedure, with deep applications, which is explained in Chapter 7.

The nilpotent almost-supplements for the Fitting subgroup in a polycyclic group are very like the Carter subgroups of a finite soluble group, and there is indeed a whole 'formation theory' for polycyclic groups analogous to that for finite soluble groups. It is mainly due to M.L. Newell, who first established the existence of the nilpotent almost-suplements. We do not pursue this aspect of the subject; full references are given in section D.

A. Generalities

Here we make some technical preparations for the later sections. First of all, we note a convenient approach to the complements of an abelian normal subgroup in a split extension:

Proposition 1   Let E be a group, A an abelian normal subgroup of E, and $G = E/A$. Let H be a complement for A in E and denote by $\mathscr{C}$ the set of all complements for A in E. Then there is a bijection

$$\Theta : \mathrm{Der}(G, A) \to \mathscr{C},$$

defined as follows. Let $\;: G \to H$ be the natural isomorphism $g = Ah \mapsto h$ ($g \in G$, $h \in H$); for $\delta \in \mathrm{Der}\,(G, A)$ and $g \in G$, put

$$g\delta^* = g' \cdot g\delta \in HA = E,$$

and define

$$\delta\Theta = \mathrm{Im}\,\delta^* = \{g\delta^* | g \in G\}.$$

Of course, A is here considered as a G-module via conjugation in E.

Proof   If $x, y \in G$ and $\delta \in \mathrm{Der}\,(G, A)$, then

$$(xy)\delta^* = x'y'(x\delta)^y(y\delta)$$
$$= x'y' \cdot y'^{-1}(x\delta)y'(y\delta)$$
$$= (x\delta^*)(y\delta^*);$$

so δ^* is a homomorphism of G into E and $\delta\Theta = \mathrm{Im}\,\delta^*$ is a subgroup of E. Using $H \cap A = 1$ and $HA = E$ one verifies easily that $\delta\Theta$ is actually a complement for A in E (note that $1_G\delta = 1_A$ always). To show that Θ is bijective, we construct the inverse map $\Psi : \mathscr{C} \to \mathrm{Der}(G, A)$ as follows. For $C \in \mathscr{C}$, let $C\Psi$ be the map sending $g \in G$ to the element $g'^{-1}c \in A$, where $c \in C \cap g$ is the unique element of C belonging to the coset g of A. It is a routine matter to verify that $C\Psi$ is a derivation; and it is then immediate from the definition that $C\Psi\Theta = C$ for all $C \in \mathscr{C}$ and that $\delta\Theta\Psi = \delta$ for all $\delta \in \mathrm{Der}(G, A)$.

Recall that for $a \in A$, the *inner derivation* on a is the map of G into A given by

$$g \mapsto a(g - 1)$$

(or $g \mapsto a^{-1}a^g$ if A is multiplicative). The set of all inner derivations $G \to A$ is a subgroup

$$\mathrm{Ider}\,(G, A) \le \mathrm{Der}\,(G, A).$$

Proposition 2 In the notation of Proposition 1, let α, $\beta \in \mathrm{Der}\,(G, A)$. Then the following are equivalent:

(a) the complements $\alpha\Theta$ and $\beta\Theta$ are conjugate in E;
(b) $\alpha\Theta$ and $\beta\Theta$ are conjugate under the action of A;
(c) α and β belong to the same coset of Ider (G, A) in $\mathrm{Der}(G, A)$.

Proof That (a) and (b) are equivalent follows from the fact that $A \cdot \alpha\Theta = E = A \cdot \beta\Theta$. Now for $a \in A$, let δ_a be the inner derivation on a, and consider the derivation $\delta_a \cdot \beta : G \to A$. In the notation of Proposition 1, for $g \in G$ we have

$$
\begin{aligned}
g(\delta_a \beta)^* &= g' \cdot g\delta_a \cdot g\beta \\
&= g' \cdot a^{-1}g'^{-1}ag' \cdot g\beta \\
&= ag' \cdot g\beta \cdot a^{-1} = a \cdot g\beta^* \cdot a^{-1}
\end{aligned}
\tag{1}
$$

(since A is abelian and $a^g = g'^{-1}ag' \in A$). Suppose (c) holds. Then $\alpha = \delta_a \beta$ for some $a \in A$, and (1) then shows that $\alpha\Theta = a \cdot \beta\Theta \cdot a^{-1}$, so (b) follows. Suppose conversely that $a^{-1} \cdot \alpha\Theta \cdot a = \beta\Theta$ with $a \in A$. From (1),

$$
\beta\Theta = a^{-1} \cdot (\delta_a \beta)\Theta \cdot a;
$$

therefore $\alpha\Theta = a \cdot \beta\Theta \cdot a^{-1} = (\delta_a \beta)\Theta$, and since Θ is bijective it follows that $\alpha = \delta_a \beta$. Thus (c) follows.

We now specialise further. For the rest of this section, G will denote a nilpotent group and $A \cong \mathbb{Z}^n$ a G-module, n being some non-negative integer. $M = \mathbb{Q} \otimes A$ will denote the G-module of fractions of A.

Lemma 1 The following are equivalent:

(a) $C_A(G) = 0$;
(b) $C_M(G) = 0$;
(c) $M = M(G - 1)$;
(d) $A/A(G - 1)$ is finite.

Here,

$$
M(G - 1) = \{\textstyle\sum \mu_i(g_i - 1) | \mu_i \in M, g_i \in G\}
$$

and analogously for $A(G - 1)$.

Proof It is easy to see that $(a) \Leftrightarrow (b)$ and $(c) \Leftrightarrow (d)$; just note that $A \leq M$ and that every element of M has some non-zero integer multiple in A. To establish the equivalence $(b) \Leftrightarrow (c)$, assume without loss of generality that

$C_G(A) = 1$. Then $C_G(M) = 1$ also. If $G = 1$ then each of (b), (c) implies that $M = 0$ and there is no more to prove; so suppose that $G \neq 1$, and choose $x \in \zeta_1(G)$ with $x \neq 1$. The map

$$\xi : M \to M; \quad \mu \mapsto \mu(x - 1)$$

is then a G-module endomorphism of M, with $\ker \xi = C_M(x)$ and $\operatorname{Im} \xi = M(x - 1)$. Therefore

$$\dim_\mathbb{Q} M = \dim_\mathbb{Q} C_M(x) + \dim_\mathbb{Q} M(x - 1). \tag{2}$$

Proof that $(b) \Rightarrow (c)$. If $C_M(x) = 0$, it follows at once from (2) that

$$M(G - 1) \geq M(x - 1) = M,$$

and we have (c). Suppose then that $C_M(x) > 0$. Then $\bar{M} = M/C_M(x)$ has smaller dimension than M; and $\bar{M} \cong M\xi \leq M$, so $C_{\bar{M}}(G) = 0$. Arguing by induction on $\dim_\mathbb{Q} M$, we may therefore assume that $\bar{M} = \bar{M}(G - 1)$. Thus

$$M = M(G - 1) + C_M(x). \tag{3}$$

Also $\dim_\mathbb{Q} C_M(x) < \dim_\mathbb{Q} M$, since $x \notin C_G(M) = 1$; so again inductively we may suppose that $C_M(x) = C_M(x)(G - 1)$. Together with (3) this gives $M = M(G - 1)$, the desired conclusion.

Proof that $(c) \Rightarrow (b)$ If $M\xi = M$, we see from (2) that

$$C_M(G) \leq C_M(x) = 0,$$

giving (b). Suppose $M\xi < M$, put $\tilde{M} = M/M\xi$, and note that $\dim_\mathbb{Q} \tilde{M} < \dim_\mathbb{Q} M$ (why?). Also $\tilde{M} = \tilde{M}(G - 1)$, so arguing inductively as before we may assume that $C_{\tilde{M}}(G) = 0$. This implies that $C_M(G) \leq C_{M\xi}(G)$. But $\dim_\mathbb{Q} M\xi < \dim_\mathbb{Q} M$ and $M\xi(G - 1) = M(G - 1)\xi = M\xi$, so again inductively we may suppose that $C_{M\xi}(G) = 0$. Thus (b) follows.

The usefulness of Lemma 1 lies mainly in its

Corollary 1 Suppose $B \cong \mathbb{Z}^m$ is a factor module of a G-submodule of A, and M' is a $\mathbb{Q}G$-factor module of a $\mathbb{Q}G$-submodule of M. If any of the conditions (a)–(d) of Lemma 1 hold, then (a) and (d) hold with B in place of A, and (b) and (c) hold with m' in place of M.

For (a) and (b) are inherited by submodules, (c) and (d) by factor modules; and the proof that $(b) \Leftrightarrow (c)$ clearly works for any finite-dimensional $\mathbb{Q}G$-module M.

Readers familiar with the cohomology of groups will have noticed that Lemma 1 relates $H^0(G, M)$ and $H_0(G, M)$ with each other and with $H^0(G, A)$ and $H_0(G, A)$. The next lemma does the same for $H^1(G, M) = \operatorname{Der}(G, M)/\operatorname{Ider}(G, M)$.

Lemma 2 Suppose that G is finitely generated. Then $\mathrm{Der}(G, A)/\mathrm{Ider}(G, A)$ is finite if and only if $\mathrm{Der}(G, M) = \mathrm{Ider}(G, M)$.

Proof Suppose $\mathrm{Der}(G, M) = \mathrm{Ider}(G, M)$, and take $\delta \in \mathrm{Der}(G, A)$. Of course δ can also be considered as a derivation into M, so by hypothesis there exists $\mu \in M$ such that $g\delta = \mu(g - 1)$ for all $g \in G$. Now $k\mu \in A$ for some non-zero $k \in \mathbb{Z}$, and then the map $k\delta$ is evidently the inner derivation on $k\mu$. Thus $\mathrm{Der}(G, A)/\mathrm{Ider}(G, A)$ is a periodic group. But $\mathrm{Der}(G, A)$ is a finitely generated abelian group; for if G is generated by $g_1, \ldots, g_r$ say, then the map

$$\delta \mapsto (g_1\delta, \ldots, g_r\delta)$$

embeds $\mathrm{Der}(G, A)$ into the direct sum of r copies of A. It follows therefore that $\mathrm{Der}(G, A)/\mathrm{Ider}(G, A)$ is finite.

Suppose conversely that $\mathrm{Der}(G, A)/\mathrm{Ider}(G, A)$ is finite. Say $G = \langle g_1, \ldots, g_r \rangle$ as above, and let $\delta \in \mathrm{Der}(G, M)$. We must show that δ is inner. Now there exists $k \in \mathbb{Z}$, $k \neq 0$, such that $k \cdot g_i \delta \in A$ for $i = 1, \ldots, r$. I claim that the derivation $k\delta$ then maps G into A. To see this, consider an element $w \in G$; w is a product of the generators g_i and their inverses, say $w = g_j^{\pm 1} \cdot v$ where v is shorter such product. Suppose inductively that $k \cdot v \in A$. Then

$$k \cdot w\delta = k \cdot g_j^{\pm 1}\delta \cdot v + k \cdot v\delta;$$

and this lies in A since $k \cdot g_j\delta \in A$ and $k \cdot g_j^{-1}\delta = - k \cdot g_j\delta \cdot g^{-1} \in A$. Thus $k\delta \in \mathrm{Der}(G, A)$, so by hypothesis $mk\delta \in \mathrm{Ider}(G, A)$ for some non-zero $m \in \mathbb{Z}$. Say $mk\delta$ is the inner derivation on $a \in A$. It is then obvious that δ is the inner derivation on the element $(mk)^{-1}a \in M$.

One could now go on to treat the second cohomology groups in a similar vein. Rather than introduce these formally however, let us work directly with group extensions. So let E be a group having A as a normal subgroup such that $E/A = G$, the G-module structure of A being realized by conjugation in E. We want to embed E into an extension $\bar{E}$ of M by G, corresponding to the given G-module structure of M, such that $\bar{E} = ME$ and $M \cap E = A$. To do this, choose a transversal

$$\{t_g | g \in G\}$$

to the cosets of A in E, with $t_1 = 1$, and define a map $\gamma : G \times G \to A$ by

$$t_g t_h = \gamma(g, h)t_{gh}, \quad g, h \in G;$$

γ is the '2-cocycle' corresponding to the extension E of A by G. Now form a new group $\bar{E}$ as follows. The elements of $\bar{E}$ are the symbols

$$\mu \cdot t_g, \quad \mu \in M, \quad g \in G,$$

and they are multiplied according to the rule

$$(\mu \cdot t_g)(\nu \cdot t_h) = \mu \nu^{g^{-1}} \gamma(g,h) \cdot t_{gh}.$$

It is a routine matter to check that this does indeed make $\bar{E}$ into a group: the formal properties of the cocycle γ, which express the fact that multiplication in E is associative, ensure that the multiplication in $\bar{E}$ is also associative; and

$$(\mu \cdot t_g)^{-1} = \mu^{-g} a_g \cdot t_{g^{-1}}$$

where $t_g^{-1} = a_g t_{g^{-1}}$ with $a_g \in A$. It is clear that by identifying each $\mu \in M$ with $\mu \cdot t_1 \in \bar{E}$, and each $t_g \in E$ with $1_M \cdot t_g \in \bar{E}$, we get $M \lhd \bar{E}$, $\bar{E}/M \cong G$, $ME = \bar{E}$ and $M \cap E = A$; of course E appears inside $\bar{E}$ as the subset $\{a \cdot t_g | a \in A, g \in G\}$.

Note that this construction works perfectly well for any group G and any G-module M with a G-submodule A. For the next result however we need G finitely generated.

Lemma 3 Let A, M and G be as before and assume G finitely generated. Let E be an extension of A by G and $\bar{E}$ the corresponding extension of M by G, described above. If $\bar{E}$ splits over M, then E has a subgroup H such that $A \cap H = 1$ and $|E:AH|$ is finite.

Thus E 'almost splits' over A provided $\bar{E}$ splits over M.

Proof M has a complement C, say, in $\bar{E}$: thus $M \cap C = 1$ and $MC = \bar{E}$. We take $H = C \cap E$. Certainly $H \cap A = 1$, so it remains to show that $|E:AH|$ is finite. Now $C \cong G$ is finitely generated, by the set

$$\{\mu_1 e_1, \ldots, \mu_s e_s\}$$

say, with each $\mu_j \in M$ and each $e_j \in E$. There exists $k \in \mathbb{Z}$, $k \neq 0$, such that $\mu_i^k \in A$ for each i (we write M and A multiplicatively for the moment); then each μ_i lies in the G-submodule

$$B = A^{1/k} = \{x \in M | x^k \in A\}$$

of M. Since B is clearly normal in $\bar{E}$, the whole of C is contained in the subgroup BE of $\bar{E}$. Since $M \cap BE = B$ and $BE \leq MC$, we have

$$BE = BC = B(C \cap BE).$$

Thus

$$|E:AH| \leq |BE:AH| = |BE:BH| \cdot |BH:AH| \leq |BE:BH| \cdot |B:A|$$

and

$$|BE:BH| \leq |(C \cap BE):H| \leq |BE:E| = |B:A|.$$

Hence

$$|E:AH| \leq |B:A|^2 = k^{2n}.$$

B. Splitting theorems

We come now to the central results of this chapter. From a more sophisticated point of view, these are really finiteness theorems for first and second cohomology groups, and in that form were systematically worked out by Derek Robinson. Details are given in section D.

Theorem 1   Let A be a finitely generated free abelian normal subgroup of a group E such that E/A is finitely generated and nilpotent. If $C_A(E) = 1$ then E has a subgroup H such that $A \cap H = 1$ and $|E:AH|$ is finite; that is, E almost splits over A.

Theorem 2   Let A be a finitely generated free abelian group and G a finitely generated nilpotent group acting on A, such that $C_A(G) = 1$. Then the complements for A in the semi-direct product $A]G$ lie in finitely many conjugacy classes.

We prove Theorem 2 first. Using Propositions 1 and 2, we reformulate it as

Theorem 2*   With A and G as in Theorem 2, if $C_A(G) = 0$ then $\mathrm{Der}(G, A)/\mathrm{Ider}(G, A)$ is finite.

This result will follow, by Lemma 2, from

Theorem 2**   Let G be a nilpotent group and M a $\mathbb{Q}G$-module of finite dimension over $\mathbb{Q}$. If $C_M(G) = 0$ then $\mathrm{Der}(G, M) = \mathrm{Ider}(G, M)$.

Proof   The proof is by induction on $\dim_{\mathbb{Q}} M$. Note that $C_{M'}(G) = 0$ for every $\mathbb{Q}G$-factor module M' of M, by Corollary 1. We consider two alternatives:

Case 1   Suppose M has a proper $\mathbb{Q}G$-submodule $N > 0$. By inductive hypothesis, we may assume that $\mathrm{Der}(G, N) = \mathrm{Ider}(G, N)$ and that $\mathrm{Der}(G, M/N) = \mathrm{Ider}(G, M/N)$. Now let $\delta : G \to M$ be a derivation. As the induced derivation $G \to M/N$ is inner, there exists $\mu \in M$ such that

$$g\delta \equiv \mu(g - 1) \bmod N \text{ for all } g \in G.$$

The map $\delta' : G \to N$; $g\delta' = g\delta - \mu(g - 1)$ is then a derivation of G into N, so there exists $\nu \in N$ such that

$$g\delta - \mu(g - 1) = g\delta' = \nu(g - 1) \text{ for all } g \in G.$$

Thus δ is the inner derivation on the element $\mu + \nu \in M$. Hence $\mathrm{Der}(G, M) = \mathrm{Ider}(G, M)$ in this case.

Case 2 Suppose M is a simple $\mathbb{Q}G$-module. Put $K = C_G(M)$, and take $\delta \in \mathrm{Der}(G, M)$. Then $\delta|_K$ is a group homomorphism, so the set $K\delta$ is an additive subgroup of M; in fact $K\delta$ is a G-submodule of M, for if $x \in K$ and $g \in G$ a simple calculation gives

$$x\delta \cdot g = (x^g)\delta \in K\delta. \tag{4}$$

This equation also shows that $\ker(\delta|_K) \lhd G$. Suppose for a moment that $\ker(\delta|_K) < K$. Since G is nilpotent, we can then find $x \in K \setminus \ker(\delta|_K)$ with $[x, G] \leq \ker(\delta|_K)$. If $g \in G$ then

$$x\delta \cdot g = (x^g)\delta = (xy)\delta = x\delta \cdot y + y\delta = x\delta$$

where $y = [x, g] \in \ker(\delta|_K)$, whence $y\delta = 0$. Therefore $x\delta \in C_M(G) = 0$ and $x \in \ker(\delta|_K)$, a contradiction. It follows that $\ker(\delta|_K) = K$. Thus $K\delta = 0$, and since $K \lhd G$ we lose no generality in assuming henceforth that

$$K = C_G(M) = 1.$$

Now let $1 \neq x \in \zeta_1(G)$. Then $C_M(x)$ is a $\mathbb{Q}G$-submodule of M; as $x \notin C_G(M) = 1$ and M is supposed simple, it follows that $C_M(x) = 0$. Similarly, $M(x - 1) = M$. Hence there exists $\mu \in M$ with

$$x\delta = \mu(x - 1).$$

Now for any $g \in G$, $gx = xg$ implies

$$g\delta \cdot x + x\delta = x\delta \cdot g + g\delta,$$

whence

$$g\delta \cdot (x - 1) = x\delta \cdot (g - 1) = \mu(x - 1)(g - 1) = \mu(g - 1)(x - 1).$$

Since $C_M(x) = 0$ it follows that $g\delta = \mu(g - 1)$. As g was an arbitrary element of G this shows that δ is the inner derivation on μ, and completes the proof.

There are various ways to prove Theorem 1. Perhaps the most perspicuous way is via an exact sequence in cohomology theory, and this is sketched in section D. Rather than introduce a lot of machinery, however, we give a simple group-theoretic proof based on Lemma 3. That lemma shows that Theorem 1 will follow once we have established

Theorem 1* Let G be a nilpotent group and M a $\mathbb{Q}G$-module of finite $\mathbb{Q}$-dimension. If $C_M(G) = 0$ then every extension of M by G splits.

Note that here, as with Theorem 2**, the group G does not have to be finitely generated: this is one advantage gained by passing from the $\mathbb{Z}G$-module A to the $\mathbb{Q}G$-module M.

Proof of Theorem 1* Assume that $M \neq 0$, and argue by induction on $\dim_{\mathbb{Q}} M$. Since $C_M(G) = 0$, $C_G(M) < G$ so we can choose an element $y \in G \setminus C_G(M)$ with y central in G modulo $C_G(M)$. Now let E be an extension of M by G and consider two alternatives.

Case 1 Suppose $C_M(y) \neq 0$. Put $B = C_M(y)$. Then B is a $\mathbb{Q}G$-submodule of M and $0 < B < M$ since $y \notin C_G(M)$. By Corollary 1, $C_{M/B}(G) = 0$, so by inductive hypothesis we may assume that the extension E/B splits over M/B. Thus E has a subgroup F with $MF = E$ and $M \cap F = B$. Then F is an extension of B by $F/B \cong E/M \cong G$; by the same inductive hypothesis this extension also splits, so F has a subgroup H with $BH = F$ and $B \cap H = 1$. Then

$$MH = M \cdot BH = MF = E$$
$$M \cap H = (M \cap F) \cap H = B \cap H = 1,$$

so E splits over M as required.

Case 2 Suppose $C_M(y) = 0$. Identifying E/M with G, we have $y = Mx$ say, with $x \in E$. Then $M \cap \langle x \rangle = 1$ since evidently $M \cap \langle x \rangle \leq C_M(x) = C_M(y)$. Now put

$$Z_i/M = \zeta_i(E/M) = \zeta_i(G)$$

for $i = 0, \ldots, c$ where $\zeta_c(G) = G$, and put $H_0 = \langle x \rangle$. Then

$$x \in H_0 \leq Z_0 \langle x \rangle$$
$$M \cap H_0 = 1$$
$$MH_0 = Z_0 \langle x \rangle.$$

Suppose inductively that for some $i \geq 0$ we have found a subgroup H_i of E such that

$$x \in H_i \leq Z_i \langle x \rangle$$
$$M \cap H_i = 1 \tag{5}$$
$$MH_i = Z_i \langle x \rangle.$$

Put

$$H_{i+1} = N_{Z_{i+1}\langle x \rangle}(H_i).$$

We shall see that for this H_{i+1}, (5) holds with $i + 1$ in place of i. Granting this for the moment, it follows by induction that there exists $\bar{H} = H_c \leq E$ with $M \cap \bar{H} = 1$ and $M\bar{H} = Z_c \langle x \rangle = E$; thus E splits over M as required.

Now write $H = H_i$, $K = H_{i+1}$. Evidently

$$x \in K \leq Z_{i+1} \langle x \rangle.$$

We show next that $M \cap K = 1$. Take $\mu \in M \cap K$. Then

$$\mu x \mu^{-1} \in \mu H \mu^{-1} = H$$

so, since $x \in H$ and $M \triangleleft E$,

$$x^{-1} \mu x \mu^{-1} \in H \cap M = 1.$$

Therefore $\mu \in C_M(x) = C_M(y) = 1$. Finally, we show that $MK = Z_{i+1}\langle x \rangle$. Note first that $Z_i \langle X \rangle \lhd Z_{i+1} \langle x \rangle$, from the definition of Z_i and Z_{i+1}. Now let $g \in Z_{i+1} \langle x \rangle$. Then

$$M \cdot H^g = (MH)^g = (Z_i \langle x \rangle)^g = Z_i \langle x \rangle = MH,$$

so H^g, like H, is a complement for M in $Z_i \langle x \rangle$. Now $Z_i \langle x \rangle / M$ is nilpotent and $C_M(Z_i \langle x \rangle) \leq C_M(x) = 1$. We can therefore apply Theorem 2** and Proposition 2 to infer that all complements for M in $Z_i \langle x \rangle$ are conjugate under the action of M. Thus in particular there exists $\mu \in M$ such that $H^g = H^\mu$. So $g\mu^{-1} \in N_{Z_{i+1}\langle x \rangle}(H) = K$, and $g \in KM = MK$. Hence $Z_{i+1} \langle x \rangle = MK$ as claimed.

Exercise 1 Let A be a finitely generated free abelian normal subgroup of a group E with E/A abelian, and suppose that A is non-central and rationally irreducible as E/A-module. Show directly that E almost splits over A.
(*Hint*: Choose $x \in E \setminus C_E(A)$ and put $H = C_E(x)$. Show that $H \cap A = 1$. Show that $|E:AH|$ is finite by showing that $AH \geq \ker \theta$, where $\theta: C_E(A/[A,x]) \to A/[A,x]$ is the map sending g to $[A,x] \cdot [x,g]$, and that $A/[A,x]$ is finite.)

Exercise 2 Prove, without recourse to the units theorem, that a soluble group is polycyclic if all its abelian subgroups are finitely generated.

(*Outline proof* (1) By induction on derived length, reduce to the case of a metabelian group, i.e. one of derived length ≤ 2.
(2) Let G be a hypothetical metabelian counterexample with $h(G')$ as small as possible, put $A = G'$ and $T = \tau(A)$. Then A/T is r.i. as G/A-module.
(3) If $A/T \not\leq \zeta_1(G/T)$, Exercise 1 provides $H \leq G$ with $|G:AH| < \infty$ and $A \cap H = T$. Show that H and hence G is polycyclic.
(4) If $A = T$, let N be a maximal abelian subgroup of $C_G(T) = C$ say. Then $N \lhd C$ and C/N embeds into $\mathrm{Hom}(N/T, T)$. Deduce that G is abelian-by-finite, hence polycyclic.
(5) If $1 \neq A/T \leq \zeta_1(G/T)$, let M/T be a maximal abelian subgroup of G/T containining A/T. Show that M is polycyclic. Also $M/T = C_{G/T}(M/T) \lhd G/T$, and G/M embeds into $\mathrm{Hom}(M/A, A/T)$. Deduce that G/M is finitely generated and hence that G is polycyclic.)

C. Nilpotent almost-supplements

A $\mathfrak{X}$-*group* is a torsion-free finitely generated nilpotent group. The main result of this section is

Theorem 3   Let G be a polycyclic group and N a normal $\mathfrak{T}$-subgroup of G such that G/N is nilpotent. Then G has a $\mathfrak{T}$-subgroup C such that $|G:NC|$ is finite.

Before proving this let us note an important corollary. Recall that an arbitrary polycyclic group G has a nilpotent-by-abelian normal subgroup G_0 of finite index (Theorem 4 of Chapter 2); and G_0 can be taken to be torsion-free, by Proposition 2 of Chapter 1. Taking G_0 for G and G_0' for N in Theorem 3 we deduce that G_0 has a nilpotent subgroup C with $|G_0:G_0'C|$ finite. Evidently $G_0' \le \text{Fitt}(G)$, so we get

Corollary 2   If G is a polycyclic group, there exists a nilpotent almost-supplement for Fitt(G) in G, that is, a nilpotent subgroup C such that $|G:\text{Fitt}(G)\,C| < \infty$.

Proof of Theorem 3   Let Z be the centre of N. Then N/Z is a $\mathfrak{T}$-group (Chapter 1, Corollary 5), so arguing by induction on the Hirsch number of G we may assume that G/Z has a $\mathfrak{T}$-subgroup H/Z such that $|G:NH|$ is finite. Now consider two alternatives:

Case 1   Suppose $Z \cap \zeta_1(H) \ne 1$. Put $K = Z \cap \zeta_1(H)$ and note that $K = C_Z(H)$. Now Z/K is torsion free (for if $z \in Z$ and $z^m \in K$, then for all $h \in H$ we have $[z,h]^m = [z^m,h] = 1$, whence $[z,h] = 1$ if $m \ne 0$); thus N/K is a $\mathfrak{T}$-group and so is its subgroup $(N \cap H)/K$. Hence by inductive hypothesis again, we may suppose that H/K has a $\mathfrak{T}$-subgroup C/K such that $|H:(N \cap H)C|$ is finite. Then C is a $\mathfrak{T}$-group, and

$$|G:NC| = |G:NH| \cdot |NH:NC| \le |G:NH| \cdot |H:(N \cap H)C| < \infty.$$

Case 2   Suppose $Z \cap \zeta_1(H) = 1$. Then Z is a free abelian normal subgroup of H, H/Z is finitely generated nilpotent, and $C_Z(H) = 1$. By Theorem 1, H has a subgroup C such that $Z \cap C = 1$ and $|H:ZC|$ is finite. Then C is a $\mathfrak{T}$-group, being isomorphic to a subgroup of H/Z; and

$$|G:NC| = |G:NH| \cdot |NH:NC| \le |G:NH| \cdot |H:ZC| < \infty.$$

This completes the proof.

In a finite soluble group, the Carter subgroups are all conjugate. It would be too much to expect the nilpotent almost-supplements for Fitt(G) in a polycyclic group G to be conjugate: for if C is one such, then so is every subgroup of finite index in C, and so is YC for every subgroup Y of $\zeta_1(G)$. However if we make suitable hypotheses to eliminate such obvious irregularities, we can at least ensure that the remaining nilpotent almost-supplements lie in only finitely many conjugacy classes. For the proof, we shall need a slight refinement of Proposition 14 of Chapter 1.

Definition   For a group N and positive integer i, the subgroup $\tau_i(N)$ is defined by

$$\tau_i(N)/\gamma_i(N) = \tau(N/\gamma_i(N)).$$

This makes sense because $N/\gamma_i(N)$ is a nilpotent group.

Lemma 4   If N is a $\mathfrak{X}$-group of class c, the series

$$N = \tau_1(N) > \tau_2(N) > \ldots > \tau_c(N) > \tau_{c+1}(N) = 1 \tag{6}$$

is a characteristic central series with torsion-free factors.

Proof   Evidently $N/\tau_{i+1}(N)$ is a $\mathfrak{X}$-group, so if $Z/\tau_{i+1}(N) = \zeta_1(N/\tau_{i+1}(N))$, then N/Z is torsion-free (Corollary 5 of Chapter 1). But $\gamma_i(N) \leq Z$ since $[\gamma_i(N), N] = \gamma_{i+1}(N) \leq \tau_{i+1}(N)$, so $\tau_i(N) \leq Z$. This shows that (6) is a central series, and the rest is clear.

Exercise 3   Let Γ be a group and A a Γ-module which is torsion-free as $\mathbb{Z}$-module. Show that if Γ acts nilpotently on a Γ-submodule B of A with A/B periodic, then Γ acts nilpotently on A.
(*Hint*: Put $A_0 = 0$ and $A_i/A_{i-1} = C_{A/A_{i-1}}(\Gamma)$ for $i = 1, 2, \ldots$. Show that A/A_i is torsion-free for each i.)

Proposition 3   Let N be a $\mathfrak{X}$-group and Γ a group acting by automorphisms on N. If Γ acts nilpotently on $N/\tau_2(N)$, then Γ acts nilpotently on N.

Proof   Write $T_j = \tau_j(N)$ for $j = 1, \ldots, c+1$, where c is the class of N. If $c = 1$ there is nothing to prove; so assuming that $c > 1$ and arguing by induction on c, we may suppose that Γ acts nilpotently on N/T_c. Now $[T_c, N] = 1$ by Lemma 4; also $[T_{c-1}, T_2] = 1$, because $N/C_N(T_{c-1})$ is isomorphic to a subgroup of the torsion-free abelian group $\operatorname{Hom}(T_{c-1}/T_c, T_c)$, by Corollary 4 of Chapter 1, whence $T_2 \leq C_N(T_{c-1})$. Commutation therefore induces a Γ-module homomorphism of $(T_{c-1}/T_c) \otimes (N/T_2)$ into T_c, with image equal to $[T_{c-1}, N]$; compare the proof of Proposition 12 of Chapter 1. Lemma 5 of Chapter 1 shows that Γ acts nilpotently on $(T_{c-1}/T_c) \otimes (N/T_2)$, consequently acts nilpotently on $[T_{c-1}, N]$. But $[T_{c-1}, N] \geq [\gamma_{c-1}(N), N] = \gamma_c(N)$ and $T_c/\gamma_c(N)$ is periodic; so by Exercise 3, Γ acts nilpotently on T_c. The result follows.

We are now ready for the next main result,

Theorem 4   Let G be a polycyclic group and N a normal $\mathfrak{X}$-subgroup of G. Then the maximal nilpotent supplements for N in G (if there are any) lie in finitely many conjugacy classes.

A *supplement* for N in G is, of course, a subgroup C of G such that $NC = G$. An argument like that leading up to Corollary 2, together with the fact that a polycyclic group has only finitely many subgroups of a given finite index (Chapter 1, Lemma 1), now gives

Corollary 3   Let G be a polycyclic group and d a positive integer. Then the maximal nilpotent subgroups C of G which satisfy $|G:\text{Fitt}(G)\cdot C| \leq d$ lie in finitely many conjugacy classes.

Proof of Theorem 4   Denote by $\mathscr{C}$ the set of all maximal nilpotent supplements for N in G; assume that $\mathscr{C}$ is non-empty, so in particular G/N is nilpotent. Then if G acts nilpotently on N, G itself is nilpotent and so $\mathscr{C} = \{G\}$ has just one member. Suppose G does not act nilpotently on N. Then by Proposition 3, G does not act nilpotently on $N/\tau_2(N)$. Let $M/\tau_2(N)$ be the maximal G-submodule of $N/\tau_2(N)$ on which G acts nilpotently. Then $C_{N/M}(G) = 1$. Also N/M is free abelian, by Exercise 3, and $h(M) < h(N)$. Now take $C \in \mathscr{C}$. Then $C_{N/M}(C) = 1$ since $G = NC$, and it follows that $MC \cap N = M$: for C acts nilpotently on $(MC \cap N)/M$ since C is nilpotent. Thus MC/M is a complement for N/M in G/M. By Theorem 2, the set of all such complements breaks up into finitely many conjugacy classes in G/M; thus there exist $C_1, \ldots, C_k \in \mathscr{C}$ such that

$$\mathscr{C} = \bigcup_{g \in G} (\mathscr{C}_1^g \cup \ldots \cup \mathscr{C}_k^g) \tag{7}$$

where for $i = 1, \ldots, k$,

$$\mathscr{C}_i = \{C \in \mathscr{C} \mid MC = MC_i\}.$$

Now each member of $\mathscr{C}_i$ is evidently a maximal nilpotent supplement for M in the group MC_i. Arguing by induction on the Hirsch number of N, we may therefore suppose that $\mathscr{C}_i$ is the union of finitely many conjugacy classes of subgroups in MC_i. Together with (7), this shows that $\mathscr{C}$ is the union of finitely many conjugacy classes in G, and completes the proof.

D. Notes

'Classical' extension theory is described in many books on group theory, for example in Huppert (1967), Chapter 1, section 14. This is the theory behind the construction of the group $\bar{E}$ in section A. A powerful and elegant framework for extension theory is provided by the *cohomology theory* of groups; Gruenberg (1970) gives a congenial introduction to this subject, and Chapter 5 deals with extension theory.

The main results of section B, Theorems 1 and 2, are special cases of theorems of D.J.S. Robinson. Using Lemma 1 and Propositions 1 and 2, one sees that Theorems 1* and 2** and both contained in Corollary 1 of Robinson (1976*b*). The rather complicated group-theoretic arguments in this paper can be replaced by smoother cohomological ones, as in Robinson (1975), (1976*a*). These papers use the rather formidable machinery of spectral sequences. For our purposes, namely Theorems 1* and 2**, it is possible to get by with a much simpler cohomological argument; it goes like this:

Theorem   Let G be a nilpotent group and M a $\mathbb{Q}G$-module with $\dim_{\mathbb{Q}} M$ finite. If $H^0(G, M) = 0$ then $H^1(G, M) = H^2(G, M) = 0$.

Proof   Recall that $H^0(G, M) = C_M(G)$. So we may assume that $C_G(M) \neq G$. Choose $x \in G \backslash C_G(M)$ with x central modulo $C_G(M)$. Put $B = C_M(x)$. The exact sequence of G-modules

$$0 \to B \to M \to M(x - 1) \to 0$$

gives the exact cohomology sequence

$$H^1(G, B) \to H^1(G, M) \to H^1(G, M(x - 1)) \to H^2(G, B)$$
$$\to H^2(G, M) \to H^2(G, M(x - 1)).$$

If $0 < B < M$, an inductive hypothesis with respect to $\dim_{\mathbb{Q}} M$ makes the 1st, 3rd, 4th and 6th terms in this sequence zero; so $H^1(G, M)$ and $H^2(G, M)$ are zero in this case. Assume then that $B = 0$. Thus $H^0(\langle x \rangle, M) = 0$, and a simple direct argument shows that $H^1(\langle x \rangle, M) = H^2(\langle x \rangle, M) = 0$. Now put $Z_i = \zeta_i(G)$, and suppose inductively that

$$H^q(Z_i \langle x \rangle, M) = 0, \quad q = 1, 2. \tag{8}$$

There are exact sequences ('inflation-restriction sequences')

$$0 \to H^1(\Gamma/N, H^0(N, M)) \to H^1(\Gamma, M) \to H^1(N, M) \tag{9}$$
$$0 \to H^2(\Gamma/N, H^0(N, M)) \to H^2(\Gamma, M) \to H^2(N, M) \tag{10}$$

for any group Γ, normal subgroup N of Γ, and Γ-module M, provided in the case of (10) that $H^1(N, M) = 0$; see Gruenberg, *op. cit.*, section 6.4, and Serre (1979), Chapter VII, section 6. Take $\Gamma = Z_{i+1} \langle x \rangle$ and $N = Z_i \langle x \rangle$. Since $x \in N$ we have $H^0(N, M) = 0$. From (8) and (9) we deduce that $H^1(\Gamma, M) = 0$, and from (8) and (10) that $H^2(\Gamma, M) = 0$. Thus (8) holds with $i + 1$ in place of i. By induction, (8) holds with $i = c$ where $Z_c = G$ and the result is proved.

The 'formation theory' for polycyclic groups has been developed by M.L.

Newell; see Newell (1973), (1975). In particular, our Corollary 2 follows from Theorem 1 of the latter paper. The conjugacy result, Corollary 3, is obviously related to Newell's Theorem 2, but I am not sure what the exact relationship is. A slightly different version of Theorem 3 appears in section 5 of Grunewald, Pickel & Segal (1980); we show there that if G is polycyclic and $G' \leq N < G$ with $N \in \mathfrak{T}$, then $\bar{N}$ has nilpotent supplements in $\bar{G}$, where $\bar{N}$ is a certain finite extension of N and $\bar{G}$ is the 'corresponding' extension of G. The connection with Theorem 3 is like that expressed in Lemma 3.

For a very neat proof of Theorem 3, see Wilson (1982), section 2.1.

A fancy version of Theorem 2 will appear in section D of Chapter 10; and in section E of that chapter a more precise, quantitative version of Theorem 1 is proved.

4
Arithmetical methods

This chapter illustrates how certain kinds of group-theoretical problem can be translated into questions of algebraic number theory. We have already seen an example of this in Chapter 2, where the Dirichlet Units Theorem played a key role; using deeper arithmetical results we shall be able to derive some more subtle properties of polycyclic groups. The basic idea is that in certain aspects of its internal structure, a polycyclic group resembles a subgroup of the semi-direct product $\mathfrak{o}^+]\mathfrak{o}^*$, where $\mathfrak{o}$ is the ring of integers in some algebraic number field: the precise sense in which this holds was stated as Proposition 1 of Chapter 2.

The first two sections, A and B, are concerned with a theorem of Baer, which says that *supersolubility* in a polycyclic group is determined by the finite quotients of the group; their main purpose is to give a simple introduction to the method. In section C we prove that polycyclic groups are *conjugacy separable*: i.e., conjugacy of elements in such a group is controlled by the finite quotients of the group. This important theorem is one of the highlights of the subject, and was only established within the last decade (by V.N. Remeslennikov and E. Formanek). As well as being aesthetically attractive because of its simplicity and depth, this result has powerful consequences for the behaviour of soluble subgroups in $GL_n(\mathbb{Z})$; these will play an important role in the later chapters.

The analogous result for conjugacy of subgroups in a polycyclic group is proved in section D. The results of both these sections depend on a theorem which relates the additive structure of a ring of algebraic integers $\mathfrak{o}$ with the multiplicative structure of its group of units $\mathfrak{o}^*$. As this theorem is so important for our subject, and seems to have escaped the usual textbooks, I include a proof of it in section E (modulo standard results in algebraic number theory).

The questions discussed in this chapter are all, ultimately, ones about the structure of $\mathfrak{o}^+$ as an $\mathfrak{o}^*$-module, and their answers give detailed information on what is happening inside a polycyclic group. Studying polycyclic groups 'from the outside' – trying to classify them up to isomorphism, or looking at their automorphism groups – one is led to arithmetical questions of a different kind: in particular, questions about 'arithmetic subgroups' of algebraic groups. But this aspect of the subject must be deferred to the later chapters.

A. Supersolubility

Definition   A group G is *supersoluble* if there is a finite series

$$1 = G_0 \leq G_1 \leq \ldots \leq G_k = G \tag{1}$$

with $G_i \triangleleft G$ and G_i/G_{i-1} cyclic for $i = 1, \ldots, k$.

This is a stronger condition than being polycyclic, since here each G_i must be normal in G.

Exercise 1   Show that

finitely generated nilpotent $\Rightarrow$ supersoluble $\Rightarrow$ polycyclic;

construct counterexamples to show that neither implication can be reversed.

Exercise 2   Show that a group G is supersoluble if and only if G has *max* and every non-identity quotient group of G has a non-identity cyclic normal subgroup. (In fact 'G has *max*' can be replaced by 'G is finitely generated' here, but this requires a little more argument.)

Exercise 3   If G is supersoluble then $1 \neq N \triangleleft G$ implies $1 \neq \langle x \rangle \triangleleft G$ for some $x \in N$.

Lemma 1   Every supersoluble group is nilpotent-by-(finite and abelian).

Proof   Suppose the group G has a series (1) of normal subgroups. Put $K_i = C_G(G_i/G_{i-1})$ for $i = 1, \ldots, k$. Then G/K_i is isomorphic to a subgroup of $\mathrm{Aut}(G_i/G_{i-1})$, and this group is finite and abelian since G_i/G_{i-1} is cyclic. It follows that if $K = \bigcap K_i$ then G/K is finite and abelian. Moreover, K is nilpotent; for it has the central series $(K \cap G_i)_{i=0,\ldots k}$.

We shall prove the following theorem of Baer:

Theorem 1   Let G be a polycyclic group. Then G is supersoluble if and only if every finite quotient of G is supersoluble.

This is analogous to Hirsch's result, Theorem 2 of Chapter 1: but in spite of the superficial resemblance, the present theorem takes us deeply into number theory. To get an idea of where exactly it will take us, let us see what it means for the 'prototype' polycyclic group

$$G = A]U,$$

where $U = \mathfrak{o}^*$ and A is the U-module $\mathfrak{o}^+$, $\mathfrak{o}$ being the ring of integers in some algebraic number field k. We assume for convenience that $\mathfrak{o}^*$ generates $\mathfrak{o}$ as a ring; this ensures that the $\mathfrak{o}^*$-submodules of $\mathfrak{o}^+$ are then exactly the *ideals* of $\mathfrak{o}$.

Suppose G is supersoluble. By Exercise 3, A then contains a non-identity cyclic normal subgroup of G. Hence $\mathfrak{o}$ has a non-zero ideal I which is cyclic as additive group, say $I = \mu\mathbb{Z} \lhd \mathfrak{o}$. But $\mathfrak{o}$ is an integral domain, so

$$\mathfrak{o} \cong \mu\mathfrak{o} = \mu\mathbb{Z} \cong \mathbb{Z}.$$

It follows that $\mathfrak{o} = \mathbb{Z}$, and hence that $k = \mathbb{Q}$. Conversely of course, if $k = \mathbb{Q}$ then G is supersoluble; for then $\mathfrak{o} = \mathbb{Z}$ and $\mathfrak{o}^* = \{1, -1\} \cong C_2$.

Now what does it mean for every finite quotient of G to be supersoluble? Suppose this is so, and let $\mathfrak{p}$ be a proper non-zero prime ideal of $\mathfrak{o}$, so $\mathfrak{p}$ is a maximal ideal since $\mathfrak{o}$ is a Dedekind domain. Write P for $\mathfrak{p}$ considered as a normal subgroup of G. Then A/P is finite, so since G/P is residually finite, G has a normal subgroup N of finite index with $N \cap A = P$ (as an exercise, show that $N = P \cdot X$ will do, where $X = \mathfrak{o}^* \cap (1 + \mathfrak{p})$). By hypothesis, G/N is supersoluble, so by Exercise 3, AN/N contains a non-identity cyclic normal subgroup of G/N. But $AN/N \cong A/(A \cap N) \cong \mathfrak{o}/\mathfrak{p}$ as $\mathfrak{o}^*$-module, so AN/N is simple as a G-module and must therefore be cyclic. Thus $\mathfrak{o}/\mathfrak{p}$ is cyclic as an additive group. As $\mathfrak{o}/\mathfrak{p}$ is a finite field, it follows that $\mathfrak{o}/\mathfrak{p} \cong GF(p)$, where p is the rational prime contained in $\mathfrak{p}$. In other words, the prime $\mathfrak{p}$ has *degree* 1. We conclude:

> *if every finite quotient of G is supersoluble, then*
> *every prime ideal of $\mathfrak{o}$ has degree* 1. (2)

Exercise 4 Prove the converse of this statement.

Thus what Baer's theorem amounts to, in the very special case under discussion, is that (2) *holds if and only* $k = \mathbb{Q}$. This is a classical, but relatively deep, theorem of algebraic number theory. Proofs can be found in many books on this subject (see section F for suitable references). We state it in a slightly stronger form, *viz.*

Theorem 2 Let k be an algebraic number field with ring of integers $\mathfrak{o}$. If all but finitely many prime ideals of $\mathfrak{o}$ have degree 1, then $k = \mathbb{Q}$.

The relevant fact about algebraic numbers having been identified, the actual proof of Theorem 1 is a fairly routine matter. Let us take the opportunity to demonstrate Noetherian induction (see Chapter 1, Section C). Thus we consider an infinite polycyclic group G such that

$$1 \neq N \lhd G \Rightarrow G/N \text{ is supersoluble,}$$

and have to show that G is supersoluble (the reader should fill in the remaining steps in the argument leading from this to Theorem 1).

Step 1   If G' is finite then G is supersoluble. For there exists $N \lhd_f G$ with $N \cap G' = 1$, and then G is isomorphic to a subgroup of $G/N \times G/G'$.

Assume henceforth that G' is infinite.

Step 2 G' is nilpotent. By Theorem 2 (and Lemma 2) of Chapter 1, it will suffice to show that G'/K is nilpotent for every $K \lhd_f G'$ with $K \lhd G$. Since G' is infinite, such a K is necessarily infinite; so by hypothesis, G/K is supersoluble, and Lemma 1 then shows that G'/K is nilpotent.

Step 3 G has an abelian normal subgroup $A \leq \zeta_1(G')$ with A rationally irreducible as G/G'-module. For by Step 2, $\zeta_1(G')$ is an infinite finitely generated abelian group (see Chapter 1, Corollary 6), so it has infinite characteristic free abelian subgroups. These are normal in G, so we may take A to be a free abelian normal subgroup of G, contained in $\zeta_1(G')$, of minimal non-zero rank.

Now apply Proposition 1 of Chapter 2. This shows that there is an algebraic number field k, with ring of integers $\mathfrak{o}$, and homomorphisms

$$^-:A \to \mathfrak{o}^+, \quad \sim :G \to \mathfrak{o}^*$$

with $\ker(^-) = 1$ and $\ker(\sim) = C_G(A)$, such that

$$(a^g)^- = \bar{a} \cdot \tilde{g} \quad \forall a \in A, \forall g \in G$$

(note that $G' \leq C_G(A)$, so we can apply the proposition to the abelian group $G/C_G(A)$ acting on A). Recall also that

$$|\mathfrak{o}^+ : \bar{A}| < \infty, \quad |\mathfrak{o}^+ : \mathbb{Z}[\tilde{G}]| < \infty$$

where $\mathbb{Z}[\tilde{G}]$ denotes the subring of $\mathfrak{o}$ generated by $\tilde{G}$. Hence there exists $t \neq 0$ in $\mathbb{Z}$ such that

$$t\mathfrak{o} \subseteq \bar{A}\mathfrak{o}$$

$$t\mathfrak{o} \subseteq \mathbb{Z}[\tilde{G}] \subseteq \mathfrak{o}.$$

Now only finitely many prime ideals of $\mathfrak{o}$ contain the integer t; for the others we can prove

Step 4   Every non-zero prime ideal of $\mathfrak{o}$ not containing t has degree 1.

Proof   Let $\mathfrak{p} \neq 0$ be a prime ideal with $t \notin \mathfrak{p}$. Then $\mathfrak{p} + t\mathfrak{o} = \mathfrak{o}$, so $\mathfrak{p} + \bar{A} = \mathfrak{o}$ and

$$\bar{A}/(\mathfrak{p} \cap \bar{A}) \cong \mathfrak{o}/\mathfrak{p}$$

(a $\mathbb{Z}[\tilde{G}]$-module isomorphism). Now since $\mathfrak{o} = \mathfrak{p} + t\mathfrak{o} = \mathfrak{p} + \mathbb{Z}[\tilde{G}]$, the $\mathbb{Z}[\tilde{G}]$-submodules of $\mathfrak{o}/\mathfrak{p}$ are just the $\mathfrak{o}$-submodules; but $\mathfrak{o}/\mathfrak{p}$ is simple as $\mathfrak{o}$-module, so $\bar{A}/(\mathfrak{p} \cap \bar{A})$ is simple as a $\mathbb{Z}[\tilde{G}]$-module. Thus taking B to be the inverse image in A of $\mathfrak{p} \cap \bar{A}$, we have (i) B is a G-submodule of A; (ii) A/B is a simple G-module; and (iii) $A/B \cong \mathfrak{o}/\mathfrak{p}$. Thus B is an infinite normal subgroup of G, so G/B is supersoluble. From (ii) and Exercise 3, A/B is a cyclic group. Hence from (iii), $\mathfrak{o}/\mathfrak{p}$ is cyclic as additive group and so $\mathfrak{p}$ has degree 1.

Conclusion   By Theorem 2, $\mathfrak{o} = \mathbb{Z}$. Therefore A is cyclic. But G/A is supersoluble by hypothesis, consequently G is supersoluble.

B. A counterexample

Let us now investigate the possibility of generalizing Theorem 1. As a temporary terminology, let us say that a polycyclic group G has *breadth* $\leq n$ if G has a series of normal subgroups

$$1 = G_0 \leq G_1 \leq \ldots \leq G_k = G$$

such that each factor G_i/G_{i-1} is an abelian group which can be generated by n elements. Thus a supersoluble group the same as a polycyclic group of breadth ≤ 1. The obvious generalization of Theorem 1 would be the following.

Conjecture   A polycyclic group has breadth $\leq n$ if and only if each of its finite quotients has breadth $\leq n$.

As before, let us try this out on the group

$$G = A]U$$

where $A = \mathfrak{o}^+$, $U = \mathfrak{o}^*$ and $\mathbb{Z}[\mathfrak{o}^*] = \mathfrak{o}$, the ring of integers in algebraic number field k. We say 'prime of $\mathfrak{o}$' for 'proper non-zero prime ideal of $\mathfrak{o}$'; a prime $\mathfrak{p}$ of $\mathfrak{o}$ is said to have *degree f* if

$$\mathfrak{o}/\mathfrak{p} \cong GF(p^f),$$

where p is the rational prime belonging to $\mathfrak{p}$.

Exercise 5 G has breadth $\leq n$ if and only if $\mathfrak{o}^+$ has rank at most n as a free abelian group.

Exercise 6   Every finite quotient of G has breadth $\leq n$ if and only if every prime of $\mathfrak{o}$ has degree at most n.

Both exercises are straightforward adaptations of the case $n = 1$, discussed in section A.

Since the additive rank of $\mathfrak{o}^+$ is the same as the degree of the field extension $[k:\mathbb{Q}]$, we see that the conjecture in our special case is equivalent to the statement: *if every prime of $\mathfrak{o}$ has degree $\leq n$, then $[k:\mathbb{Q}] \leq n$.* (The 'only if' parts are of course always true, and need not concern us.) Knowing a certain amount of algebraic number theory, one can easily find counterexamples to this.

It is easiest to work with an algebraic number field k which is Galois over $\mathbb{Q}$. If p is a rational prime, then

$$p\mathfrak{o} = \mathfrak{p}_1^e \ldots \mathfrak{p}_r^e$$

with $\mathfrak{p}_1, \ldots, \mathfrak{p}_r$ distinct primes of $\mathfrak{o}$, all of the same degree f say, and $efr = [k:\mathbb{Q}] = d$, say.

Exercise 7   Show that if d is prime, then there exist primes of degree d in $\mathfrak{o}$. (*Hint*: recall Theorem 2.)

So for a counterexample we must avoid fields of prime degree. We want to find a field k, of degree d over $\mathbb{Q}$, such that for every rational prime p the corresponding f is strictly less than d. Now if $f = d$ we have $e = r = 1$ and $p\mathfrak{o} = \mathfrak{p}_1$ is a prime of $\mathfrak{o}$: one says that p is *inertial* in k. There is an obvious natural homomorphism

$$\pi:\mathrm{Gal}(k/\mathbb{Q}) \to \mathrm{Gal}(\mathfrak{o}/\mathfrak{p}_1/\mathbb{Z}/p\mathbb{Z}),$$

and it is known that π is *surjective*. Since the field degrees $[k:\mathbb{Q}]$ and $[\mathfrak{o}/\mathfrak{p}_1:\mathbb{Z}/p\mathbb{Z}]$ are equal, the Galois groups have the same order and so π is an *isomorphism*. As the Galois group of a finite field is cyclic, we may improve Exercise 7 to

Lemma 2   If k is a Galois extension of $\mathbb{Q}$ such that some prime of $\mathbb{Z}$ is inertial in k, then $\mathrm{Gal}(k/\mathbb{Q})$ is cyclic.

Thus every non-cyclic Galois extension of $\mathbb{Q}$ provides a counterexample of the required kind. For this to make a counterexample to the original conjecture, we also need to ensure that $\mathbb{Z}[\mathfrak{o}^*] = \mathfrak{o}$; but this is often the case.

Exercise 8   Put $k = \mathbb{Q}(\sqrt{2}, \sqrt{3})$. Then $[k:\mathbb{Q}] = 4$. Show that every prime of $\mathfrak{o}$ has degree ≤ 2, and that $\mathbb{Z}[\mathfrak{o}^*] = \mathfrak{o}$.
(*Hint for the last part*: $1 + \sqrt{2}$ and $2 + \sqrt{3}$ belong to $\mathfrak{o}^*$.)

Exercise 9 (for readers who are familiar with the *Frobenius automorphism* belonging to a prime of o) Construct polycyclic groups of arbitrarily large breadth, whose finite quotients all have breadth ≤ 2. (*Hint*: try $k = \mathbb{Q}(\sqrt{2}, \sqrt{3}, \sqrt{5}, \ldots)$.)

All the above-quoted facts about number fields can be found in most books on algebraic number theory: see section F for references.

C. Conjugacy

A moment's reflection shows that a group G is residually finite if and only if it has the following property: two elements of G are equal if they have equal images in every finite quotient of G. The reader who has come across the subject of decision problems in group theory will recognise that this observation forms the basis for an algorithmic solution of the *word problem* in a finitely presented residually finite group. If the word 'equal' is replaced, at each appearance, by the word 'conjugate', the group G is said to be 'residually finite w.r.t. conjugacy', or *conjugacy separable*. Thus G is conjugacy separable if and only if for any two elements a and b of G,

$$aN \text{ conjugate to } bN \text{ in } G/N \qquad \forall N \lhd_f G$$

$$\Rightarrow a \text{ conjugate to } b \text{ in } G.$$

Again, a finitely presented conjugacy separable group has an algorithmically soluble *conjugacy problem*. We do not pursue this aspect of the matter here; instead, we shall discuss at some length the proof of the following theorem, and some of its ramifications:

Theorem 3 Every polycyclic-by-finite group is conjugacy separable.

This theorem resisted attempts at a proof for some time. The reason was that, concealed within, it contains a gem of algebraic number theory: the latter had to be recognised before the former could be proved. To dig out this gem, let us proceed in the usual manner and see what Theorem 3 has to say for a polycyclic group

$$\Gamma = A]U,$$

where A is the additive group of a ring of algebraic integers o, and U is a subgroup of o*. We take two elements α and β of A and consider what it means for them (i) to be conjugate in Γ, and (ii) to have conjugate images in every finite quotient of Γ.

Assume for convenience that $\alpha \neq 0$.

(i) Evidently, α and β are conjugate in Γ if and only if these exists $x \in U$

with $\alpha x = \beta$ (using additive 'module' notation here, so αx represents the product in the ring $\mathfrak{o}$).

(ii) Suppose α and β have conjugate images in all finite quotients of Γ. For $0 \neq m \in \mathbb{Z}$, put

$$U_m = U \cap (1 + m\mathfrak{o}),$$
$$N_m = mA \cdot U_m \subseteq \Gamma.$$

Then U_m is a subgroup of finite index in U, and it acts trivially on A/mA; so N_m is a normal subgroup of finite index in Γ. Therefore αN_m and βN_m are conjugate in Γ/N_m, and this simply means that

$$\alpha x_m - \beta \in mA \qquad\qquad (3)$$

for some $x_m \in U$. Thus for every non-zero integer m there exists $x_m \in U$ such that (3) holds. (If this is so, then conversely it is easy to see that α and β do have conjugate images in all finite quotients of Γ.)

Thus Theorem 3 for the group Γ implies the following:

> let $\alpha, \beta \in \mathfrak{o}$, with $\alpha \neq 0$. If for each $m \in \mathbb{N}$ there exists $x_m \in U$
> with $\alpha x_m - \beta \in m\mathfrak{o}$, then there exists $x \in U$ with $\alpha x = \beta$. $\qquad$ (4)

As a first step towards establishing (4), observe that there certainly exists $v \in \mathfrak{o}^*$ with $\alpha v = \beta$ (this is why it would not have got us very far to consider the case where $U = \mathfrak{o}^*$, as we did in sections A and B). The argument is as follows. Since $\alpha \neq 0$, the ideal $\alpha \mathfrak{o}$ has finite index in $\mathfrak{o}$, so $r\mathfrak{o} \subseteq \alpha\mathfrak{o}$ for some $r \in \mathbb{N}$. From (3),

$$\alpha x_r - \beta \in r\mathfrak{o} \subseteq \alpha\mathfrak{o},$$

whence $\beta \in \alpha\mathfrak{o}$. Similarly, $\alpha \in \beta\mathfrak{o}$ (**Exercise**: show that $\beta \neq 0$); and it follows that $\alpha^{-1}\beta = v$ say belongs to $\mathfrak{o}^*$.

Now eliminating β from (3) we get

$$\alpha(x_m - v) \in m\mathfrak{o} \quad \forall m \in \mathbb{N}.$$

Taking r as above, this gives

$$\alpha(x_{mr} - v) \in mr\mathfrak{o} \subseteq \alpha \cdot m\mathfrak{o},$$

and since $\mathfrak{o}$ is an integral domain this implies that

$$x_{mr} - v \in m\mathfrak{o} \quad \forall m \in \mathbb{N}.$$

Thus writing $\mathfrak{o}^*(m) = (1 + m\mathfrak{o}) \cap \mathfrak{o}^*$, we have

$$v \in \bigcap_{m \in \mathbb{N}} (U + m\mathfrak{o}) \cap \mathfrak{o}^*$$
$$= \bigcap_{m \in \mathbb{N}} U \cdot \mathfrak{o}^*(m). \qquad\qquad (5)$$

The essential content of (4) is now evident: if $v \in \mathfrak{o}^*$ belongs to the group (5), it must belong to U. In other words, *an arbitrary subgroup $U \leq \mathfrak{o}^*$ is equal to* $\bigcap_{m \in \mathbb{N}} U \cdot \mathfrak{o}^*(m)$. Now each $\mathfrak{o}^*(m)$ is a subgroup of finite index in $\mathfrak{o}^*$, so (5) is the intersection of a certain family of subgroups of finite index in $\mathfrak{o}^*$ containing U. Since $\mathfrak{o}^*/U$ is a finitely generated abelian group, it is residually finite. So to show that the given intersection is equal to U (and not bigger), it would suffice to establish that *every subgroup of finite index in $\mathfrak{o}^*$ contains a suitable $\mathfrak{o}^*(m)$*. This is indeed the case, and we have isolated the key fact which we were seeking:

Theorem 4   Let $\mathfrak{o}$ be the ring of integers in an algebraic number field. Then every subgroup of finite index in $\mathfrak{o}^*$ contains a subgroup of the form $(1 + m\mathfrak{o}) \cap \mathfrak{o}^*$ for some $m \in \mathbb{N}$.

The proof of this, an instructive exercise in algebraic number theory, is deferred until section E. But to make sure that we have really hit on the 'right' number-theoretic fact, here is

Exercise 10   Deduce Theorem 4 from Theorem 3.
(*Hint*: consider $\Gamma = A]U$ as above, where U is a hypothetical subgroup of finite index in $\mathfrak{o}^*$ but not containing $\mathfrak{o}^*(m)$ for any $m \in \mathbb{N}$.)

The deduction of Theorem 3 from Theorem 4 is not quite such a simple matter. Before giving it, let us mention an essentially equivalent result, and discuss some of its consequences.

Theorem 5   Let n be a positive integer. For $0 \neq m \in \mathbb{Z}$, write

$$K_m = \{g \in GL_n(\mathbb{Z}) | g \equiv 1_n \bmod m\}.$$

Let G be a soluble-by-finite subgroup of $GL_n(\mathbb{Z})$. Then

(i) $G = \bigcap_{m \in \mathbb{N}} G \cdot K_m;$

(ii) every subgroup of finite index in G contains $G \cap K_m$ for some $m \neq 0$.

This looks rather like Theorem 4; and the similarity is not misleading. For if $\mathfrak{o}$ in Theorem 4 is given a $\mathbb{Z}$-basis, where $\mathfrak{o} \cong \mathbb{Z}^n$, and the action of $\mathfrak{o}^*$ on $\mathfrak{o}$ is represented by matrices according to this basis, then $\mathfrak{o}^*$ becomes identified with an abelian subgroup G of $GL_n(\mathbb{Z})$, and $\mathfrak{o}^*(m)$ is exactly $G \cap K_m$. So Theorem 4 is a special case of Theorem 5(ii). As we are going to deduce Theorem 5 directly from Theorem 3, this gives an alternative approach to Exercise 10.

There is a suggestive topological formulation for Theorem 5.

Definition   The *profinite topology* on a group H is the topology in which a base for the open sets is the set of all cosets of normal subgroups of finite index in H.

Exercise 11   (i) If $K, L \vartriangleleft_f H$ and $xK \cap yL \neq 0$ then $xK \cap yL$ is a coset of $K \cap L$; so the given family of cosets does form a base for the open sets of a topology. (ii) The group operations are continuous. (iii) The closure of a subset X of H is equal to

$$\bigcap \{NX \mid N \vartriangleleft_f H\}.$$

(iv) Every subgroup of finite index is both open and closed. (v) The topology is Hausdorff if and only if H is residually finite. (vi) The group H is conjugacy separable if and only if every conjugacy class in H is closed. (vii) If H is polycyclic-by-finite, then every subgroup of H is closed.

Definition   The *congruence topology* in $GL_n(\mathbb{Z})$ is the subspace topology induced from the profinite topology on the additive group $M_n(\mathbb{Z})$.

Exercise 12   (i) The group operations are continuous w.r.t. the congruence topology. (ii) A base for the neighbourhoods of 1 in the congruence topology on $GL_n(\mathbb{Z})$ is given by the set of all subgroups K_m, $0 \neq m \in \mathbb{Z}$ (K_m as in Theorem 5). (iii) On any subgroup of $GL_n(\mathbb{Z})$, the profinite topology is finer than the congruence topology.

It is now clear that Theorem 5 can be formulated as follows:

(i) Every soluble-by-finite subgroup of $GL_n(\mathbb{Z})$ is closed in the congruence topology on $GL_n(\mathbb{Z})$;

(ii) If $G \leq GL_n(\mathbb{Z})$ is soluble-by-finite, then the congruence topology induces the profinite topology on G.

The convenience of this topological approach is amply illustrated by the

Proof of Theorem 5   Let M denote the additive group of $M_n(\mathbb{Z})$, and make the group $G \leq GL_n(\mathbb{Z})$ act on M by matrix multiplication from the right, giving an injective homomorphism $\rho : G \to \mathrm{Aut}(M)$. By Corollary 1 of Chapter 2, G is polycyclic-by-finite; so the semi-direct product $H = M]G\rho$ is also polycyclic-by-finite. By Theorem 3 and Exercise 11(vi), each conjugacy class in H is closed in the profinite topology of H. It is easy to see that this topology induces the profinite topology on M (check this!); the conjugacy class in H of the identity matrix in M is exactly the subset $G \subseteq M$, so G is closed w.r.t. the profinite topology of $M = M_n(\mathbb{Z})$. By definition, then, G is closed w.r.t. the congruence topology on $GL_n(\mathbb{Z})$, and part (i) is established. For part (ii), it will suffice, by Exercise 12, to show that if K is a subgroup of finite index in G then K is relatively open w.r.t. the congruence topology on G. But applying part (i) to K we see that K is closed w.r.t. the congruence topology on $GL_n(\mathbb{Z})$, hence relatively closed w.r.t. that topology in G. Then each right coset of K in G is also relatively closed, so K, being the complement of the union of finitely many such cosets, is itself relatively open.

Exercise 13 Prove the following theorem (Lennox and Wilson): if G is polycyclic-by-finite and A and B are subgroups of G, then the set $AB = \{ab \mid a \in A, b \in B\}$ is closed w.r.t. the profinite topology on G.
(*Hint*: Assume $G \leq GL_n(\mathbb{Z})$ (see Chapter 5). Take M as in the proof of Theorem 5 and let $\rho: G \to \operatorname{Aut} M$, $\lambda: G \to \operatorname{Aut} M$ be given by right and left matrix multiplication respectively. Now consider the conjugacy class in the group $M] (A\lambda \times B\rho)$ of the identity matrix in M.)

Exercise 14 Deduce that for G, A and B as above, the set AB is a subgroup of G if and only if $A\theta \cdot B\theta$ is a subgroup of $G\theta$ for every homomorphism θ of G onto a finite group.

We now press ahead with proving Theorem 3. Having isolated the relevant piece of algebraic number theory, we are left with the problem of constructing a group-theoretic argument which will carry us from the special case – a subgroup of $o^+]o^*$ – to the case of a general polycyclic group. In the special case, 'the 'group-theoretic' structure was essentially trivial, and number theory was paramount; it is often helpful, when faced with the present kind of problem, to go to the opposite extreme and see what happens for a polycyclic group whose 'arithmetical' structure is trivial. Now the algebraic number fields come into the structure of a polycyclic group G in connection with free abelian factors of G which are rationally irreducible as G-modules. To make sure that all such factors are trivial as G-modules, one has to assume that G is nilpotent. So let us, slightly more generally, establish

Proposition 1 Every (finitely generated nilpotent)-by-finite group is conjugacy separable.

Proof Let G_1 be a nilpotent normal subgroup of finite index in a group G, and put $Z = \zeta_1(G_1)$. Consider two elements a and b of G which have conjugate images in every finite quotient of G. We must show that a and b are conjugate in G; arguing by Noetherian induction, we assume that

G is infinite, and that a and b have conjugate images in every proper quotient of G. Then Z is infinite, so b is conjugate to a modulo Z. Replacing b by a suitable conjugate we may then assume that

$$x = b^{-1}a \in Z.$$

Now for every $m \in \mathbb{N}$ there exists $g_m \in G$ such that

$$b^{g_m} \equiv a \bmod Z^m;$$

since $a \equiv b \bmod Z$, each g_m must belong to the group C defined by

$$C/Z = C_{G/Z}(bZ).$$

Thus since

$$x = b^{-1}a \equiv b^{-1}b^{g_m} \bmod Z^m \ \forall m \in \mathbb{N},$$

we have

$$x \in \bigcap_{m \in \mathbb{N}} [b, C] \cdot Z^m, \tag{6}$$

where $[b, C] = \{[b, c] \mid c \in C\}$. Put $C_1 = C \cap G_1$. Since $Z = \zeta_1(G_1)$, the map

$$\Delta : C_1 \to Z$$
$$c \mapsto [b, c]$$

is a homomorphism. The set $[b, C_1] = C_1 \Delta$ is therefore a subgroup of Z. The group $Z/C_1 \Delta$ is then residually finite, so $C_1 \Delta$ is closed in the profinite topology on Z (see Exercise 11). If we could show that the larger set $[b, C]$ is also closed in Z, it would follow from (6) that $x \in [b, C]$. But then $x = [b, c]$ for some $c \in C$ and

$$b^{-1}a = x = b^{-1}b^c$$

then gives $a = b^c$, concluding the proof.

It remains to see that $[b, C]$ is closed w.r.t. the profinite topology on Z. Let $\{t_1, \ldots, t_n\}$ be a transversal to the right cosets of C_1 in C. Then

$$[b, C] = \bigcup_{i=1}^{n} [b, C_1 t_i]$$

$$= \bigcup_{i=1}^{n} [b, t_i] \cdot [b, C_1]^{t_i}.$$

$$= \bigcup_{i=1}^{n} y_i \cdot (C_i \Delta)^{t_i} \tag{7}$$

say. The maps $z \mapsto y_i z$ and $z \mapsto z^{t_i} (z \in Z)$ are easily seen to be homeomorphisms of Z w.r.t. the profinite topology; since $C_1 \Delta$ is closed, each term in (7) is also a closed set. It follows that $[b, C]$ is closed.

We would like to mimic this argument with a polycyclic-by-finite group in the role of G. Instead of Z, central in some subgroup of finite index in G, we shall have to make do with an abelian normal subgroup of G, call it A. If $b\in G$ and $C/A = C_{G/A}(bA)$, the map $\Delta:C \to A$ given by

$$c\Delta = [b,c], \quad c\in C,$$

while perhaps not a homomorphism, will certainly be a *derivation*:

$$(c_1c_2)\Delta = [b,c_1c_2] = [b,c_2][b,c_1]^{c_2} = (c_1\Delta)^{c_2}\cdot(c_2\Delta)$$

(as A is abelian). So to make the argument work, what we need is

Theorem 6 Let C be a polycyclic-by-finite group and A a C-module, finitely generated as a $\mathbb{Z}$-module. If Δ is any derivation of C into A, the image

$$C\Delta = \{c\Delta | c\in C\}$$

is closed w.r.t. the profinite topology on A.

Recall that a derivation $\Delta:C \to A$ is *inner* if there exists $a\in A$ such that $c\Delta = a(c-1)$ for all $c\in C$.

Exercise 15 Assuming Theorem 3 deduce Theorem 6 for the case of an inner derivation Δ. (*Hint*: consider $G = A]C$.)

In fact a simple trick will reduce Theorem 6 in general to the special case of an inner derivation; in any case the last exercise and the one below show that Theorem 6 is a not unreasonable reformulation of Theorem 3:

Exercise 16 Deduce Theorem 3 from Theorem 6.
(*Hint*: copy the proof of Proposition 1.)

It remains, then, to prove Theorem 6. Knowing as we do that Theorem 4 must be the key to this, the construction of the proof is now almost a routine matter. To apply Theorem 4 at all, we shall evidently have to use Proposition 1 of Chapter 2 at some stage, and this means finding a rationally irreducible module. So the idea is to break the module A up into pieces, and accordingly we must deal with three aspects of the problem: (i) the case where the module is trivial; (ii) the case where the module is r.i. and non-trivial; and (iii) the business of fitting the steps together.

Rather than formalise this division into three aspects, we shall argue in the usual way by induction on the Hirsch number of A: the point of mentioning the division is to clarify the underlying structure of the argument. It will be convenient to isolate some of the reduction steps, in the form of

Lemma 3   Let G be a group and A a G-module, finitely generated as a $\mathbb{Z}$-module. Let $\Delta : G \to A$ be a derivation.

(i) If $H \leq_f G$ and $H\Delta$ is closed in A (w.r.t. the profinite topology), then $G\Delta$ is closed in A.

(ii) If B is a G-submodule of A then

$$B\Delta^{-1} = \{g \in G | g\Delta \in B\}$$

is a subgroup of G; if $|A:B|$ is finite then $|G:B\Delta^{-1}|$ is finite.

(iii) If B is a $\mathbb{Z}$-submodule of A with $G\Delta \subseteq B$, then $G\Delta$ is closed in B if and only if $G\Delta$ is closed in A.

(iv) If G is abelian, A is $\mathbb{Z}$-free, and $C_A(G) = 0$, then $l\Delta$ is an inner derivation for some non-zero integer l.

(v) If A is $\mathbb{Z}$-free and $l \cdot G\Delta$ is closed in A for some non-zero integer l, then $G\Delta$ is closed in A.

Proof (i) $G\Delta$ is the union of finitely many sets of the form

$$H\Delta \cdot t + t\Delta$$

with $t \in G$; each of these sets is closed. Compare the last paragraph in the proof of Proposition 1.

(ii) The first claim is a simple exercise; for the second, note that if $C = C_G(A/B)$, then $|G:C|$ is finite, and $C \cap B\Delta^{-1}$ is the kernel of the homomorphism $c \mapsto c\Delta + B$ of C into the finite group A/B.

(iii) holds because B is closed in A and the profinite topology of A induces that of B, since A/Y is residually finite for every $Y \leq_f B$.

(iv) follows from Theorem 2* of Chapter 3.

(v) $l \cdot G\Delta$ is closed in lA, and multiplication by l^{-1} is a homeomorphism of lA onto A.

Proof of Theorem 6   We have a polycyclic-by-finite group C, a finitely $\mathbb{Z}$-generated C-module A, and a derivation $\Delta : C \to A$; we must show that $C\Delta$ is closed in A (w.r.t. the profinite topology). If A is finite there is nothing to prove (why?), so assume A infinite and argue by induction on $h(A)$. Now A certainly contains non-zero $\mathbb{Z}$-free C-submodules, so among all pairs (H, B), with H a polycyclic subgroup of finite index in C and B a non-zero $\mathbb{Z}$-free H-submodule of A, we can choose one so that the $\mathbb{Z}$-rank of B is as small as possible. Then B is r.i. as a G-module for every subgroup G of finite index in H. By Corollary 6 of Chapter 2, H has a normal subgroup G of finite index such that $G/C_G(B)$ is abelian.

In view of Lemma 3 (i), it will be enough to show that $G\Delta$ is closed in A.

The next part of the argument is to reduce the problem to consideration of G and B. For each $m \in \mathbb{N}$, composing $\Delta|_G$ with the natural homomorphism of A onto A/mB gives a derivation of G into A/mB. By inductive hypothesis, the image of this derivation is closed in A/mB: this means that the set

$$G\Delta + mB = \{g\Delta + mb | g \in G, b \in B\} \tag{8}$$

is closed in A. Now let a be an arbitrary element of the closure of $G\Delta$ in A. For each $m \in \mathbb{N}$, then, a belongs to the set (8), so there exist $g_m \in G$ and $b_m \in B$ with

$$a = g_m\Delta + mb_m.$$

If $m > 1$,

$$b_1 g_1^{-1} = a g_1^{-1} - (g_1\Delta)g_1^{-1}$$
$$= g_m\Delta \cdot g_1^{-1} + mb_m g_1^{-1} - (g_1\Delta)g_1^{-1}$$
$$= mb_m g_1^{-1} + (g_m g_1^{-1})\Delta.$$

Thus for each m, $g_m g_1^{-1} \in B\Delta^{-1} \cap G = K$ say; and so

$$b_1 g_1^{-1} \in \bigcap_{m \in \mathbb{N}} (K\Delta + mB). \tag{9}$$

Now suppose $h(B) < h(A)$. By inductive hypothesis, the image of the derivation $\Delta|_K : K \to B$ is closed in B. Hence from (9), $b_1 g_1^{-1} \in K\Delta$; say $b_1 g_1^{-1} = x\Delta$ with $x \in K$. Then

$$a = g_1\Delta + b_1 = g_1\Delta + (x\Delta)g_1 = (xg_1)\Delta \in G\Delta.$$

Thus in this case, it follows that $G\Delta$ is indeed closed in A.

We are left with the 'essential' case, where $h(B) = h(A)$. Then B has finite index in A, so by Lemma 3(i, ii, iii), $|G:K|$ is finite, and it will suffice to prove that $K\Delta$ is closed in B. Note that from the way it was chosen, B is still r.i. as a K-module. Write $X = C_K(B)$. One verifies easily that $X\Delta$ is actually a K-submodule of B, so either $X\Delta = 0$ or $X\Delta$ has finite index in B. In the latter case, $K\Delta$ is certainly closed in B, as it is a union of cosets of $X\Delta$. So assume that $X\Delta = 0$. Then $\Delta|_K$ factors through K/X and we lose no generality in assuming that $X = 1$. Then K is abelian and acts faithfully on its r.i. module B. If $K = 1$ there is nothing to prove (why?), so assume that $K \neq 1$. Then $C_B(K) = 0$. In view of Lemma 3(iv, v), we can replace $\Delta|_K$ by $l \cdot \Delta|_K$ for some non-zero integer l, and assume that there exists $b \in B$ such that

$$x\Delta = b(x - 1) \quad \forall x \in K. \tag{10}$$

By Proposition 1 of Chapter 2, there exist an algebraic number field, with ring of integers $\mathfrak{o}$, and embeddings

$$^- : B \to \mathfrak{o}^+, \quad ^- : K \to \mathfrak{o}^*$$

such that

$$(bx)^- = \bar{b} \cdot \bar{x} \quad \forall b \in B, \quad \forall x \in K.$$

Now consider an element a lying in the closure of $K\Delta$ in B. Then for each $m \in \mathbb{N}$ there exists $x_m \in K$ with $x_m \Delta - a \in mB$; using (10), this gives

$$\bar{b}\bar{x}_m - (a + b)^- \in m\bar{B} \subseteq m\mathfrak{v}. \tag{11}$$

This brings us right back to the situation discussed at the beginning of this section. Specifically, put $\alpha = \bar{b}, \beta = \bar{a} + \bar{b}$, and $U = \bar{K}$: then (11) is nothing other than formula (3). The whole point of that initial discussion was to show that, given Theorem 4, one can infer the existence of $x \in U$ with $\alpha x = \beta$. In the present context, we have $x = \bar{g}$ for some $g \in K$, and

$$\bar{b}\bar{g} = \bar{a} + \bar{b}.$$

Dropping the bars gives

$$a = b(g - 1) = g\Delta \in K\Delta.$$

Thus we have shown that $K\Delta$ is closed in B, and come to the end of the proof.

All the theorems of this section have now been established, modulo Theorem 4: this will be dealt with in section E. One might say that the 'transition point' from arithmetic to algebra in the whole argument was where we showed that Theorem 4 implies the statement (4); in fact (4) itself can be generalised:

Corollary 1 Let G be a soluble-by-finite group and M a G-module, finitely generated as a $\mathbb{Z}$-module. Let $a, b \in M$ and suppose that for every positive integer m there exists $g_m \in G$ such that $ag_m - b \in mM$. Then there exists $g \in G$ with $ag = b$.

Proof By Theorem 1 of Chapter 2, we may replace G by $G/C_G(M)$ and assume that G is polycyclic-by-finite. Let $\delta: G \to M$ be the inner derivation $g \mapsto a(g - 1)$. Then

$$b - a \in \bigcap_{m \in \mathbb{N}} (G\delta + mM).$$

Hence by Theorem 6 there exists $g \in G$ such that $b - a = g\delta$. Then $b = a + a(g - 1) = ag$.

If we are prepared to accept a result from the next chapter, we can now derive Theorem 3 in a different way. If G is a polycyclic-by-finite group, Theorem 5 of Chapter 5 shows that G may be embedded as a subgroup into $GL_n(\mathbb{Z})$, for some n. Suppose $a, b \in G$ have conjugate images in every finite quotient of G. Then for each $m \in \mathbb{N}$, since the normal subgroup

$$G(m) = (1 + m M_n(\mathbb{Z})) \cap G$$

has finite index in G, there exists $g_m \in G$ with $a^{g_m} b^{-1} \in G(m)$. Now G acts by conjugation on the $\mathbb{Z}$-module $M = M_n(\mathbb{Z})$. For each m we have

$$a^{g_m} - b = (a^{g_m} b^{-1} - 1) b \in mM \,;$$

so Corollary 1 shows that there exists $g \in G$ with $a^g = b$, and Theorem 3 is proved anew.

Results similar to Corollary 1, for groups which may be far from soluble, will be discussed in Chapter 9.

D. Conjugate subgroups

We have seen that the conjugacy of elements in a polycyclic group G is determined by the finite quotients of G. In this section we establish the analogous result for conjugacy of subgroups:

Theorem 7   Let G be a polycyclic-by-finite group and X, Y two subgroups of G. If X and Y have conjugate images in every finite quotient of G then X and Y are conjugate in G.

The proof depends on various other results, which I shall state as we come to need them. Let G, X and Y be as in the theorem. As usual, we argue by induction on the Hirsch number of G. If G is finite there is nothing to prove, so suppose G infinite, and let M be an infinite free abelian normal subgroup of G. By inductive hypothesis, MX/M and MY/M are conjugate in G/M; so replacing Y by a suitable conjugate in G, we may assume that $MX = MY = Q$, say. For each positive integer m, the inductive hypothesis gives an element $g_m \in G$ such that $M^m X^{g_m} = M^m Y$, and it is easy to see that each such g_m normalizes the group Q; replacing G by $N_G(Q)$, we may therefore assume that $Q \lhd G$.

Put $A = X \cap M$ and $B = Y \cap M$. Then for each m we have

$$M^m A^{g_m} = M^m X^{g_m} \cap M = M^m Y \cap M = M^m B.$$

We want to conclude, of course, that A and B are actually conjugate in G; to do so, we appeal to the following result, which is a 'relative' version of the theorem:

Proposition 2   Let H be a polycyclic-by-finite group and $M \cong \mathbb{Z}^n$ an H-module. Let A and B be $\mathbb{Z}$-submodules of M. Suppose that for every positive integer m there exists $h_m \in H$ such that

$$A h_m + mM = B + mM.$$

Then there exists $h \in H$ with $Ah = B$.

This will be proved later. We now apply it with $H = G$ to give an element $g \in G$ such that $A^g = B$. Replacing X by X^g, we can then assume that

$$X \cap M = Y \cap M = D, \quad \text{say.}$$

There are now two cases to consider, requiring different methods.

Case 1   Where $D = 1$. Then $Q = M]X = M]Y$, so X and Y are complements for M in Q. Now we described a technique for dealing with complements in Chapter 3: Proposition 1 of that chapter gives

Lemma 4   Let $\mathscr{C}$ be the set of all complements for M in $Q = M]X$. Then there is a bijective map

$$\Psi = \Psi(M, X) : \mathscr{C} \to \operatorname{Der}(Q/M, M);$$

if $C \in \mathscr{C}$ and $Mc = Mx$ with $c \in C$ and $x \in X$, then the derivation $C\Psi$ maps the coset $Mc \in Q/M$ to the element $x^{-1}c \in M$.

We also have to consider the action of G on the set $\mathscr{C}$; the reader may easily verify

Lemma 5   (i) If G is a group acting on $Q = M]X$ and fixing M, then G acts on $\operatorname{Der}(Q/M, M)$ by

$$\delta^g = g_1^{-1} \cdot \delta \cdot g_2, \quad \delta \in \operatorname{Der}(Q/M, M), \quad g \in G$$

where g_1 denotes the action of g on Q/M and g_2 its action on M.
   (ii) The map

$$\Delta = \Delta(M, X) : G \to \operatorname{Der}(Q/M, M)$$
$$g\Delta = X^g \Psi$$

is a derivation of G into the G-module $\operatorname{Der}(Q/M, M)$.

Case 1 of Theorem 7 can now be finished off. Let m be a positive integer, and apply the two lemmas with M/M^m in place of M and $Q/M^m = (M/M^m)](M^m X/M^m)$ in place of Q; writing Ψ_m and Δ_m for the corresponding mappings, we see that

$$(M^m Y)\Psi_m = (M^m X)^{g_m}\Psi_m = g_m \Delta_m.$$

A quick checking of the definitions shows that this now implies that

$$Y\Psi \equiv g_m \Delta \bmod m \operatorname{Der}(Q/M, M).$$

As m was an arbitrary positive integer, we see that $Y\Psi$ belongs to the profinite closure of $G\Delta$ in $\operatorname{Der}(Q/M, M)$. But the latter is a free $\mathbb{Z}$-module of finite rank, so Theorem 6 shows that

$$Y\Psi = g\Delta$$

for some $g \in G$. Thus $Y\Psi = X^g\Psi$, and as Ψ is a bijective map it follows that $Y = X^g$. This completes the proof of Case 1.

Case 2   Where $D \neq 1$. For each $m \in \mathbb{N}$, we have an element $g_m \in G$ such that $M^m X^{g_m} = M^m Y$. Suppose we could choose each g_m to lie in $N_G(D)$. Then X/D and Y/D would have conjugate images in every finite quotient of $N_G(D)/D$; since D is infinite, the inductive hypothesis would ensure that X/D and Y/D were conjugate in $N_G(D)/D$. Thus X and Y would be conjugate in G, and the proof would be finished.

The rest of the argument is to show that we can, in fact, choose each g_m in $N_G(D)$. Let m be a positive integer. There exist $r \in \mathbb{N}$ such that $Q^r \cap M \leq M^m$, and $s \in \mathbb{N}$ such that $G^s \cap Q \leq Q^r$. Now I claim that

$$N_G(M^k D) \leq N_G(D) G^s \text{ for some } k \in \mathbb{N}. \tag{12}$$

Accepting (12) for the moment, put $l = mk$. Then since $M^l \leq M^k \leq M$ and $M^l X^{g_l} = M^l Y$, we obtain

$$(M^k D)^{g_l} = M^k(M \cap M^l X^{g_l}) = M^k(M \cap M^l Y) = M^k D.$$

Thus $g_l \in N_G(M^k D)$, so from (12) we have $g_l = hf^{-1}$, say, with $h \in N_G(D)$ and $f \in G^s$. Now G^s normalizes $M^m Y$: for by the choice of s we know that G^s centralizes Q/Q^r, and since $Q = MY$, $Q^r \leq MY^r$ which implies that $Q^r = (M \cap Q^r)Y^r \leq M^m Y$. Thus

$$M^m X^h = (M^m X^{g_l})^f = (M^m Y)^f = M^m Y.$$

So we may replace g_m by $h \in N_G(D)$, which is what we had to show.

We still have to prove (12). Put $H = N_G(D)G^s$ and let $\{g_1 = 1, g_2, \ldots, g_d\}$ be a transversal to the left cosets of H in G. Suppose by way of contradiction that $N_G(M^k D) \not\leq H$ for every positive integer k. Then for each k there exists $i(k) \in \{2, \ldots, d\}$ such that

$$N_G(M^{k!}D) \cap g_{i(k)}H \neq \varnothing. \tag{13}$$

At least one value of $i(k)$ will occur for infinitely many values of k; let i_0 be such a value. Since $N_G(M^{l!}D) \geq N_G(M^{k!}D)$ whenever $k \geq l$, it follows that (13) will hold with $i(k) = i_0$ for every k. Putting $x = g_{i_0}$, we thus have for each positive integer k some element $h_k \in H$ with

$$xh_k \in N_G(M^{k!}D) \leq N_G(M^k D).$$

So

$$M^k D^x = M^k D^{h_k^{-1}} \quad \forall k \in \mathbb{N}.$$

By Proposition 2, with D^x for A and D for B, it follows that $D^{xh} = D$ for some $h \in H$. But then

$$g_{i_0} = x \in N_G(D)h^{-1} \leq H,$$

contradicting the choice of $i_0 \neq 1$. This final contradiction now establishes (12) and finishes the proof of Theorem 7.

The rest of this section is devoted to the proof of Proposition 2. As in the previous section, the essential case concerns a ring of algebraic integers:

Lemma 6   Let $\mathfrak{o}$ be the ring of integers in an algebraic number field k, and let U be a subgroup of $\mathfrak{o}^*$. Let $a, b \in \mathfrak{o}$, and suppose that for every positive integer m there exists $g_m \in U$ with

$$ag_m \mathbb{Z} + m\mathfrak{o} = b\mathbb{Z} + m\mathfrak{o}.$$

Then there exists $g \in U$ with $ag\mathbb{Z} = b\mathbb{Z}$.

Proof   This is similar to the statement (4) in section C; only here the cyclic groups $a\mathbb{Z}$ and $b\mathbb{Z}$ are playing the roles of the elements α and β. The first step is to show that

$$b = av \tag{14}$$

for some unit v of $\mathfrak{o}$. If either a or b is zero the result is easy, so suppose that $ab \neq 0$. There is a positive integer t with $t\mathfrak{o} \subseteq a\mathfrak{o}$ and $t\mathfrak{o} \subseteq b\mathfrak{o}$. Then

$$b \in ag_t \mathbb{Z} + t\mathfrak{o} \subseteq a\mathfrak{o}, \quad a \in bg_t^{-1} \mathbb{Z} + t\mathfrak{o} \subseteq b\mathfrak{o},$$

so $a\mathfrak{o} = b\mathfrak{o}$ and (14) holds for some $v \in \mathfrak{o}^*$.

Eliminating b from the hypothesis, we get

$$ag_m \mathbb{Z} + m\mathfrak{o} = av\mathbb{Z} + m\mathfrak{o}$$

for each m; so for each m there exists $r_m \in \mathbb{Z}$ with $av - ag_m r_m \in m\mathfrak{o}$. Taking t as above and putting $r = r_{mt}$, this gives

$$a(v - g_{mt} r) \in mt\mathfrak{o} \subseteq a \cdot m\mathfrak{o}$$

whence

$$r \equiv g_{mt}^{-1} v \bmod m\mathfrak{o}. \tag{15}$$

Now we take *norms* from $\mathfrak{o}$ to $\mathbb{Z}$: the norm of an algebraic number is the product of all its conjugates over $\mathbb{Q}$. The norm of a unit is always ± 1, while the norm of $r \in \mathbb{Z}$ is r^n, where $n = [k : \mathbb{Q}]$. The conclusion from (15) is that

$$r^n \equiv \pm 1 \bmod m,$$

and plugging this back into (15) gives

$$(g_{mt}^{-1} v)^{2n} \equiv 1 \bmod m\mathfrak{o}. \tag{16}$$

Thus $v^{2n} \equiv g_{mt}^{2n} \bmod \mathfrak{o}^*(m)$, where $\mathfrak{o}^*(m) = (1 + m\mathfrak{o}) \cap \mathfrak{o}^*$.

Now the subgroup U^{2n} is closed in the congruence topology on $\mathfrak{o}^*$, by the results of section C. So it follows from the last paragraph that $v^{2n} \in U^{2n}$. Hence there exist $x \in U$ and a $2n$th root of unity ζ in $\mathfrak{o}^*$ such that $v = x\zeta$. To finish the proof, it will suffice to show that $\pm \zeta \in U$.

We need a simple arithmetic fact:

Lemma 7 There exists a positive integer s such that the only roots of unity lying in $s\mathfrak{o} + \mathbb{Z}$ are 1 and -1.

I shall prove this below. Let s be as given by Lemma 7, put $H = \mathfrak{o}^*(s) \cap U$ and $K = H^{2n}$. By Theorem 4, there exists $m \in \mathbb{N}$ with $\mathfrak{o}^*(m) \cap U \leq K$, and we may take m to be a multiple of s. Put $g = g_{mt}$. Then from (16),

$$(g^{-1}x)^{2n} = (g^{-1}x\zeta)^{2n} = (g^{-1}v)^{2n} \in \mathfrak{o}^*(m) \cap U \leq K,$$

so $(g^{-1}x)^{2n} = h^{2n}$ for some $h \in H$, and $g^{-1}x = \lambda h$ for some $2n$th root of unity λ. Since $h \in \mathfrak{o}^*(s)$ and $s \mid m$, (15) now gives

$$\zeta\lambda \equiv \zeta\lambda h = g^{-1}\zeta x = g^{-1}v \equiv r \in \mathbb{Z} \bmod s\mathfrak{o}.$$

Hence, by the choice of s, $\zeta\lambda = \pm 1$; and we get

$$\zeta = \pm\lambda^{-1} = \pm gx^{-1}h \in \pm U$$

as required.

Proof of Lemma 7 The ring $\mathfrak{o}$ only contains a finite number of roots of unity (by the Units Theorem); let $\rho_1, \ldots, \rho_w$ be those not equal to ± 1. If for each i we have $\rho_i \notin s_i\mathfrak{o} + \mathbb{Z}$, then clearly $s = \text{l.c.m.} (s_1, \ldots, s_w)$ will do for Lemma 7. So let ρ be a root of unity, and suppose that $\rho \in m\mathfrak{o} + \mathbb{Z}$ for every positive integer m. Since $\mathfrak{o} \cong \mathbb{Z}^n$, it follows that $\rho \in \mathbb{Z}$, hence that $\rho = \pm 1$. Thus if $\rho = \rho_i$ with $1 \leq i \leq w$, then a positive integer s_i exists for which $\rho \notin s_i\mathfrak{o} + \mathbb{Z}$. This establishes the lemma.

Before proceeding with the proof of Proposition 2, I want to isolate a reduction step:

Lemma 8 Let $K \leq_f H$ be groups and $M \cong \mathbb{Z}^n$ an H-module. If Proposition 2 holds for the pair (M, K) then it holds for the pair (M, H).

Proof Let $A, B \leq M$ and suppose that for each $m \in \mathbb{N}$ there exists $h_m \in H$ with $Ah_m + mM = B + mM$. Let $\{t_1, \ldots, t_s\}$ be a transversal to the left cosets of K in H, and put

$$h_m = t_{i(m)}k_m,$$

with $k_m \in K$, for each m. There exists i_0 such that $i(m!) = i_0$ for infinitely many values of m. Putting $t = t_{i_0}$ we then have

$$At \cdot k'_m + mM = B + mM$$

for every m, where $k'_m = k_{r!}$ for a suitable value of $r \geq m$. By hypothesis, it follows that there exists $k \in K$ with $Atk = B$; and the lemma is proved.

Proof of Proposition 2 *Step 1* Suppose A and B are cyclic. Just as in the proof of Theorem 6 in section C, we can find a subgroup G of finite index in H and a rationally irreducible G-submodule N of M such that $G/C_G(N)$ is abelian; replacing N by its isolator in M, we may assume also that M/N is free abelian. By Lemma 8, it will suffice to prove the proposition for the pair (M, G). There are two cases to consider.

Split case   Where $A \cap N = B \cap N = 0$. Here we argue as in Case 1 of Theorem 7 – the argument did not depend on Proposition 2 (take $M]G$ for G, N for M, A for X and B for Y).

Non-split case   Where $A \cap N \neq 0$. Then $A \leq N$, since A is cyclic and M/N is torsion-free. For every $m \in \mathbb{N}$, we assume that there exists $g_m \in G$ with $Ag_m + mM = B + mM$. Then

$$B \leq \bigcap_{m \in \mathbb{N}} (Ag_m + mM) \leq \bigcap_{m \in \mathbb{N}} (N + mM) = N,$$

so $B \leq N$ also. Since N is r.i. as a G-module and $G/C_G(N)$ is abelian, Proposition 1 of Chapter 2 reduces this case to Lemma 6, above.

Step 2   Now let A and B have arbitrary rank. We show next that there exists $h \in H$ such that

$$\overline{Ah} = \bar{B},$$

where $^-$ denotes isolator in M (i.e. $\bar{A}/A = \tau(M/A)$).

Say $A \cong \mathbb{Z}^r$ and $B \cong \mathbb{Z}^s$. Take $m \in \mathbb{N}$ such that $mM \cap A \subseteq 2A$ and $mM \cap B \subseteq 2B$. Since

$$A/(mM \cap A) \cong (A + mM)/mM \cong (B + mM)/mM \cong B/(mM \cap B),$$

it follows that $A/2A \cong B/2B$, and hence that $r = s$. Now we need a trick to reduce this case to the previous one. The way to do this is via *exterior powers*: see for example Lang's 'Algebra', Chapter XVI, section 6. The sth exterior power of M is a $\mathbb{Z}$-module

$$M^* = \wedge^s M \cong \mathbb{Z}^{\binom{n}{s}};$$

this has cyclic submodules

$$A^* = \wedge^s A = (a_1 \wedge \ldots \wedge a_s)\mathbb{Z}, \quad B^* = \wedge^s B = (b_1 \wedge \ldots \wedge b_s)\mathbb{Z},$$

where $\{a_1, \ldots, a_s\}$ is a basis for A and $\{b_1, \ldots, b_s\}$ is a basis for B. One makes M^* into an H-module via

$$(x_1 \wedge \ldots \wedge x_s)h = (x_1 h) \wedge \ldots \wedge (x_s h), \quad h \in H, x_j \in M.$$

Now take $m \in \mathbb{N}$. Since $Ah_m \equiv B \bmod mM$, there is a matrix $\mu \in M_s(\mathbb{Z})$ such that

$$a_i h_m \equiv \sum_j \mu_{ij} b_j \bmod mM, \quad i = 1, \ldots, s.$$

Then

$$A^* h_m = (a_1 h_m \wedge \ldots \wedge a_s h_m)\mathbb{Z}$$
$$\equiv \det \mu \cdot (b_1 \wedge \ldots \wedge b_s)\mathbb{Z} \bmod mM^*$$
$$= \det \mu \cdot B^*.$$

Thus $A^*h_m \leq B^* + mM^*$, and similarly $B^* \leq A^*h_m + mM^*$; consequently
$$A^*h_m + mM^* = B^* + mM^*.$$

This holds for each m; since A^* and B^* are cyclic, Step 1 (with N^* for M, A^* for A and B^* for B) shows that $A^*h = B^*$ for some $h \in H$.

This means that $a_1 h \wedge \ldots \wedge a_s h = \pm b_1 \wedge \ldots \wedge b_s$; so if $1 \leq i \leq s$ we have
$$b_i \wedge a_1 h \wedge \ldots \wedge a_s h = \pm b_i \wedge b_1 \wedge \ldots \wedge b_i \wedge \ldots \wedge b_s = 0.$$

This can only happen if b_i is linearly dependent on $a_1 h, \ldots, a_s h$, i.e. if $b_i \in \overline{Ah}$. Therefore $B \leq \overline{Ah}$; similarly $Ah \leq \bar{B}$, and so $\bar{B} = \overline{Ah}$ as required.

Step 3 Conclusion. In view of Step 2, we may replace A by Ah, for a suitable $h \in H$, and assume that
$$\bar{A} = \bar{B} = C, \text{ say.}$$

Then $|C : A|$ and $|C : B|$ are finite, so for some positive integer f we have
$$fM \cap C \leq A \cap B.$$

Then
$$fM \cap A = fM \cap C = fM \cap B = D, \text{ say;}$$

and D is isolated in fM (i.e. $fM \cap \bar{D} = D$). Put $G = C_H(M/fM)$. Then G has finite index in H, and a glance at the proof of Lemma 8 shows that we can find an element $t \in H$ such that for every positive integer m there exists $g_m \in G$ with
$$Atg_m + mM = B + mM.$$

Then
$$\begin{aligned} Dtg_{fm} + mfM &= (fM \cap Atg_{fm}) + mfM \\ &= fM \cap (Atg_{fm} + mfM) \\ &= fM \cap (B + mfM) \quad = D + mfM. \end{aligned}$$

Now Dt, like D, is isolated in fM; so applying Step 2 with fM for M, G for H, Dt for A and D for B, we conclude that
$$Dtg = D$$

for some $g \in G$. Now evidently C is the isolator of D in M, and Ctg is the isolator of Dtg. Therefore $Ctg = C$. Since G centralizes M/fM, we have
$$B + fM = Atg_f + fM = Atg + fM.$$

Intersecting both sides with $C = Ctg$ we get
$$\begin{aligned} B = (B + fM) \cap C &= (Atg + fM) \cap Ctg \\ &= ((A + fM) \cap C)tg = Atg, \end{aligned}$$

and the proof is finished.

E. Congruences and algebraic integer units.

Everything we have done in sections C and D rested squarely on Theorem 4, the proof of which we shall now discuss. Consider an algebraic number field k with ring of integers $\mathfrak{o}$. Slightly reformulated, Theorem 4 says the following:

$$\text{for every } t \in \mathbb{N} \text{ there exists } m \in \mathbb{N} \text{ such that} \tag{17}$$
$$x \in \mathfrak{o}^*, \; x \equiv 1 \bmod m\mathfrak{o} \Rightarrow x \in \mathfrak{o}^{*t}.$$

Exercise 17   Let K be a finite extension field of k, with ring of integers $\bar{\mathfrak{o}}$. Show that if (17) holds with $\bar{\mathfrak{o}}$ in place of $\mathfrak{o}$, then (17) holds for $\mathfrak{o}$.

Thus to prove (17), we may replace k by any finite extension field. In particular, we may – and henceforth shall – assume that

$$\sqrt{-1} \in k$$

(this is to obviate some unpleasant complications later on). Now if h.c.f. $(s, t) = 1$ then $\mathfrak{o}^{*s} \cap \mathfrak{o}^{*t} = \mathfrak{o}^{*st}$; so if (17) holds whenever t is a prime-power, it will hold in general. We fix a prime-power p^e say, with $e \geq 1$, and seek to establish:

$$\exists m \in \mathbb{N} \text{ such that } x \in \mathfrak{o}^*, \; x \equiv 1 \bmod m\mathfrak{o} \Rightarrow x \in \mathfrak{o}^{*p^e}. \tag{18}$$

How does one set about finding a suitable m? The ideal $m\mathfrak{o}$ is a product of primes (i.e. non-zero proper prime ideals) of $\mathfrak{o}$, so let us see what can be concluded about a prime $\mathfrak{p}$ of $\mathfrak{o}$ if for some $x \in \mathfrak{o}^*$, $x \equiv 1 \bmod \mathfrak{p}$ but $x \notin \mathfrak{o}^{*p^e}$. One can then hope to build $m\mathfrak{o}$ out of primes which do not satisfy that conclusion.

So let $\mathfrak{p}$ be a prime of $\mathfrak{o}$, and suppose there is an $x \in \mathfrak{o}^*$ with $x \equiv 1 \bmod \mathfrak{p}$ and $x \notin \mathfrak{o}^{*p^e}$. For some $j \leq e$ then, $x \in \mathfrak{o}^{*p^{j-1}} \setminus \mathfrak{o}^{*p^j}$; so $x = y^{p^{j-1}}$ with $y \in \mathfrak{o}^*$ and $y \notin \mathfrak{o}^{*p}$. Adjoin a pth root z of y to the field k, getting $F = k(z)$ say, and write $\mathfrak{o}_F$ for the ring of integers of F. As Galois extensions are nicer to deal with, let us make the

First simplifying assumption: k contains a primitive pth root of unity.

Then F is Galois over k, with cyclic Galois group of order p. Consider now a prime $\mathfrak{P}$ of $\mathfrak{o}_F$ containing $\mathfrak{p}\mathfrak{o}_F$; then $\mathfrak{P} \cap \mathfrak{o} = \mathfrak{p}$. Write

$$^{-} : \mathfrak{o}_F \to \mathfrak{o}_F / \mathfrak{P}$$

for the natural map. Since

$$z^{p^j} = y^{p^{j-1}} = x \equiv 1 \bmod \mathfrak{p},$$

we see that $\bar{z}$ is a p^jth root of unity in $\bar{\mathfrak{o}}_F$. Now what we are trying to do is to eliminate x, and hence z, from the scene: so let us make a

Second simplifying assumption: k contains a primitive p^eth of unity, ζ say.

We can then get rid of $\bar{z}$ by supposing that

the distinct powers of ζ remain distinct modulo $\mathfrak{p}$; $\qquad$ (19)

for then the p^j distinct powers of $\zeta^{p^{e-j}}$ give p^j distinct zeros of the polynomial $X^{p^j} - 1$ in $\bar{\mathfrak{o}}_F$, and so, for some l,

$$\bar{z} = \bar{\zeta}^{p^{e-jl}} \in \bar{\mathfrak{o}}.$$

Hence

$$(\mathfrak{o}[z] + \mathfrak{P})/\mathfrak{P} = \bar{\mathfrak{o}}[\bar{z}] = \bar{\mathfrak{o}} \cong \mathfrak{o}/\mathfrak{p}.$$

As $F = k(z)$, one can expect the subring $\mathfrak{o}[z]$ to be rather close to $\mathfrak{o}_F$; in any case, if we have

$$\mathfrak{o}[z] + \mathfrak{P} = \mathfrak{o}_F \qquad (20)$$

we can deduce that $\mathfrak{o}_F/\mathfrak{P} \cong \mathfrak{o}/\mathfrak{p}$. Since $[F:k] = p > 1$, this shows that $\mathfrak{p}$ is not inertial in F (i.e. $\mathfrak{p}\mathfrak{o}_F$ is not prime in $\mathfrak{o}_F$).

Summarizing the conclusions so far, we have established

Lemma 9 Assume that k contains a primitive p^eth root of unity ζ. Let $\mathfrak{p}$ be a prime of $\mathfrak{o}$, and suppose an element $x \in \mathfrak{o}^*$ exists such that $x \equiv 1 \bmod \mathfrak{p}$ and $x \notin \mathfrak{o}^{*p^e}$. Then one of the following obtains:

(*a*) (19) is false;

(*b*) there exists a cyclic Galois extension F of degree p over k, generated by a pth root z of some element in $\mathfrak{o}^*$, such that *either* (20) is false *or* $\mathfrak{p}$ is not inertial in F.

We have still not quite got rid of the element z; to do so, we want a condition on $\mathfrak{p}$ which will imply (20).

Definition The relative discriminant of z over k is

$$D(z) = \prod (z^\tau - z^\sigma),$$

the product being over all τ and σ in $\mathrm{Gal}(F/k)$ with $\tau \neq \sigma$.

Lemma 10 If $F = k(z)$ and $z \in \mathfrak{o}_F$ then $D(z)\mathfrak{o}_F \subseteq \mathfrak{o}[z]$.

This is an elementary fact; see section F for a reference. To apply it, we must compute $D(z)$. Since $z^p = y \in \mathfrak{o}^*$, the distinct conjugates of z over k are $z\zeta_1^r$, $r = 0, 1, \dots p - 1$, where $\zeta_1 = \zeta^{p^{e-1}}$ is a primitive pth root of 1. Thus

$$D(z) = \prod (z\zeta_1^r - z\zeta_1^s) = z^{p(p-1)} \prod (\zeta_1^r - \zeta_1^s) = y^{p-1} \prod (\zeta_1^r - \zeta_1^s).$$

Thus – most conveniently – the hypothesis (19) implies that $D(z) \notin \mathfrak{p}$, in which case we have $D(z)\mathfrak{o} + \mathfrak{p} = \mathfrak{o}$, and hence, from Lemma 10,

$$\mathfrak{o}_F = D(z)\mathfrak{o}_F + \mathfrak{p}\mathfrak{o}_F \subseteq \mathfrak{o}[z] + \mathfrak{P}.$$

In other words, (19) implies (20). Turning Lemma 9 around, then, gives

Lemma 11   Assume that k contains a primitive p^e th root of unity ζ. Suppose that for every proper extension F of the form $F = k(y^{1/p})$ with $y \in \mathfrak{o}^*$ there is a prime $\mathfrak{p}_F$ of $\mathfrak{o}$ which satisfies (19) and is inertial in F. Then any $x \in \mathfrak{o}^*$ which is $\equiv 1 \bmod \mathfrak{p}_F$ for every such F must lie in $\mathfrak{o}^{*p^e}$.

Now $\mathfrak{o}^*$ is a finitely generated group, say $\mathfrak{o}^* = \langle u_1, \ldots, u_d \rangle$. The field $E = k(u_1^{1/p}, \ldots, u_d^{1/p})$ is a finite extension of k, and provided k contains a primitive pth root of unity, E contains all pth roots of all elements of $\mathfrak{o}^*$. Hence there are only finitely many fields F of the kind mentioned in Lemma 11; call them $F_1, \ldots, F_n$. We now need the existence of primes of $\mathfrak{o}$ which are inertial in the $F_i s$; it is guaranteed by

Theorem 8   Let k be an algebraic number field and F a finite Galois extension of k with cyclic Galois group. Then infinitely many primes of $\mathfrak{o}$ (the ring of integers of k) are inertial in F.

The reader will recognise this as a strong converse to Lemma 2 of section B. For the proof, look up *Čebotarev's Theorem* or the *Frobenius Density Theorem* in any book on algebraic number theory: see section F for references.

By Theorem 8, we can choose for each $i = 1, \ldots, n$, a prime $\mathfrak{p}_i$ of $\mathfrak{o}$ which is inertial in F_i and satisfies (19). Let $m \in \mathbb{N}$ be such that

$$m\mathfrak{o} \subseteq \bigcap_{i=1}^{n} \mathfrak{p}_i.$$

Then – provided we have $\zeta \in K$ – Lemma 11 shows that

$$x \in \mathfrak{o}^*, \quad x \equiv 1 \bmod m\mathfrak{o} \Rightarrow x \in \mathfrak{o}^{*p^e}.$$

So we have established

Proposition 3   If k contains a primitive p^eth root of unity, then (18) holds.

The proof of (18) in general is now quite elementary, though rather fiddly. The reader can readily convince himself that the following result will finish the job:

Proposition 4   Let ζ be a primitive p^eth root of unity. If $x \in k(\zeta)$ and $x^{p^e} \in \mathfrak{o}^*$, then $x^{p^e} \in \mathfrak{o}^{*p^e}$.

(Recall that $\sqrt{-1} \in k$; as an exercise, show that the proposition may fail if $p = 2$ and $\sqrt{-1} \notin k$.)

Write

$$\zeta_i = \zeta^{p^{e-i}}$$

for $i = 0, 1, \ldots, e$, so $\zeta_e = \zeta$, ζ_1 is a primitive pth root of 1, and $\zeta_0 = 1$. Let $\mathfrak{o}_i$ be the ring of integers in $k(\zeta_i)$. For Proposition 4, it will suffice to establish, for $1 \leq i \leq e$, that

$$x \in \mathfrak{o}_i^* \text{ and } x^{p^e} \in \mathfrak{o}^* \Rightarrow x \in \mathfrak{o}_{i-1}^* \langle \zeta \rangle. \tag{21}$$

Thus suppose $x \in k(\zeta)$ satisfies $x^{p^e} \in \mathfrak{o}^*$. Then $x \in \mathfrak{o}_e^*$, and (21) shows that $x = x_1 \zeta^{s_1}$ for some $x_1 \in \mathfrak{o}_{e-1}^*$ and $s_1 \in \mathbb{Z}$; and $x_1^{p^e} = x^{p^e}$. Repeating with x_1 in place of x, and so on, gives after e steps an element $x_e \in \mathfrak{o}_0^* = \mathfrak{o}^*$ such that $x_e^{p^e} = x^{p^e}$. Thus $x^{p^e} \in \mathfrak{o}^{*p^e}$ and Proposition 4 follows.

Proof of (21) *for* $i = 1$. This is trivial if $p = 2$, so suppose p odd. The degree $d = [k(\zeta_1):k]$ divides $p - 1$, since $[\mathbb{Q}(\zeta_1):\mathbb{Q}] = p - 1$. Therefore h.c.f. $(p, d) = 1$. Now let $y \in k$ be the relative norm of x, that is the product of all conjugates of x over k. Thus writing $\Gamma = \mathrm{Gal}\,(k(\zeta_1)/k)$, we get

$$y^{p^e} = \left(\prod_{\sigma \in \Gamma} x^\sigma \right)^{p^e} = \prod_{\sigma \in \Gamma} (x^{p^e})^\sigma = x^{p^e d} = (x^d)^{p^e},$$

since $x^{p^e} \in \mathfrak{o}^*$ is fixed by Γ. Therefore

$$x^d = y \zeta^r, \text{ for some } r \in \mathbb{Z}.$$

Since h.c.f. $(p, d) = 1$, $\zeta^r = \zeta^{sd}$ for some $s \in \mathbb{Z}$, and then

$$(x\zeta^{-s})^d = y \in \mathfrak{o}^*.$$

But $(x\zeta^{-s})^{p^e} = x^{p^e} \in \mathfrak{o}^*$ also; since h.c.f. $(p, d) = 1$ it follows that $x\zeta^{-s} \in \mathfrak{o}^*$, giving the result.

For the general case we shall need

Lemma 12 *If* $\zeta_i \notin k(\zeta_{i-1})$ *then* $\zeta_{i+1} \notin k(\zeta_i)$, *provided in case* $p = 2$ *that* $i \geq 3$, *and otherwise that* $i \geq 2$.

Proof Suppose first that $p = 2$. Then $k(\zeta_i) = k(\zeta_{i-1}) \oplus \zeta_i k(\zeta_{i-1})$. If $\zeta_{i+1} \in k(\zeta_i)$ we have $\zeta_{i+1} = a + \zeta_i b$ with $a, b \in k(\zeta_{i-1})$ and $b \neq 0$. Then $\zeta_i = \zeta_{i+1}^2 = a^2 + \zeta_{i-1} b^2 + 2ab\zeta_i$, so $a^2 + \zeta_{i-1} b^2 = 0$. Since $i \geq 3$, $\zeta_2 \in k(\zeta_{i-1})$; and

$$\zeta_{i-1} = -(b^{-1}a)^2 = (\zeta_2 b^{-1}a)^2.$$

Therefore $\zeta_i = \pm \zeta_2 b^{-1} a \in k(\zeta_{i-1})$, contradicting hypothesis.

Now consider p an odd prime. Since $\zeta_1 \in k(\zeta_{i-1})$, the Galois group of $k(\zeta_i)$

over $k(\zeta_{i-1})$ has order p and is generated by an automorphism σ such that $\zeta_i^\sigma = \zeta_i\zeta_1$. Suppose $\zeta_{i+1} \in k(\zeta_i)$. Then for $0 \le j \le p-1$,

$$(\zeta_{i+1}^{\sigma^j})^p = \zeta_i^{\sigma^j} = \zeta_i\zeta_1^j. \tag{22}$$

Now $\prod_{j=0}^{p-1} \zeta_{i+1}^{\sigma^j} = \eta$ say lies in $k(\zeta_{i-1})$. From (22),

$$\eta^p = \prod_j (\zeta_i\zeta_1^j) = \zeta_i^p\zeta_1^{1+2+\cdots+(p-1)} = \zeta_{i-1},$$

since $1 + 2 + \ldots + (p-1) \equiv 0 \bmod p$ for any odd prime p. Therefore $\zeta_i = \zeta_1^r\eta \in k(\zeta_{i-1})$, for some r, contradicting hypothesis.

Proof of (21) *for $i \ge 2$ ($i \ge 3$ if $p = 2$).* Take $x \in \mathfrak{o}^*$ with $x^{p^e} \in \mathfrak{o}^*$, where $i \le e$. If $\zeta_i \in k(\zeta_{i-1})$, $\mathfrak{o}_{i-1} = \mathfrak{o}_i$ and there is nothing to prove, so assume that $\zeta_i \notin k(\zeta_{i-1})$. Then as in Lemma 12, $\mathrm{Gal}(k(\zeta_i)/k(\zeta_{i-1}))$ is generated by an automorphism σ sending ζ_i to $\zeta_i\zeta_1 = \zeta_i^g$ where $g = 1 + p^{i-1}$. Now x^{p^e} is fixed by σ, so $(x^\sigma)^{p^e} = x^{p^e}$ and

$$x^\sigma = x\zeta^r, \text{ for some } r \in \mathbb{Z}.$$

By Lemma 12, $\zeta_{i+1} \notin k(\zeta_i)$, so r must be divisible at least by p^{e-i} and $\zeta^r = \zeta_i^s$ for some $s \in \mathbb{Z}$. A simple calculation then gives

$$x^{\sigma^p} = x\zeta_i^{s(1+g+\ldots+g^{p-1})}.$$

But $x^{\sigma^p} = x$, so

$$s(1 + g + \ldots + g^{p-1}) \equiv 0 \bmod p^i. \tag{23}$$

We must now separate two cases.

Case 1 p odd, $i \ge 2$. Then one can verify that $g = 1 + p^{i-1}$ gives

$$1 + g + \ldots + g^{p-1} \equiv p \bmod p^i.$$

Then from (23), $s \equiv 0 \bmod p^{i-1}$. Hence $\zeta_i^s = \zeta_1^t$ for some t, and

$$\begin{aligned}(x\zeta_i^{-t})^\sigma &= x^\sigma\zeta_i^{-t}\zeta_1^{-t}\\ &= x\zeta_1^t\zeta_i^{-t}\zeta_1^{-t}\\ &= x\zeta_i^{-t};\end{aligned}$$

therefore $x\zeta_i^{-t} \in k(\zeta_{i-1})$ and the result follows.

Case 2 $p = 2$, $i \ge 3$. In this case, (23) reads

$$s(1 + (1 + 2^{i-1})) \equiv 0 \bmod 2^i.$$

Hence $2^{i-1}|s(1 + 2^{i-2})$, and since $i \ge 3$ it follows that $2^{i-1}|s$. The argument is then concluded as in Case 1.

Proof of (21) *for $i = p = 2$:* this is trivial, since $\zeta_2 = \sqrt{-1} \in k$.

All cases of (21) having been dealt with, Proposition 4 is established; and the proof of Theorem 4 is complete.

Exercise 18 Deduce Theorem 2 (section A), for the case of a Galois extension of $\mathbb{Q}$, from Theorem 8.

F. Notes

For the basic facts about number fields and their rings of integers which I have used more or less tacitly throughout the chapter, see Lang (1965), Chapters VII, VIII, IX; Lang (1970), Chapter 1; Janusz (1973), Chapters I, III, or any book on algebraic number theory. For a more 'concrete' approach, one should also look at Cohn (1978), Chapters 6, 7, 9 and 10. In particular, our Lemma 10 (section E) is formula (7.26) on page 59 of Cohn's book. The deeper results we used, Theorems 2 and 8, are both consequences of Čebotarev's Theorem, or of its precursor the Frobenius Density Theorem. An accessible proof of the latter is described in the book of Janusz, Chapter IV. Our Theorem 8 is Corollary 5.4 of that chapter. For our Theorem 2, one can embed the given number field k in its normal closure K over $\mathbb{Q}$. Then all but finitely many primes of o_K also have degree 1 over $\mathbb{Q}$ (if p splits completely in k and is unramified in K, then the decomposition field of every prime $\mathfrak{p}$ of o_K lying over p contains k; the conjugates of k generate K, from which it follows that the decomposition field of each such $\mathfrak{p}$ is actually K) — this reduces the problem to our Exercise 18. Alternatively, one can deduce Theorem 2 from a purely 'rational' result:

Theorem 2* Let $f \in \mathbb{Z}[X]$ be a monic, irreducible, non-linear polynomial. Then for infinitely many primes p, the image f_p of f in $(\mathbb{Z}/p\mathbb{Z})[X]$ has no zero in $\mathbb{Z}/p\mathbb{Z}$.

This is Exercise 5 of Chapter IV in Janusz's book. To deduce Theorem 2, take f to be the minimal polynomial of an integral primitive element for k over $\mathbb{Q}$, assuming that $k \neq \mathbb{Q}$; it is easy to see that if $\mathfrak{p}$ is a prime of o of degree 1, containing p, then f_p does have a zero in $\mathbb{Z}/p\mathbb{Z}$. Hence, by Theorem 2*, infinitely many primes of o do not have degree 1.

Now some notes on the other results of the chapter. Theorem 1 was published in Baer (1957); generalizations have been given in Wehrfritz (1971a), Wehrfritz (1976), and Segal (1975). Related to the discussion in section B is the article Baer (1974).

The connection of residual finiteness with the word problem is explained in Lyndon & Schupp (1977): see Theorem 4.6 on page 195. The argument given there can easily be adapted to show that a finitely presented conjugacy separable group has soluble conjugacy problem. Exercise 14 and Theorem 7 also imply the solubility of corresponding decision problems, as does the following theorem (Kegel (1966)): *a subgroup of a polycyclic group is subnormal if and only if its image in every finite quotient of the group is subnormal.* Exercises 13 and 14 are due to J.C. Lennox and J.S. Wilson; Exercise 14 is from Lennox & Wilson (1977).

The conjugacy separability of finitely generated nilpotent groups was first established in Blackburn (1965). The extension to polycyclic groups was achieved independently (and simultaneously, despite appearances) by V.N. Remeslennikov and E. Formanek; see Remeslennikov (1969) and Formanek (1976). There are several at least superficially different proofs of this theorem. Formanek proves it by deducing it from Theorem 5; the latter he reduces to Theorem 4 by matrix-group arguments. The proof of Remeslennikov (which was not flawless, see *Math. Reviews* **43**, # 6313) works directly within the polycyclic group, again reducing the result to Theorem 4.

The ingenious deduction of Theorem 5 from Theorem 3 is due to Wehrfritz (1973*a*). The proof of Theorem 3 I have given comes from Grunewald & Segal (1978*a*); Theorem 6 is taken from that paper, and so is the whole of section D.

Theorem 4 was proved in Schmidt (1930) and, independently, in Chevalley (1951). Generalizations to wider classes of rings are given in Wehrfritz (1978) and Segal (1979). A more elaborate version of Theorem 4 appears in Grunewald & Segal (1979*a*), a paper we shall refer to again in Chapter 9.

5
Faithful representations

We saw in Chapter 2 that soluble linear groups over $\mathbb{Z}$ are always polycyclic. The converse of this statement – that every polycyclic group is isomorphic to a subgroup of some $GL_n(\mathbb{Z})$ – was conjectured by Philip Hall in the 1950s, and proved a decade later by Louis Auslander. In a sense this would seem to have finished the subject of polycyclic groups; my feeling is rather that one should view it as the characterization of soluble $\mathbb{Z}$-linear groups, and as such, something which gives added relevance to the whole subject.

This chapter (and in a way the next two also) are about the embedding of a polycyclic group into a matrix group over $\mathbb{Z}$. The first step is to embed $\mathfrak{T}$-groups – torsion-free finitely generated nilpotent groups – into $\mathrm{Tr}_1(n,\mathbb{Z})$. There is a natural way to do this, and sections A and B are devoted to it. Section C shows how to go from there to the case of an arbitrary polycyclic group G; in fact not only G but its holomorph $G]\mathrm{Aut}\,G$ can be embedded into some $GL_n(\mathbb{Z})$. The embedding described here is rather un-canonical; a more canonical, but more complicated, approach will be explained in Chapter 7.

A. Dimension subgroups

Let G be a group. The constant map $\varepsilon: G \to \{1\}$ extends to a ring homomorphism $\varepsilon: \mathbb{Z}G \to \mathbb{Z}$, the *augmentation*, given by

$$(\textstyle\sum n_g g)\varepsilon = \sum n_g.$$

The kernel of ε is the *augmentation ideal*

$$\mathfrak{g} = \{\textstyle\sum n_g g \,|\, \sum n_g = 0\}$$

$$= \bigoplus_{1 \neq x \in G} \mathbb{Z}(x-1).$$

Exercise 1 Establish the above equality.

For a positive integer j, the jth *dimension subgroup* of G is the subgroup

$$\Delta_j(G) = (1 + \mathfrak{g}^j) \cap G.$$

This is exactly the kernel of the action of G by right multiplication on the factor ring $\mathbb{Z}G/\mathfrak{g}^j$. Putting $\mathfrak{g}^0 = \mathbb{Z}G$ as usual, observe that we have a series of G-submodules

$$\mathbb{Z}G/\mathfrak{g}^j = \mathfrak{g}^0/\mathfrak{g}^j \supseteq \mathfrak{g}/\mathfrak{g}^j \supseteq \ldots \supseteq \mathfrak{g}^i/\mathfrak{g}^j \supseteq \ldots \supseteq \mathfrak{g}^j/\mathfrak{g}^j = 0$$

and that G acts trivially on each of the factors of this series, since $\mathfrak{g}^{i-1}(G-1) \subseteq \mathfrak{g}^i$ for each i. Hence $G/\Delta_j(G)$ is nilpotent of class at most $j-1$ (Chapter 1, Proposition 11). Thus

$$\gamma_j(G) \le \Delta_j(G). \tag{1}$$

(Another way to see this is to embed $G/\Delta_j(G)$ into the subgroup $(\mathfrak{g}/\mathfrak{g}^j) + 1$ of the units in the ring $\mathbb{Z}G/\mathfrak{g}^j$, and observe that Proposition 9 of Chapter 1 applies.) For some time it was conjectured that $\gamma_j(G) = \Delta_j(G)$ for all groups G and all j; this conjecture was refuted by E. Rips in 1972, and much effort has since gone into elucidating the precise relationship which obtains in general between Δ_j and γ_j. That need not concern us here, however. All that we need is

Theorem 1   If G is a $\mathfrak{X}$-group of class c, then

$$\Delta_{c+1}(G) = \gamma_{c+1}(G) = 1.$$

This theorem is due to S.A. Jennings. The proof is rather long and rather computational, and I cannot better the exposition given in P. Hall's lecture notes *Nilpotent Groups*, so I shall refer the reader to Chapter 7 of that work for the full proof of Theorem 1 (see also section D below). It will be instructive however to prove the qualitative result,

Theorem 1*   If G is a $\mathfrak{X}$-group, then $\Delta_m(G) = 1$ for some positive integer m.

It is obvious from (1) above that $m \ge c+1$ here, if c is the class of G, so Theorem 1 is the best possible quantitative version of this. As far as our applications are concerned, we could in fact do without Theorem 1 altogether, by replacing '$\mathfrak{g}^{c+1}$' wherever it occurs by '$\mathfrak{g}^m$', where m is minimal subject to $\Delta_m(G) = 1$'; but this would be somewhat artificial, not to say cumbersome.

The proof of Theorem 1* occupies the rest of this section. Many of the techniques and arguments will be needed again later on. Until further notice, G will denote an arbitrary group.

Lemma 1   Let M be a $\mathbb{Z}G$-module. Then G stabilizes a series of length d in M if and only if $M\mathfrak{g}^d = 0$.

This is evident.

Lemma 2 The map $\theta: G/G' \to \mathfrak{g}/\mathfrak{g}^2$ which sends gG' to $(g-1)+\mathfrak{g}^2$ is an isomorphism (of the multiplicative group G/G' to the additive group $\mathfrak{g}/\mathfrak{g}^2$).

Proof For g and h in G,

$$gh - 1 = (g-1) + (h-1) + (g-1)(h-1), \tag{2}$$

so $g \mapsto (g-1) + \mathfrak{g}^2$ maps G homomorphically into $\mathfrak{g}/\mathfrak{g}^2$. Since the latter is an abelian group, this map induces a homomorphism $\theta: G/G' \to \mathfrak{g}/\mathfrak{g}^2$. To show that θ is bijective we construct its inverse. As $\mathfrak{g}$ is $\mathbb{Z}$-free on the basis $\{g - 1 | 1 \neq g \in G\}$, there is a homomorphism, ψ say, of $\mathfrak{g}$ into G/G' such that $(g-1)\psi = gG'$ for $1 \neq g \in G$. Now $\mathfrak{g}^2$ is additively generated by elements of the form $(g-1)(h-1)$, and

$$(g-1)(h-1)\psi = ((gh-1) - (g-1) - (h-1))\psi$$
$$= (gh)G' \cdot (gG')^{-1} \cdot (hG')^{-1}$$
$$= 1 \in G/G';$$

so $\mathfrak{g}^2 \leq \ker\psi$ and ψ induces a map $\varphi: \mathfrak{g}/\mathfrak{g}^2 \to G/G'$. It is immediate that $\theta\varphi$ is the identity map on G/G', and evaluating $\varphi\theta$ on the generators $(g-1) + \mathfrak{g}^2$ shows that $\varphi\theta$ is the identity map on $\mathfrak{g}/\mathfrak{g}^2$.

Since $\ker\theta = \Delta_2(G)/G'$, we infer:

Corollary 1 $\Delta_2(G) = \gamma_2(G)$ for all groups G.

Lemma 3 If G is finitely generated then each $\mathfrak{g}^i/\mathfrak{g}^{i+1}$ is finitely generated as a $\mathbb{Z}$-module.

Proof Lemma 2 shows that $\mathfrak{g}/\mathfrak{g}^2$ is finitely generated. If $i \geq 1$, there is a $\mathbb{Z}$-module epimorphism

$$\pi_i: \mathfrak{g}^{i-1}/\mathfrak{g}^i \otimes \mathfrak{g}/\mathfrak{g}^2 \to \mathfrak{g}^i/\mathfrak{g}^{i+1}$$

given by

$$(r + \mathfrak{g}^i) \otimes (s + \mathfrak{g}^2) \to rs + \mathfrak{g}^{i+1}.$$

Since the tensor product of finitely generated $\mathbb{Z}$-modules is again finitely generated, the lemma follows by induction on i.

Lemma 4 Suppose a group Γ acts by automorphisms on G, and Γ acts nilpotently on G/G'. Then Γ acts nilpotently on $\mathfrak{g}/\mathfrak{g}^m$ for each m.

Proof Here Γ is supposed to act on $\mathbb{Z}G$ by linear extension; it clearly leaves invariant all the ideals $\mathfrak{g}^i$. Now the map $\theta: G/G' \to \mathfrak{g}/\mathfrak{g}^2$ given in Lemma 2 is evidently a Γ-module isomorphism, so Γ acts nilpotently on $\mathfrak{g}/\mathfrak{g}^2$. Also the map π_i defined in the proof of Lemma 3 is a Γ-module epimorphism; it follows then from Lemma 5 of Chapter 1 that Γ acts nilpotently on each factor $\mathfrak{g}^i/\mathfrak{g}^{i+1}$, and this gives the result.

Exercise 2   Let $G = NX$ be a group which is the product of a normal subgroup N and a subgroup X, and let M be a G-module. Show that if both N and X act nilpotently on M, then so does G.
(*Hint*: X acts nilpotently on each factor of a suitable N-central series in M.)

The next result plays an important role in the embedding theorems:

Proposition 1   Let $G = H]\,X$ be a semi-direct product. Then the action of H, by right multiplication, and the action of X, by conjugation, on $\mathbb{Z}H$ extend simultaneously to give an action of G on $\mathbb{Z}H$. Thus there is a homomorphism

$$* : G \to \mathrm{Aut}_\mathbb{Z} \mathbb{Z}H$$

such that for $h \in H$, $x \in X$ and $a = \sum_{c \in H} a_c c \in \mathbb{Z}H$,

$$a^{h^*} = ah, \quad a^{x^*} = a^x = \sum_{c \in H} a_c c^x.$$

Proof   What has to be checked is this: if $h_i \in H$ and $x_i \in X$ satisfy

$$(h_1 x_1)(h_2 x_2) = h_3 x_3,$$

then

$$h_1^* x_1^* \cdot h_2^* x_2^* = h_3^* x_3^*.$$

Try it out on an element $a \in \mathbb{Z}H$. We get

$$a^{h_1^* x_1^* h_2^* x_2^*} = (((ah_1)^{x_1})h_2)^{x_2}$$
$$= a^{x_1 x_2} h_1^{x_1 x_2} h_2^{x_2}$$
$$= (ah_1 h_2^{x_1^{-1}})^{x_1 x_2}$$
$$= (ah_3)^{x_3} = a^{h_3^* x_3^*}$$

as required, since $h_3 = h_1 h_2^{x_1^{-1}}$ and $x_3 = x_1 x_2$.

Proof of Theorem 1* G is a $\mathfrak{X}$-group of class c. If $c = 1$ the theorem follows from Corollary 1, so we assume that $c > 1$. Now argue by induction on the Hirsch number of G; if $h(G) = 1$ then $c = 1$, so the first step is clear. By Corollary 5 of Chapter 1, $G/\zeta_{c-1}(G)$ is free abelian, so G has a normal subgroup $H \geq \zeta_{c-1}(G)$ with G/H infinite cyclic. Say $G = H]\langle x \rangle$. By inductive hypothesis there is a positive integer r such that $\Delta_r(H) = 1$, i.e.

$$(1 + \mathfrak{h}^r) \cap H = 1 \qquad (3)$$

where $\mathfrak{h}$ is the augmentation ideal of $\mathbb{Z}H$. Taking $X = \langle x \rangle$ in Proposition 1, we get an action $*$ of G on $\mathbb{Z}H$. It is easy to see that each power of $\mathfrak{h}$ is invariant under G^*. Now H^* acts nilpotently on $\mathbb{Z}H/\mathfrak{h}^r$ (Lemma 1), and X^* acts nilpotently on $\mathbb{Z}H/\mathfrak{h}^r$ by Lemma 4, since X acts nilpotently on H/H' because G is a nilpotent group. Hence, by Exercise 2, G^* acts nilpotently on $\mathbb{Z}H/\mathfrak{h}^r$. Thus if we extend $*: G \to \operatorname{Aut}_{\mathbb{Z}} \mathbb{Z}H$ to a ring homomorphism $*: \mathbb{Z}G \to \operatorname{End}_{\mathbb{Z}} \mathbb{Z}H$, we have

$$\mathbb{Z}H^{(\mathfrak{g}^m)^*} \subseteq \mathfrak{h}^r$$

for some positive integer m. I claim now that $\Delta_m(G) = 1$; this will finish the proof. Suppose then that $\Delta_m(G) \neq 1$. Then $\Delta_m(G) \cap \zeta_1(G) \neq 1$ (Chapter 1, Exercise 4); since $H \geq \zeta_{c-1}(G) \geq \zeta_1(G)$ it follows that $\Delta_m(G) \cap H \neq 1$. So there exists $h \in H$ with $h \neq 1$ and $h - 1 \in \mathfrak{g}^m$. Then

$$h - 1 \in \mathbb{Z}H \cdot (h - 1) = \mathbb{Z}H^{(h-1)^*} \subseteq \mathbb{Z}H^{(\mathfrak{g}^m)^*} \subseteq \mathfrak{h}^r.$$

But then (3) above forces $h = 1$, a contradiction.

B. The embedding

Until further notice, G will denote a $\mathfrak{T}$-group of class c. Define $I(G) \subseteq \mathbb{Z}G$ by

$$I(G)/\mathfrak{g}^{c+1} = \tau(\mathbb{Z}G/\mathfrak{g}^{c+1}),$$

the torsion subgroup of the additive group $\mathbb{Z}G/\mathfrak{g}^{c+1}$.

Lemma 5 $I = I(G)$ is an ideal of $\mathbb{Z}G$, $\mathbb{Z}G/I$ is finitely generated free as a $\mathbb{Z}$-module, and $(I + 1) \cap G = 1$.

Proof   The first statement is almost obvious, and the second follows from Lemma 3. To show that $(I + 1) \cap G = 1$, it will suffice to show that $(I + 1) \cap G = K$ say is finite, since G is supposed torsion-free. Let N be the kernel of the action of G, by right multiplication, on $I(G)/\mathfrak{g}^{c+1}$. Since, by Theorem 1, G acts faithfully on $\mathbb{Z}G/\mathfrak{g}^{c+1}$, it follows that $K \cap N$ is isomorphic to a subgroup of $\operatorname{Hom}_{\mathbb{Z}}(\mathbb{Z}G/I(G), \, I(G)/\mathfrak{g}^{c+1})$ (Chapter 1, Proposition 11). Therefore $K \cap N$ is finite. But N has finite index in G, so K is finite as required.

Now define the invariant $n(G)$ to be *the rank of the free $\mathbb{Z}$-module $\mathbb{Z}G/I(G)$,* i.e.

$$n(G) = h(\mathbb{Z}G/\mathfrak{g}^{c+1})$$

where c is the class of G. Lemma 5 shows that G acts faithfully on $\mathbb{Z}G/I(G)$, so we have our canonical embedding $G \to GL_n(\mathbb{Z})$ where $n = n(G)$. But we can do better than this:

Lemma 6 Put $X = \text{Aut } G$ and define the map $*:G]X \to \text{Aut}_\mathbb{Z} \mathbb{Z}G$ as in Proposition 1 (with G for H). Then $I(G)$ is invariant under $(G]X)^*$, and $G]X$ acts faithfully via $*$ on $\mathbb{Z}G/I(G)$.

Proof It is clear from the definition that $I(G)$ is left fixed by automorphisms of G, and this establishes the first claim. For the second claim, suppose $g \in G$ and $x \in X$ are such that $(gx)^*$ acts trivially on $\mathbb{Z}G/I(G)$. We have to show that $g = 1$ and $x = 1$. Now

$$g^x - 1 = 1_{\mathbb{Z}G}^{(gx)^*} - 1 \in I(G),$$

so $g^x = 1$ by Lemma 5. Therefore $g = 1$. Let $h \in G$ be arbitrary. Then

$$h^{-1}h^x - 1 = h^{-1}(h^{x^*} - h) \in h^{-1} \cdot I(G) \subseteq I(G),$$

so again by Lemma 5, $h^{-1}h^x = 1$. It follows that x is the identity automorphism of G, as required.

Theorem 2 Let G be a $\mathfrak{T}$-group. There exists a canonical injective homomorphism

$$\beta_G : G] \text{Aut } G \to GL_n(\mathbb{Z}),$$

with $n = n(G)$, such that $G\beta_G \leq \text{Tr}_1(n, \mathbb{Z})$.

To say that β_G is *canonical*, here, means that it is uniquely determined up to the choice of basis in $\mathbb{Z}^n$: a different choice has the effect of composing β_G with some inner automorphism of $GL_n(\mathbb{Z})$. Theorem 2 follows from Lemma 6: identify $\text{Aut } \mathbb{Z}G/I(G)$ with $GL_n(\mathbb{Z})$ by choosing a $\mathbb{Z}$-basis for $\mathbb{Z}G/I(G)$, and define $(gx)\beta_G$, for $g \in G$ and $x \in \text{Aut } G$, to be the automorphism induced on $\mathbb{Z}G/I(G)$ by $(gx)^*$. Since G^* acts nilpotently on $\mathbb{Z}G/I(G)$, by Lemma 1, the basis can be so chosen that $G\beta_G$ is represented by unitriangular matrices (this is discussed more fully in Lemma 7, below). The canonical nature of β_G is made precise in

Exercise 3 Let G and H be $\mathfrak{T}$-groups and $\alpha:G \to H$ an isomorphism. Show that there exists an inner automorphism γ of $GL_n(\mathbb{Z})$, where $n = n(G) = n(H)$, such that the following diagram commutes,

$$
\begin{array}{ccc}
G] \text{ Aut } G & \xrightarrow{\ \beta_G\ } & GL_n(\mathbb{Z}) \\
{\scriptstyle \alpha^*}\big\downarrow & & \big\downarrow{\scriptstyle \gamma} \\
H] \text{ Aut } H & \xrightarrow[\ \beta_H\ ]{} & GL_n(\mathbb{Z})
\end{array}
$$

where α^* is the obvious induced isomorphism.

(*Hint*: α induces an isomorphism $\bar{\alpha}:\mathbb{Z}G/I(G)\to\mathbb{Z}H/I(H)$. Let μ be the matrix of $\bar{\alpha}$ w.r.t. the chosen bases of these two modules and consider the inner automorphism of $GL_n(\mathbb{Z})$ given by μ.)

It is interesting to push the result of this exercise a bit further. Of course the exercise was really no more than an exercise on notation; but the relationship between $G\beta_G$ and $H\beta_H$ when H is a subgroup of G (not necessarily isomorphic to G) is not such a trivial question. When H has finite index in G, however, the relationship is as close as one might hope for:

Theorem 3 Let H be a subgroup of finite index in the $\mathfrak{X}$-group G. Then $n(H) = n(G)$ and there exists $\mu \in GL_n(\mathbb{Q})$, where $n = n(G)$, such that the following diagram commutes (here μ^* denotes conjugation by μ):

$$
\begin{array}{ccccc}
G & \xrightarrow{\beta_G} & GL_n(\mathbb{Z}) & \hookrightarrow & GL_n(\mathbb{Q}) \\
\big\downarrow & & & & \big\uparrow{\scriptstyle\mu^*} \\
H & \xrightarrow[\beta_H]{} & GL_n(\mathbb{Z}) & \hookrightarrow & GL_n(\mathbb{Q})
\end{array}
\tag{4}
$$

To prove this we shall have to take a closer look at the augmentation powers $\mathfrak{g}^i$. Recall (Chapter 3, section C) that for each i, $\tau_i(G)/\gamma_i(G)$ is the torsion subgroup of $G/\gamma_i(G)$. Denote by $m_i(G)$ the rank of the free abelian group $\tau_i(G)/\tau_{i+1}(G)$; thus $m_i(G)$ is the Hirsch number of $\gamma_i(G)/\gamma_{i+1}(G)$. Let $d_i(G)$ denote the Hirsch number of the additive group $\mathfrak{g}^i/\mathfrak{g}^{i+1}$; then $d_i(G)$ is equal to the dimension of the $\mathbb{Q}$-vector space $(\mathbb{Q}\mathfrak{g})^i/(\mathbb{Q}\mathfrak{g})^{i+1}$, where $\mathbb{Q}\mathfrak{g}$ is the rational augmentation ideal of G, i.e. the kernel of the obvious augmentation map from $\mathbb{Q}G$ to $\mathbb{Q}$. We must now take on trust the following fact, which arises in the proof of Theorem 1; see P. Hall, *Nilpotent Groups*, Chapter 7:

Proposition 2 For a $\mathfrak{X}$-group G of class c and for $1 \le i \le c$, the number $d_i(G)$ depends only on i, c, and the numbers $m_1(G), \ldots, m_c(G)$.

In fact, $d_i(G)$ is equal to the coefficient of t^i in the power series expansion of

$$f(t) = (1 - t)^{-m_1}(1 - t^2)^{-m_2}\ldots(1 - t^c)^{-m_c},$$

but we do not need to use this. Since

$$n(G) = 1 + d_1(G) + \ldots + d_c(G),$$

it will follow that for $H \le G$, $n(H) = n(G)$ provided that H also has class c and that $m_i(H) = m_i(G)$ for each i.

Exercise 4 Let G be a $\mathfrak{X}$-group of class c and let $H \le_f G$. Show that H has class c and that for each i, $|\gamma_i(G):\gamma_i(H)|$ is finite.

(*Hint*: Assume that $H \triangleleft G$ (see Chapter 1, Exercise 4(iv)). To show that H has class c, it suffices to show that $\gamma_c(H) \neq 1$. Put $K = \gamma_{c-1}(H), Z = \zeta_1(G)$ and suppose inductively that $K \nleq Z$. Then $G/C_G(K)$ is not periodic (Chapter 1, Exercise 13), so $H \nleq C_G(K)$ and the result follows. For the second part, put $T/\gamma_i(H) = \tau(G/\gamma_i(H))$ and apply the first part to find the nilpotency class of G/T.)

Exercise 5   Suppose $H \leq_f G$ and $G^e \leq H$ (i.e. $g^e \in H$ for every $g \in G$). Show that

$$eg \subseteq \mathfrak{h} + \mathfrak{g}^2.$$

Deduce that for every $m > 0$,

$$e^{m(m+1)/2} \mathbb{Z}G \subseteq \mathbb{Z}H + \mathfrak{g}^m.$$

(*Hint*: For the first part, use Lemma 2. Then argue by induction on m.)

We can now prove Theorem 3. We have a $\mathfrak{X}$-group G of class c and a subgroup H of finite index in G. By Exercise 4, H has class c, so $I(H)/\mathfrak{h}^{c+1}$ is the torsion subgroup of $\mathbb{Z}H/\mathfrak{h}^{c+1}$. Now in the picture

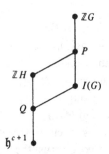

we have $\mathbb{Z}G/P$ finite by Exercise 5 (P, Q being the join, meet respectively of $\mathbb{Z}H$ and $I(G)$). Therefore

$$n(G) = h(\mathbb{Z}G/I(G)) = h(P/I(G)) = h(\mathbb{Z}H/Q).$$

But Proposition 2 together with Exercise 4 shows that $n(H) = n(G)$; and $n(H) = h(\mathbb{Z}H/\mathfrak{h}^{c+1})$. It follows that $Q/\mathfrak{h}^{c+1}$ is also finite, hence that $Q = I(H)$. Now choose bases for $\mathbb{Z}H/I(H)$ and $\mathbb{Z}G/I(G)$, to define the maps β_H and β_G. Let μ be the matrix, w.r.t. these bases, of the natural map of $\mathbb{Z}H/I(H)$ into $\mathbb{Z}G/I(G)$ suggested by the picture (i.e. the map induced by the inclusion $\mathbb{Z}H \hookrightarrow \mathbb{Z}G$). What we have established so far shows that μ is non-singular, so $\mu \in GL_n(\mathbb{Q})$ where $n = n(G) = n(H)$; and it is more or less obvious that the diagram (4) does commute (the reader can convince himself of this better than I can).

To conclude this section, let us briefly discuss *unipotent groups of* matrices. A subgroup G of $GL_n(k)$, k a field, is called *unipotent* if each element of G acts nilpotently on the module k^n. A classical theorem says that such a group G can be put into *unitriangular form*, i.e. that G is conjugate in $GL_n(k)$ to a subgroup of $\text{Tr}_1(n,k)$. In particular, it follows that the group G as a whole acts nilpotently on k^n. We do not need this result here (and when, in later chapters, unipotent groups are discussed, it would do no harm to take the latter property as the definition); but one ought to know this fact. What we do need, and have used above, is the fact that a subgroup of $GL_n(\mathbb{Z})$ which acts nilpotently on $\mathbb{Z}^n$ can be put into unitriangular form *over* $\mathbb{Z}$. Slightly more generally, one has

Lemma 7 Let R be a commutative principal ideal domain and G a group acting nilpotently by R-automorphisms on a finitely generated free R-module M. Then M has an R-basis with respect to which the action of G is represented by unitriangular matrices.

In other words, G can be put into unitriangular form over R.

Proof Define the 'central flag' of G in M,

$$0 = M_0 \subset M_1 \subset \ldots \subset M_k = M,$$

by taking M_i/M_{i-1} to be the set of fixed points of G in M/M_{i-1}, for $i = 1, 2, \ldots$. Then each M_{i-1} is an R-submodule of M and M/M_{i-1} is R-torsion-free. Since R is a P.I.D., M_i/M_{i-1} is a free R-module and therefore M_i splits over M_{i-1}:

$$M_i = M_{i-1} \oplus \bigoplus_{j=1}^{r_i} u_{ij} R$$

say. It is then easy to see that

$$(u_{k1}, \ldots, u_{kr_k}, \ldots, u_{21}, \ldots, u_{2r_2}, u_{11}, \ldots, u_{1r_1})$$

is a basis of the required kind.

This result is false for other kinds of coefficient ring! The reader might like to do

Exercise 6 Let R be a Dedekind Domain which is not a P.I.D.. Show that there is a unipotent matrix in $GL_2(R)$ which cannot be put into unitriangular form over R. Give such a matrix for $R = \mathbb{Z}[\sqrt{-6}]$.
(*Hint*: Let I be a non-zero ideal of R. Then $I \oplus J \cong R \oplus R$ as R-module for a certain fractional ideal J of R with $R \subseteq J$ (see P. Cohn's *Algebra* vol. 2, Chapter 11). Taking $M = I \oplus J$ we can identify $\text{Aut}_R M$ with $GL_2(R)$ by choosing a basis. Let $g: M \to M$ send (x, y) to $(x, y + x)$ for $x \in I$, $y \in J$. Show

that if g is upper-triangular w.r.t. some basis of M, then I is principal. For the second part, try $I = 2R + \sqrt{-6R}$.)

C. Embedding polycyclic groups

Theorem 2 which we proved in the previous section shows that a group of the special form $G = N]X$, where N is a $\mathfrak{X}$-group and $X \leq \operatorname{Aut} N$, can be embedded into a suitable $GL_n(\mathbb{Z})$. Not all polycyclic groups have this form, as we shall see in Chapter 11; there are two ways round this difficulty. One is to embed the given group into one of the form $N]X$: this needs some new ideas, and will be done in Chapter 7. The other way is based on the techniques of section A above, and we describe it now. Using an inductive argument, we shall establish

Theorem 4 Let G be a torsion-free polycyclic group such that $G/\operatorname{Fitt} G$ is free abelian. Then there exist a positive integer n and an injective homomorphism $\psi : G \to GL_n(\mathbb{Z})$ such that $(\operatorname{Fitt} G)\psi \leq \operatorname{Tr}_1(n, \mathbb{Z})$.

It is an easy step from this, which is proved below, to the full embedding theorem. To make the step, use

Exercise 7 Every polycyclic-by-finite group has a characteristic subgroup G of finite index satisfying the hypotheses of Theorem 4; and the following general principle:

Lemma 8 Let G be a subgroup of finite index d in a group H. If G is isomorphic to a subgroup of $GL_n(\mathbb{Z})$, then H is isomorphic to a subgroup of $GL_{nd}(\mathbb{Z})$.

Proof Identify G with a subgroup of $GL_n(\mathbb{Z})$. Thus $\mathbb{Z}^n$ becomes a G-module, which we call M. Then $M \otimes_{\mathbb{Z}G} \mathbb{Z}H$ is an H-module, isomorphic to $\mathbb{Z}^{nd}$, and we leave it to the reader to verify that H acts faithfully on this module. (Taking a transversal $\{x_1, \ldots, x_d\}$ to the right cosets of G in H, we have

$$M \otimes_{\mathbb{Z}G} \mathbb{Z}H = M \otimes x_1 \oplus \ldots \oplus M \otimes x_d.$$

If $h \in H$ and $x_i h = g x_j$ with $g \in G$, then for $\mu \in M$,

$$(\mu \otimes x_i)h = \mu g \otimes x_j.)$$

It is clear now that Theorem 4 implies the main result,

Theorem 5 Every polycyclic-by-finite group is isomorphic to a subgroup of some $GL_n(\mathbb{Z})$.

We need one more fact for the proof of Theorem 4:

Lemma 9 Let H be a polycyclic group and Q an ideal of $\mathbb{Z}H$. If $\mathbb{Z}H/Q$ is finitely generated as a $\mathbb{Z}$-module, then so is $\mathbb{Z}H/Q^m$ for every positive integer m.

Proof Every right ideal of $\mathbb{Z}H$ is finitely generated, i.e. $\mathbb{Z}H$ is a *right Noetherian* ring. To see this, let $K \lhd H$ with H/K cyclic, and assume inductively that $\mathbb{Z}K$ is right Noetherian. If H/K is finite, $\mathbb{Z}H$ is finitely generated as a right $\mathbb{Z}K$-module, so every $\mathbb{Z}K$-submodule of $\mathbb{Z}H$ is finitely generated. In particular every right ideal of $\mathbb{Z}H$ is finitely generated as a right $\mathbb{Z}K$-module, hence, *a fortiori*, as a right ideal of $\mathbb{Z}H$. If H/K is infinite we have $H = K]\langle x \rangle$ for some infinite cycle $\langle x \rangle$, and

$$\mathbb{Z}H = \mathbb{Z}K[x, x^{-1}],$$

a sort of polynomial ring. The 'variable' x need not commute with all the elements of $\mathbb{Z}K$, but it does 'normalize' $\mathbb{Z}K$ in an obvious sense. A careful look at the usual proof of Hilbert's basis theorem now shows that under these circumstances, '$\mathbb{Z}K$ right Noetherian' implies '$\mathbb{Z}H$ right Noetherian.' This result is due to Philip Hall, see section D.

The proof of the Lemma is now easily finished. For each i, Q^i is finitely generated as a right ideal of $\mathbb{Z}H$, so Q^i/Q^{i+1} is finitely generated as a right $\mathbb{Z}H$-module. But Q annihilates this module, so Q^i/Q^{i+1} is a finitely generated $\mathbb{Z}H/Q$-module. By hypothesis this ring is finitely generated as a $\mathbb{Z}$-module, therefore Q^i/Q^{i+1} is also finitely generated as a $\mathbb{Z}$-module. As this holds for $i = 0, 1, \ldots, m-1$, the result follows.

Remark In fact, Lemma 9 holds more generally for every finitely generated group H. The group ring $\mathbb{Z}H$ will not then in general be right Noetherian and a different argument is needed; see section D for a reference.

Proof of Theorem 4 G is a torsion-free polycyclic group with Fitting subgroup N, say, such that G/N is free abelian; N is a $\mathfrak{X}$-group. Let d be the rank of G/N. If $d = 0$ then Theorem 2 gives the required embedding

$$\psi = \beta_G : N = G \to \mathrm{Tr}_1(n, \mathbb{Z}).$$

Suppose $d > 0$. Then G has a normal subgroup H with $N \le H$ and G/H infinite cyclic, so

$$G = H]X$$

with $X = \langle x \rangle$, say, infinite cyclic. Note that the action of X on H (by conjugation) is faithful, for $C_X(H) \le \zeta_1(G) \le N$ but $X \cap N \le X \cap H = 1$. Also $N = \mathrm{Fitt}\, H$, since $H \lhd G$, and H/N is free abelian of rank $d-1$. So arguing by induction on d, we may assume the existence of an injective homomorphism $\varphi : H \to GL_n(\mathbb{Z})$, for some n, such that $N\varphi \le \mathrm{Tr}_1(n, \mathbb{Z})$. Then φ extends to a ring homomorphism

$$\bar{\varphi} : \mathbb{Z}H \to M_n(\mathbb{Z}).$$

Put $J = \ker \bar{\varphi}$. Note that $\mathbb{Z}H/J$ is finitely generated as a $\mathbb{Z}$-module, since $M_n(\mathbb{Z})$ is. Now for each element $y \in N$, $(y - 1)\bar{\varphi} = y\varphi - 1$ is an upper-triangular matrix with zero diagonal. Hence for $y_1, \ldots, y_n \in N$,

$$(y_1 - 1) \ldots (y_n - 1)\bar{\varphi} = 0,$$

and it follows that the augmentation ideal $\mathfrak{n}$ of $\mathbb{Z}N$ satisfies

$$\mathfrak{n}^n \le \ker \bar{\varphi} = J.$$

Since $N \lhd H$, this implies that the ideal

$$\bar{\mathfrak{n}} = \mathfrak{n}\mathbb{Z}H = \mathbb{Z}H\mathfrak{n}$$

of $\mathbb{Z}H$ also satisfies

$$\bar{\mathfrak{n}}^n \le J. \tag{5}$$

Now put

$$L = (\bar{\mathfrak{n}} + J)^n$$

and let T/L be the torsion subgroup of $\mathbb{Z}H/L$. We are going to make G act on the module $\mathbb{Z}H/T$.

Note first that $\mathbb{Z}H/T \cong \mathbb{Z}^l$ for some finite l. For $\mathbb{Z}H/(\bar{\mathfrak{n}} + J)$ is finitely generated as a $\mathbb{Z}$-module, hence, by Lemma 9, so is $\mathbb{Z}H/L$; factoring out the torsion leaves us with a finitely generated free $\mathbb{Z}$-module.

Observe next that T is invariant under right multiplication by H and under conjugation by X. Both will follow from the corresponding statements for L. Now L is clearly invariant under multiplication by H since it is an ideal of $\mathbb{Z}H$. It is not immediately apparent that $L^x = L$: this is the main trick of the proof. If $h \in H$ then

$$h^x - h = h([h, x] - 1) \in h(G' - 1) \subseteq h(N - 1) \subseteq \bar{\mathfrak{n}} \tag{6}$$

(recall that G/N is abelian); thus x acts trivially on the factor module $\mathbb{Z}H/\bar{\mathfrak{n}}$, and therefore leaves invariant the ideal $\bar{\mathfrak{n}} + J$. It follows that $L^x = L$ as required.

Now apply Proposition 1: this give a homomorphism

$$*: G = H]X \to \mathrm{Aut}_{\mathbb{Z}} \mathbb{Z}H.$$

We have just seen that T is invariant under the action of G^*, so there is an induced homomorphism

$$\psi : G \to \mathrm{Aut}_{\mathbb{Z}} \mathbb{Z}H/T \cong GL_l(\mathbb{Z}).$$

Since $N \le H, N$ acts by right multiplication, via $*$, on $\mathbb{Z}H$; so

$$\mathbb{Z}H(N^* - 1)^n = \mathbb{Z}H\mathfrak{n}^n = \bar{\mathfrak{n}}^n \le L \le T.$$

Thus N^* acts nilpotently on $\mathbb{Z}H/T$, so by choosing a suitable $\mathbb{Z}$-basis in $\mathbb{Z}H/T$ we can ensure that $N\psi \leq \mathrm{Tr}_1(l, \mathbb{Z})$.

All that remains now is to check that ψ is injective. Now ψ gives the action of G on $\mathbb{Z}H/T$ induced by the map $*: G = H]X \to \mathrm{Aut}_{\mathbb{Z}} \mathbb{Z}H$, and T/L is the torsion subgroup of $\mathbb{Z}H/L$. Suppose we can show that

$$(1 + T) \cap H = 1. \tag{7}$$

Then, just as in Lemma 6, it will follow that ψ is injective, since X acts faithfully on H.

To prove (7), recall that

$$L = (\bar{n} + J)^n \subseteq \bar{n}^n + J = J,$$

by (5). Since $\mathbb{Z}H/J$ is torsion-free, it follows that $T \subseteq J$, and so

$$(1 + T) \cap H \subseteq (1 + J) \cap H = \ker \varphi = 1.$$

Thus (7) is true, and the proof is finished.

Let us take another look at this proof. The main part of the argument served to show that an embedding $\varphi: H \to GL_n(\mathbb{Z})$ such that $N\varphi \leq \mathrm{Tr}_1(n, \mathbb{Z})$ gives rise to an embedding of $H]X$ into $GL_l(\mathbb{Z})$ for some l. Now we never actually exploited the precise nature of the group X: we used the fact that X acts faithfully on H, and, in order to establish (6), we used the fact that $[H, X] \leq N$. Thus we can state, as a corollary of the above proof,

Corollary 2 Let H be a torsion-free polycyclic group, N a nilpotent normal subgroup of H, and X a group acting faithfully on H so that $[H, X] \leq N$. If there exists an injective homomorphism $\varphi: H \to GL_n(\mathbb{Z})$ such that $N\varphi \leq \mathrm{Tr}_1(n, \mathbb{Z})$, then for some l there exists an injective homomorphism $\psi: H]X \to GL_l(\mathbb{Z})$.

The beauty of this is that there is no need for X to be polycylic here. The following lemmas will enable us to improve quite dramatically on Theorem 5.

Lemma 10 Let G be a polycyclic-by-finite group and let $N \leq H$ be characteristic subgroups of G such that G/H is finite, H/N is free abelian, and $N = \mathrm{Fitt}\, H$. Then $C_{\mathrm{Aut}\,G}(H/N)$ has finite index in $\mathrm{Aut}\, G$.

Proof Let π denote the natural map of $\mathrm{Aut}\, G$ into $\mathrm{Aut}\, H/N$. We must show that $\mathrm{Im}\, \pi$ is finite. Now $\mathrm{Aut}\, H/N \cong GL_k(\mathbb{Z})$ for some k, and therefore has a torsion-free subgroup of finite index (see Exercise 8 below). Hence every periodic subgroup of $\mathrm{Aut}\, H/N$ is finite, and it suffices to show that $\mathrm{Im}\, \pi$ is periodic. In other words, we must show that for each $x \in \mathrm{Aut}\, G$

there is a positive integer m with $[H, x^m] \leq N$. Now the group $G]\langle x \rangle$ is polycyclic-by-finite, hence (nilpotent-by-abelian)-by-finite (Chapter 2, Theorem 4); so if $K = \text{Fitt}(G\langle x \rangle)$, there exist $m \neq 0$ and $P \lhd_f G$ such that $[P, x^m] < K$. Then $[P \cap H, x^m] \leq K \cap H = N$. Since H/N is free abelian and $P \cap H$ has finite index in H it follows that $[H, x^m] \leq N$ as required.

Lemma 11   Let H be a characteristic subgroup of finite index d in a group G. Then there is an injective homomorphism

$$\sigma : C_{\text{Aut}\,G}(G/H) \to \prod_{d\,\text{copies}} H]\text{Aut}\,H.$$

Proof   Let $\{t_1, \ldots, t_d\}$ be a transversal to the cosets of H in G. For $\gamma \in C_{\text{Aut}\,G}(G/H)$ put $\gamma\sigma$ equal to

$$((t_1^{-1} t_1^{\gamma^{-1}}) \cdot \gamma|_H, \ldots, (t_d^{-1} t_d^{\gamma^{-1}}) \cdot \gamma|_H).$$

Then σ is clearly injective, since H, $t_1, \ldots, t_d$ together generate G. One computes directly that σ is also a homomorphism.

We have now assembled all the pieces required for

Theorem 6   If G is polycyclic-by-finite then Aut G is isomorphic to a subgroup of some $GL_n(\mathbb{Z})$.

Proof G has a characteristic torsion-free subgroup H of finite index such that $H/\text{Fitt}\,H$ is free abelian, by Exercise 7. Put $N = \text{Fitt}\,H$. By Lemma 10, the group

$$\Gamma = C_{\text{Aut}\,G}(H/N) \cap C_{\text{Aut}\,G}(G/H)$$

has finite index in Aut G, so by Lemma 8 it will suffice to embed Γ into some $GL_n(\mathbb{Z})$. If we put $X = C_{\text{Aut}\,H}(H/N)$, then Lemma 11 and its proof show that Γ can be embedded in the direct product of finitely many copies of $H]X$. But now Theorem 4 and Corollary 2 together show that $H]X$ has an embedding into some $GL_l(\mathbb{Z})$; the result follows.

As a final flourish, let us stick Theorems 5 and 6 together in

Theorem 7   If G is polycyclic-by-finite then the holomorph $G]$ Aut G can be embedded into some $GL_n(\mathbb{Z})$.

Proof   Consider the *wreath product*

$$G \wr C_2 = (G \times G)]\langle t \rangle$$

where t is the automorphism, of order 2, of $G \times G$ such that

$$(a, b)^t = (b, a).$$

Evidently $G \wr C_2$ is polycyclic-by-finite if G is, so by Theorem 6 it will be enough to show that $G] \operatorname{Aut} G$ can be embedded in $\operatorname{Aut}(G \wr C_2)$. The reader can verify directly that the following recipe gives a faithful action of $G] \operatorname{Aut} G$ on $G \wr C_2$: for $g \in G$ and $\gamma \in \operatorname{Aut} G$,

$$((a, b) \cdot 1)^{g\gamma} = ((g^{-1}ag)^\gamma, b^\gamma) \cdot 1 \quad \text{for} \quad (a, b) \in G \times G$$

$$((a, b) \cdot t)^{g\gamma} = ((g^{-1}a)^\gamma, (bg)^\gamma) \cdot t \quad \text{for} \quad (a, b) \in G \times G$$

(you might reflect on the true reason why this works).

The following exercises give an alternative approach to some of the theorems of Chapter 1.

Exercise 8 Fix a positive integer n. For each positive integer m let K_m be the subgroup of $GL_n(\mathbb{Z})$ consisting of all matrices congruent to 1_n modulo m (the 'principal congruence subgroup mod m'). Thus $K_1 = GL_n(\mathbb{Z})$ and $K_m \le K_s$ if $s \mid m$. (i) For each m, $K_m \lhd_f K_1$; (ii) Let M be $\mathbb{Z}^n$ considered as K_1-module. Then K_1/K_{m^e} acts faithfully on $M/m^e M$, and K_m/K_{m^e} stabilizes the series

$$M/m^e M > mM/m^e M > \ldots > m^e M/m^e M = 0;$$

(iii) K_m/K_{m^e} is a nilpotent group of exponent dividing m^{e-1}; (iv) If $m > 1$ then $\bigcap_{e=1}^{\infty} K_{m^e} = 1$; (v) For every prime p, K_p is residually a finite p-group, and $K_p \lhd_f GL_n(\mathbb{Z})$; (vi) If $p \ne q$ are primes, then $K_p \cap K_q$ is torsion-free.

Corollary 3 If G is polycyclic-by-finite, then for every prime p the holomorph of G has a subgroup of finite index which is residually a finite p-group. The holomorph of G is (torsion-free)-by-finite.

Exercise 9 Fix n as above and for each m put

$$U_m = \{g \in \operatorname{Tr}_1(n, \mathbb{Z}) \mid g \equiv 1_n \mod m\}.$$

(i) For each $m \ne 0$, U_1/U_m is a finite group, with order dividing a power of m; (ii) For each prime p, $\bigcap_{k=1}^{\infty} U_{p^k} = 1$; (iii) $\operatorname{Tr}_1(n, \mathbb{Z})$ is residually a finite p-group for every prime p.

Corollary 4 A $\mathfrak{X}$-group is residually finite-p for every prime p.

D. Notes

Dimension subgroups For the latest state of play on the 'dimension subgroup problem', i.e. the identification of $\Delta_j(G)$ in group-theoretic terms, see Passi (1979). The famous counterexample of Rips appeared in Rips (1972).

Theorem 1 was stated in Jennings (1955); for the proof, see Hall (1969), Chapter 7. The result is actually more general than Theorem 1 as stated. What Jennings and Hall prove is that for every group G and positive integer m,

$$((\mathbb{Q}\mathfrak{g})^m + 1) \cap G = \tau_m(G). \tag{8}$$

It is easy to see that $(\mathbb{Q}\mathfrak{g})^m$ is the space spanned over $\mathbb{Q}$ by $\mathfrak{g}^m$, and our Theorem 1 follows from (8) with $m = c + 1$. This material is also presented very fully in Passman (1977), Chapter 4, section 3. Passman's book is also a good reference for Proposition 2.

Embedding $\mathfrak{X}$-groups   That Theorem 2 is a consequence of Theorem 1 was pointed out by Hall (*op. cit.*). The natural approach to Theorem 3 is really to consider, as well as $\mathfrak{X}$-groups, also torsion-free nilpotent groups of finite rank which are radicable, i.e. every element has an nth root for every positive integer n. If G is a $\mathfrak{X}$-group, there is an essentially unique radicable torsion-free nilpotent group $\bar{G}$ containing G such that every element of $\bar{G}$ has some positive power in G (if G were abelian, $\bar{G}$ would be $\mathbb{Q} \otimes G$). The group $\bar{G}$ is the *radicable hull* or *Mal'cev completion* of G. Now if $\bar{G}$ has class c, then $((\mathbb{Q}\bar{\mathfrak{g}})^{c+1} + 1) \cap \bar{G} = 1$, and we get a canonical embedding $\beta_{\bar{G}} : \bar{G} \to \mathrm{Tr}_1(n, \mathbb{Q})$. The point of Theorem 3 is that β_G is just $\beta_{\bar{G}|G}$; if $H \leq_f G$ then $\bar{H} = \bar{G}$, and the transition from β_G to β_H is effected simply by choosing a different basis for $\mathbb{Q}^n$. A discussion of radicable hulls is given in the next chapter, based on Theorems 2 and 3 above. For other ways to approach this topic, see the references given in Chapter 6, section E.

There is a more direct way to embed a $\mathfrak{X}$-group into $\mathrm{Tr}_1(n, \mathbb{Z})$ than by way of Theorem 1. One argues by induction on the Hirsch number of the $\mathfrak{X}$-group G, and copies the proof of Theorem 4. We have $H \lhd G$ with $G = H]\langle x \rangle$ say, and, inductively, an embedding $\varphi : H \to \mathrm{Tr}_1(n, \mathbb{Z})$. Taking $N = H$ in the said proof, we have x^* acting nilpotently on $\mathbb{Z}H/T$, by Lemma 4; so the homomorphism $\psi : G = H\langle x \rangle \to GL_l(\mathbb{Z})$ will send G into $\mathrm{Tr}_1(l, \mathbb{Z})$, provided we choose a suitable basis for $\mathbb{Z}H/T \cong \mathbb{Z}^l$. It is no longer clear that ψ is injective now; but certainly $\psi|_H$ is injective, and to make ψ injective we can 'blow it up' as follows: define $\psi_1 : G \to \mathrm{Tr}_1(l + 2, \mathbb{Z})$ by

$$(hx^i)\psi_1 = \begin{bmatrix} (hx^i)\psi & & 0 \\ & & \\ 0 & 1 & i \\ & 0 & 1 \end{bmatrix}.$$

Unipotent matrix groups   For more about these, including their triangularizability, see Wehrfritz (1973*b*), Chapter 1.

Embedding polycyclic groups   Theorem 5 is due to Auslander (1967). The proof given above is essentially that of Swan (1967). The fact that $\mathbb{Z}H$ is Noetherian if H is polycyclic was proved in Hall (1954); see also Passman (1977), Chapter 10, section 2. For a proof of Lemma 9 applicable to any finitely generated group see Wehrfritz (1973*b*), Chapter 2, or Robinson (1972), vol. 2, section 10.1. These last two references also give proofs of Theorem 5, similar to Swan's.

Theorem 7, on embedding the holomorph of a polycyclic group, is due to Merzljakov (1970). The proof given here is a great simplification of Merzljakov's argument, discovered by Wehrfritz (1974); see Wehrfritz (1973*c*) for a simple account. Theorem 7 will find a surprising application in Chapter 8.

6
On unipotent groups

We saw in the last chapter that a finitely generated torsion-free nilpotent group has a natural representation as a unipotent group of matrices over $\mathbb{Q}$. A good way to study a unipotent group is to apply the 'logarithm' map, which embeds the group into a certain Lie algebra of matrices. The group-theoretic operations are then reflected in the Lie algebra operations; moreover, if the group is finitely generated, its logarithm will be almost (but not quite) a lattice in the Lie algebra. In this way, certain questions about unipotent groups get translated into questions about lattices in a Lie algebra (which is, in particular, a finite-dimensional vector space over $\mathbb{Q}$), and these things are usually easier to deal with.

In section A we develop the necessary formal properties of the logarithm operation, use them to construct the Lie algebra of a unipotent matrix group over $\mathbb{Q}$, and as an application construct the radicable hull (or 'Mal'cev completion') of such a group. Section B explores the connection between finitely generated unipotent groups and lattices: we shall see that such a group is only 'a finite distance away' from a *lattice group*, that is a group whose logarithm is actually a lattice. These results are applied in section C to show that the automorphism group of a finitely generated nilpotent group is in a natural way isomorphic to an arithmetic group: what this means, and some of its implications, will be discussed when we get there. Section D, finally, explains in outline how one can use these methods to solve the so-called *isomorphism problem* for nilpotent groups.

The technique introduced in this chapter is vital for the deeper study of finitely generated nilpotent groups. It also plays an essential role in the theory of (non-nilpotent) polycyclic groups, as we shall see in the following chapter.

A. log and exp

We keep fixed a positive integer n. An $n \times n$ matrix g is *unipotent* if $(g - 1)^n = 0$, and a group of matrices is called unipotent if each element in it is unipotent. Since such a group can always be put into unitriangular form (at least over a field, or over $\mathbb{Z}$, see section B of Chapter 5), we will lose no

100

generality and gain some simplicity by restricting attention to unitriangular groups. For any ring R, we denote by

$$\mathrm{Tr}_1(n, R)$$

the group of all upper-triangular $n \times n$ matrices over R with each diagonal entry equal to 1, and by

$$\mathrm{Tr}_0(n, R)$$

the set (it is a ring without 1) of upper-triangular $n \times n$ matrices over R with each diagonal entry equal to 0.

For the whole of this section, k will denote an arbitrary field of characteristic zero.

Definition For $x = 1 + u \in \mathrm{Tr}_1(n, k)$,

$$\log\ x = u - \tfrac{1}{2}u^2 + \ldots + (-1)^n (n-1)^{-1} u^{n-1}.$$

For $v \in \mathrm{Tr}_0(n, k)$,

$$\exp\ v = 1 + (2!)^{-1} v^2 + \ldots + (n-1)!^{-1} v^{n-1}.$$

Observe that these expressions for log and exp are really the same, on their respective domains, as the usual power series for logarithm and exponential, since $u^n = v^n = 0$ here. It follows therefore from the usual identities that

$$\log : \mathrm{Tr}_1(n, k) \to \mathrm{Tr}_0(n, k)$$

and

$$\exp : \mathrm{Tr}_0(n, k) \to \mathrm{Tr}_1(n, k)$$

are mutually inverse bijections; similarly, for commuting matrices $x, y \in \mathrm{Tr}_1(n, k)$ we have

$$\log xy = \log x + \log y \tag{1}$$

and for commuting matrices $u, v \in \mathrm{Tr}_0(n, k)$ we have

$$\exp u + v = (\exp u) \cdot (\exp v). \tag{2}$$

Formulae (1) and (2) cease to be valid if the matrices in question do not commute: 'correction terms' have to be introduced. The fact that this can be done, in a universal way, is embodied in the *Baker–Campbell–Hausdorff* formula. To express this we need, in addition to the multiplicative commutators

$$[x, y] = x^{-1} y^{-1} xy,$$

the additive commutators or *Lie brackets*

$$(u, v) = uv - vu.$$

Note that the map sending the pair (u, v) to the Lie bracket (u, v) is bilinear, a fact we shall often use without special mention. For $r > 2$, we write

$$[x_1, x_2, \ldots, x_r] = [[x_1, \ldots, x_{r-1}], x_r]$$
$$(v_1, v_2, \ldots, v_r) = ((v_1, \ldots, v_{r-1}), v_r);$$

these things are called repeated commutators (Lie brackets) of length r. For a vector $\mathbf{e} = (e_1, \ldots, e_r)$ of positive integers, write

$$[x, y]_\mathbf{e} = [x, \underbrace{y, \ldots, y}_{e_1}, \underbrace{x, \ldots, x}_{e_2}, \ldots]$$

$$(u, v)_\mathbf{e} = (u, \underbrace{v, \ldots, v}_{e_1}, \underbrace{u, \ldots, u}_{e_2}, \ldots).$$

Theorem 1   There exists constants $q_\mathbf{e} \in \mathbb{Q}$, one for each vector $\mathbf{e}$ of positive integers, such that $q_{(1)} = \frac{1}{2}$ and such that for any two matrices u, $v \in \mathrm{Tr}_0(n, k)$, the matrix

$$u * v = u + v + \sum_\mathbf{e} q_\mathbf{e}(u, v)_\mathbf{e} \tag{3}$$

has the property

$$(\exp u) \cdot (\exp v) = \exp u * v. \tag{4}$$

Since a repeated Lie bracket in $\mathrm{Tr}_0(n, k)$ of length n or more is easily seen to be zero, the summation in (3) needs only to be taken over those vectors $\mathbf{e} = (e_1, \ldots, e_s)$ such that $e_1 + \ldots + e_s < n$. In fact (3) and (4) represent an identity of power series (in non-commuting variables), if we allow formal infinite sums in (3) and in the definitions of log and exp; the coefficients $q_\mathbf{e}$ are independent of n and of the field k. We shall not prove Theorem 1; the proof, as well as an explict formula for the constants $q_\mathbf{e}$, can be found in books on Lie algebras (see section E for references). The theorem has a number of rather technical consequences which we shall be needing.

Corollary 1   For x and $y \in \mathrm{Tr}_1(n, k)$,

$$\log xy = (\log x) * (\log y).$$

This is immediate from Theorem 1. Less immediately, we also have

Corollary 2   For $x_1, \ldots, x_s \in \mathrm{Tr}_1(n, k)$, we have

$$\log[x_1, \ldots, x_s] - (\log x_1, \ldots, \log x_s) = \sum_i r_i c_i \tag{5}$$

where each c_i is a repeated Lie bracket of length at least $s + 1$ in the arguments $\log x_1, \ldots, \log x_s$, each of which appears at least once; and the coefficients r_i are universal constants lying in $\mathbb{Q}$.

Proof To see that the r_i are 'universal', we can suppose k to be the field of rational functions in many variables over $\mathbb{Q}$; having proved a formula like (5) for matrices over k, we can deduce that the same formula holds for an arbitrary s-tuple of matrices in $\mathrm{Tr}_1(n, k')$, for any field k' of characteristic zero, by specializing s matrices with distinct indeterminate entries. This argument also applies to some other identities we shall meet, and I shall not bother to repeat it.

It remains, then, to show that (5) holds. Let $x, y \in \mathrm{Tr}_1(n, k)$, put $\log x = a$, $\log y = b$, and let L be the $\mathbb{Q}$-space spanned by all repeated Lie brackets of length at least 3 in a and b. Using the fact that

$$a * b \equiv a + b + \tfrac{1}{2}(a, b) \bmod L$$
$$= b + a - \tfrac{1}{2}(b, a),$$

and the fact that $u * (a, b) \equiv u + (a, b) \bmod L$ for any u in the algebra generated by a and b, it is easy to see that there exists $\mu \in L$ with

$$(b * a) * (a, b) = (a * b) * \mu,$$

whence

$$(\exp b)(\exp a)(\exp (a, b)) = (\exp a)(\exp b)(\exp \mu).$$

Thus

$$\begin{aligned}\log[x, y] &= \log((\exp a)^{-1}(\exp b)^{-1}(\exp a)(\exp b)) \\ &= \log((\exp(a, b))(\exp \mu)^{-1}) \\ &= (a, b) * (-\mu) \\ &= (a, b) + v \quad \text{for some } v \in L \\ &= (\log x, \log y) + v.\end{aligned} \tag{6}$$

This establishes (5) for $s = 2$. Now argue by induction on s and suppose $s > 2$. Applying (6) with $[x_1, \ldots, x_{s-1}]$ for x and x_s for y gives

$$\log[x_1, \ldots, x_s] = (\log[x_1, \ldots, x_{s-1}], \log x_s) + v \tag{7}$$

where v is now a $\mathbb{Q}$-linear combination of repeated Lie brackets of length at least 3 in $\log[x_1, \ldots, x_{s-1}]$ and $\log x_s$. By inductive hypothesis,

$$\log[x_1, \ldots, x_{s-1}] = (\log x_1, \ldots, \log x_{s-1}) + \sigma \tag{8}$$

where σ is a $\mathbb{Q}$-linear combination of repeated Lie brackets of length at least s in $\log x_1, \ldots, \log x_{s-1}$, involving each of these arguments at least once. Since repeated Lie brackets are multilinear, putting (7) and (8) together now gives (5).

Corollary 3 For $x_1, \ldots, x_s \in \mathrm{Tr}_1(n, k)$, we have
$$(\log x_1, \ldots, \log x_s) = \log[x_1, \ldots, x_s] + \sum_i s_i \log v_i \tag{9}$$

where each v_i is a repeated group-theoretic commutator of length at least $s + 1$ in $x_1, \ldots, x_s$, each of which appears at least once, and the coefficients s_i are universal constants lying in $\mathbb{Q}$ (but depending now on n).

Proof Products of n elements in $\mathrm{Tr}_0(n, k)$ are zero, and repeated commutators of length n in $\mathrm{Tr}_1(n, k)$ are equal to 1. So (9) holds trivially for $s \geq n$. The result for $s \leq n$ then follows easily from Corollary 2, by induction on $n - s$.

Let us put these technical results to work. The algebra $\mathrm{Tr}_0(n, \mathbb{Q})$ has a natural structure as a Lie algebra over $\mathbb{Q}$, given by the Lie bracket operation. The operation $*$ given by (3) above is defined in terms of the Lie algebra operations only, but it is associative and indeed gives $\mathrm{Tr}_0(n, \mathbb{Q})$ the structure of a group, isomorphic via exp with $\mathrm{Tr}_1(n, \mathbb{Q})$. Conversely, the log mapping identifies subgroups of $\mathrm{Tr}_1(n, \mathbb{Q})$ with certain subsets of $\mathrm{Tr}_0(n, \mathbb{Q})$. The next theorem sums up the main features of this procedure. For a subset S of $\mathrm{Tr}_0(n, \mathbb{Q})$, we denote by $\mathbb{Q}S$ (respectively $\mathbb{Z}S$) the $\mathbb{Q}$-subspace (respectively $\mathbb{Z}$-submodule) of $\mathrm{Tr}_0(n, \mathbb{Q})$ generated by S.

Theorem 2   Let G be a subgroup of $\mathrm{Tr}_1(n, \mathbb{Q})$ and put

$$\mathscr{L}(G) = \mathbb{Q} \log G.$$

(i) $\mathscr{L}(G)$ is a Lie subalgebra of $\mathrm{Tr}_0(n, \mathbb{Q})$;
(ii) $\exp \mathscr{L}(G)$ is a subgroup of $\mathrm{Tr}_1(n, \mathbb{Q})$;
(iii) $\exp \mathscr{L}(G)$ is torsion-free, nilpotent and radicable;
(iv) $G \leq \exp \mathscr{L}(G)$ and every element of $\exp \mathscr{L}(G)$ has some positive power lying in G.

(A group H is *radicable* if for every $x \in H$ and every $m \in \mathbb{N}$ there exists $y \in H$ with $y^m = x$.)

Proof (i) Since the Lie bracket is bilinear, it will suffice to show that for x and $y \in G$ we have $(\log x, \log y) \in \mathscr{L}(G)$. This is immediate from Corollary 3.

(ii) follows from part (i) and Theorem 1, since for λ and μ in $\mathscr{L}(G)$,

$$(\exp \lambda)^{-1} = \exp(-\lambda) \quad \text{and} \quad (\exp \lambda)(\exp \mu) = \exp(\lambda * \mu).$$

(iii) The group $\mathrm{Tr}_1(n, \mathbb{Q})$ is torsion-free and nilpotent (see Proposition 10 of Chapter 1), hence so is its subgroup $\exp \mathscr{L}(G)$. If $\lambda \in \mathscr{L}(G)$ then also $m^{-1}\lambda \in \mathscr{L}(G)$ for any $m \in \mathbb{N}$, and

$$(\exp m^{-1}\lambda)^m = \exp \lambda.$$

So $\exp \mathscr{L}(G)$ is radicable.

(iv) Evidently $G = \exp \log G \subseteq \exp \mathscr{L}(G)$. For the final claim we must work harder:

Lemma 1   There exists a positive integer m, depending on n, such that

$$m\mathbb{Z} \log G \subseteq \log G \tag{10}$$

for every subgroup G of $\mathrm{Tr}_1(n,\mathbb{Q})$.

Before proving this let us deduce (iv) of Theorem 2. Suppose $\lambda \in \mathscr{L}(G)$; then $r\lambda \in \mathbb{Z} \log G$ for some $r \in \mathbb{N}$, so $mr\lambda \in \log G$, and

$$(\exp \lambda)^{mr} = \exp (mr\lambda) \in G,$$

justifying the claim.

Proof of Lemma 1 For x and $y \in \mathrm{Tr}_1(n,\mathbb{Q})$,

$$\log xy - \log x - \log y = (\log x) * (\log y) - (\log x + \log y)$$
$$= \sum q_e (\log x, \log y)_e$$
$$= \sum t_f \log [x, y]_f, \tag{11}$$

where the coefficients t_f are rational numbers depending only on the vector **f** and on n, not on x and y: they come from applying (9) to each of the brackets $(\log x, \log y)_e$, multiplying by q_e, and adding up the results. Let $r \in \mathbb{N}$ be a common denominator for all the coefficients t_f appearing in (11). We shall show that

$$m = r^{2^{n-1}} - 1$$

gives (10). For $0 \le s < n$, let G_s be the subgroup of G consisting of the matrices of the form

$$\tag{12}$$

i.e. those $g \in G$ with

$$g_{ij} = 0 \quad \text{for} \quad 0 < j - i < n - s.$$

Then

$$1 = G_0 \le G_1 \le \ldots \le G_{n-1} = G$$

is a central series of G. Now if $g_1, \ldots, g_s \in G_t$ then

$$\log g_1 + \ldots + \log g_s - \log(g_1 \ldots g_s) \in r^{-1} \mathbb{Z} \log G_{t-1}, \tag{13}$$

as we see by repeated applications of (11) (since $[G_t, G_t] \le G_{t-1}$). As G_1 is abelian, we certainly have

$$r^{2^0 - 1} \mathbb{Z} \log G_1 = \mathbb{Z} \log G_1 = \log G_1.$$

Fix $t > 1$ and suppose inductively that for every $q < t$,

$$r^{2^{q-1} - 1} \mathbb{Z} \log G_q \subseteq \log G_q. \tag{14}$$

Let $\lambda \in \mathbb{Z} \log G_t$. Then $\lambda = \log g_1 + \ldots + \log g_s$ for certain elements $g_1, \ldots, g_s$ in G_t, and (13) gives

$$r^{2^{t-2}} \lambda - \log w \in r^{2^{t-2} - 1} \mathbb{Z} \log G_{t-1}$$

where w is the $r^{2^{t-2}}$th power of $g_1 g_2 \ldots g_s$. By (14) there is then some $h_{t-1} \in G_{t-1}$ with

$$r^{2^{t-2}} \lambda = \log w + \log h_{t-1}.$$

Now apply (11) to get

$$r^{2^{t-2} + 1} \lambda = r \log w h_{t-1} - r \sum t_f \log [w, h_{t-1}]_f$$
$$= \log w' + \mu$$

say, with $w' = (w h_{t-1})^r$ and $\mu \in \mathbb{Z} \log G_{t-1}$. By (14) again, there exists $h_{t-2} \in G_{t-2}$ such that

$$r^{2^{t-3} - 1} \mu = \log h_{t-2};$$

and then

$$r^{2^{t-2} + 2^{t-3}} \lambda = \log w'' + \log h_{t-2}$$

with w'' a suitable power of w'. Repeating the last two steps in the obvious way, we end up with

$$r^{2^{t-2} + 2^{t-3} + \ldots + 2^0} \lambda = \log z + \log h_1$$

with $h_1 \in G_1$ and $z \in G_t$. But G_1 is central in G, so $\log z + \log h_1 = \log z h_1 \in \log G_t$. This establishes the induction step, since

$$2^{t-2} + 2^{t-3} + \ldots + 2^0 = 2^{t-1} - 1;$$

and it proves the lemma, since $G_{n-1} = G$.

By way of a digression, let us use Theorem 2 to construct the *radicable hull* or Mal'cev completion of a $\mathfrak{X}$-group G.

Definition   A radicable hull of a $\mathfrak{X}$-group G is a torsion-free radicable nilpotent group G^* containing G as a subgroup, such that every element of G^* has some positive power lying in G.

If G is a subgroup of $\mathrm{Tr}_1(n, \mathbb{Q})$, then Theorem 2 tells us that $\exp \mathscr{L}(G)$ is a radicable hull of G. We saw in Chapter 5 that for any $\mathfrak{X}$-group G there is a canonical embedding $\beta_G : G \to \mathrm{Tr}_1(n, \mathbb{Z})$ for a suitable n. If we think of β_G as embedding G in the group $\exp \mathscr{L}(G\beta_G)$, we see that the arbitrary $\mathfrak{X}$-group G has a radicable hull also. Of course, to deserve its name such a hull ought to be essentially unique; this will follow from

Exercise 1   Let G be a $\mathfrak{X}$-group and let $\beta_G : G \to \bar{G} = \exp \mathscr{L}(G\beta_G)$ be as above. Suppose M is a torsion-free nilpotent group and $\theta : G \to M$ is an injective homomorphism such that every element of M has some positive power lying in $G\theta$. Show that there exists a unique homomorphism $\psi : M \to \bar{G}$ making the following diagram commute:

and that ψ is injective. (In other words, the map $\theta^{-1} : G\theta \to G$ extends uniquely to an embedding of M into $\bar{G}$.)

(*Hint*: Since $\bar{G}$ is uniquely radicable – i.e. for $x \in \bar{G}$ and $m \in \mathbb{N}$ there exists a unique $y \in \bar{G}$, namely $y = \exp(m^{-1} \log g)$, with $y^m = x$ – it is obvious how to define ψ, and equally obvious that if ψ exists, then it is unique and injective. The difficult part is to show that the obvious map ψ is a group homomorphism. For this it will suffice to show that $\psi|_L$ is a homomorphism for every finitely generated subgroup L of M containing $G\theta$. Now for such an L, the index $|L : G\theta|$ is finite (why?). Apply Theorem 3 and Exercise 3 of Chapter 5 to find $\lambda, \mu \in GL_n(\mathbb{Q})$ such that

$$g\beta_G^{\lambda\mu} = g\theta\beta_{G\theta}^{\mu} = g\theta\beta_L \quad \text{for all } g \in G,$$

and verify that then

$$x\psi = \lambda\mu \cdot x\beta_L \cdot \mu^{-1}\lambda^{-1} \quad \text{for all } x \in L.$$

This shows that $\psi|_L$ is the homomorphism β_L composed with the homomorphism 'conjugation by $\mu^{-1}\lambda^{-1}$'.)

Corollary 4   For a $\mathfrak{X}$-group G, any two radicable hulls of G are isomorphic by a map fixing G elementwise.

Because of this it is usual to talk of 'the radicable hull' of a $\mathfrak{X}$-group (just as one talks of 'the field of fractions' of an integral domain).

Corollary 5   If G^* is the radicable hull of the $\mathfrak{X}$-group G then every automorphism of G extends uniquely to an automorphism of G^*.

Corollary 6   If $G \leq_f H$ for a $\mathfrak{X}$-group H and G^* is the radicable hull of G, then H embeds into G^* by a map fixing G elementwise, and then G^* is also a radicable hull for H.

As a worthwhile extended exercise, the reader might like to reprove the results of section C of Chapter 3 by embedding a polycyclic group G with a normal $\mathfrak{X}$-subgroup N into a group $\bar{G} = \bar{N}G$, with $\bar{N}$ a radicable hull of N, and then using Theorems 1* and 2** of that chapter directly.

B. Lattices

For a radicable subgroup G of $\mathrm{Tr}_1(n, \mathbb{Q})$, Theorem 2 shows that the set $\log G$ is a $\mathbb{Q}$-vector space. But our primary concern is with polycyclic groups, and particularly, here, with subgroups of $\mathrm{Tr}_1(n, \mathbb{Z})$. What does $\log G$ look like for such a group G? We have already seen in Lemma 1 that $\log G$ is not very far removed from being a $\mathbb{Z}$-module. However, it does not always have to be a $\mathbb{Z}$-module:

Exercise 2   Show that the set $\log \mathrm{Tr}_1(3, \mathbb{Z})$ is not closed under addition.

In this section we show that a slight adjustment to a unipotent group suffices to make its log into a $\mathbb{Z}$-module.

Definition   A unipotent matrix group G over $\mathbb{Q}$ is a *lattice group* if $\log G$ is an additive subgroup of the matrix ring.

The reason for the name is that if $G \leq \mathrm{Tr}_1(n, \mathbb{Q})$ is finitely generated, then $\mathbb{Z} \log G$ is actually a *lattice* in the $\mathbb{Q}$-space $\mathscr{L}(G)$, i.e. a free $\mathbb{Z}$-submodule which spans $\mathscr{L}(G)$ over $\mathbb{Q}$. To see this, do

Exercise 3   Let $G \leq \mathrm{Tr}_1(n, \mathbb{Q})$ be finitely generated. Show that $\mathbb{Z} \log G$ is a finitely generated $\mathbb{Z}$-module. Deduce that $\mathbb{Z} \log G$ is a lattice in $\mathscr{L}(G)$.
(*Hint*: Put $H = G \cap \mathrm{Tr}_1(n, \mathbb{Z})$. Show that $|G : H|$ is finite, and deduce that for some $s \in \mathbb{N}$

$$s \log G \subseteq \log H \subseteq (n-1)!^{-1} \mathrm{Tr}_0(n, \mathbb{Z}).$$

Then use Lemma 1.)

Having done Exercise 2, you will no doubt have guessed that $\mathrm{Tr}_1(3,\mathbb{Z})$ is the 'wrong group': it is easy to see that the slightly larger group

$$\begin{pmatrix} 1 & \mathbb{Z} & \tfrac{1}{2}\mathbb{Z} \\ & 1 & \mathbb{Z} \\ & & 1 \end{pmatrix} \tag{15}$$

is indeed a lattice group. In fact the same sort of adjustment works for every degree. For positive integers $c_1,\ldots,c_{n-1}$ define

$$\Gamma_n(c_1,\ldots,c_{n-1})$$

to be the set of all matrices $g\in\mathrm{Tr}_1(n,\mathbb{Z})$ such that

$$c_{j-i}\,g_{ij}\in\mathbb{Z} \quad \text{for } 1\le i<j\le n.$$

(Thus the group (15) is $\Gamma_3(1,2)$, for example.) We want to choose $c_1,\ldots,c_{n-1}$ in such a way that $\Gamma_n(c_1,\ldots,c_{n-1})$ becomes a lattice group. First of all, it had better be a group:

Exercise 4   Show that $\Gamma_n(c_1,\ldots,c_{n-1})$ is a subgroup of $\mathrm{Tr}_1(n,\mathbb{Q})$ provided that

$$c_i^2\,|\,c_{i+1} \quad \text{for } i=1,\ldots,n-2.$$

The lattice condition is a little more complicated. Let A_n (just for now) denote the algebra of all upper-triangular $n\times n$ matrices over $\mathbb{Q}$, and define algebra homomorphisms

$$\pi_1,\pi_2:A_n\to A_{n-1}$$

as follows. For $g\in A_n$, $g\pi_1$ is the matrix that is left when the last column of g is deleted; $g\pi_2$ is the matrix left when the first row of g is deleted. We shall also use the notation

$$\hat{g}=\begin{bmatrix} g_{11} & g_{12} & \cdots & g_{1,n-1} & 0 \\ & g_{22} & \cdots & g_{2,n-1} & g_{2n} \\ & & \ddots & & \vdots \\ & & & & g_{nn} \end{bmatrix} \qquad \bar{g}=\begin{bmatrix} g_{11} & 0 & \cdots & 0 & g_{1n} \\ & g_{22} & 0 & \cdots & 0 \\ & & \ddots & & \vdots \\ & & & & 0 \\ & & & & g_{nn} \end{bmatrix}$$

(to get $\hat{g}$ put the top right-hand corner equal to 0, to get $\bar{g}$ put all the above-diagonal entries except the corner one equal to 0). One verifies at once that for $g \in \mathrm{Tr}_1(n, \mathbb{Q})$,

$$\bar{g} = \hat{g}^{-1}g, \quad \log \bar{g} = \bar{g} - 1_n = \log g - \log \hat{g};$$

and a trivial calculation shows that for $g, h \in \mathrm{Tr}_1(n, \mathbb{Q})$,

$$\exp(\log g + \log h) = \exp(\log \hat{g} + \log \hat{h}) + \begin{bmatrix} 0 \dots 0 & g_{1n} + h_{1n} \\ & \cdot & 0 \\ & \cdot & \vdots \\ & \cdot & 0 \end{bmatrix}. \qquad (16)$$

Now suppose we have positive integers $c_1, \dots, c_{n-2}$ such that $H = \Gamma_{n-1}(c_1, \dots, c_{n-2})$ is a lattice group, and such that $c_i^2 | c_{i+1}$ for $i = 1, \dots, n-3$. We take a positive multiple c_{n-1} of c_{n-2}^2, and want to ensure that $G = \Gamma_n(c_1, \dots, c_{n-1})$ is also a lattice group. In view of Exercise 4 this simply means that $\log G$ should be closed under addition. Let $g, h \in G$ and put $w = \exp(\log g + \log h)$; we want w to lie in G. Now $g\pi_1$ and $h\pi_1$ belong to the group H, so by hypothesis

$$w\pi_1 = \exp(\log g\pi_1 + \log h\pi_1) \in \exp \mathbb{Z} \log H = H$$

(the first equality because π_1 is an algebra homomorphism). Thus

$$c_{j-i} w_{ij} \in \mathbb{Z} \text{ for } 1 \le i < j \le n-1.$$

Exactly the same argument using π_2 instead of π_1 shows that

$$c_{j-i} w_{ij} \in \mathbb{Z} \text{ for } 2 \le i < j \le n.$$

So the matrix w will belong to G provided its top right-hand entry w_{1n} satisfies

$$c_{n-1} w_{1n} \in \mathbb{Z}. \qquad (17)$$

From (16) above we know that $w_{1n} - g_{1n} - h_{1n}$ is equal to the $(1, n)$-entry of the matrix $\exp(\log \hat{g} + \log \hat{h})$. But the entries of $\hat{g}$ and of $\hat{h}$ all lie in $c_{n-2}^{-1} \mathbb{Z}$, so from the expressions for log and exp it follows that the entries of $\exp(\log \hat{g} + \log \hat{h})$ all lie in

$$((n-1)!((n-1)! c_{n-2}^{n-1})^{n-1})^{-1} \mathbb{Z} = (n-1)!^{-n} c_{n-2}^{-(n-1)^2} \mathbb{Z}.$$

Since g_{1n} and h_{1n} are in $c_{n-1}^{-1} \mathbb{Z}$, we see that (17) will hold provided

$$(n-1)!^n c_{n-2}^{(n-1)^2} | c_{n-1}.$$

We have now established

Proposition 1 Define integers $c_1, c_2, \dots$ recursively by

$$c_1 = 1, \quad c_i = i!^{i+1} c_{i-1}^{i^2} \text{ for } i > 1.$$

Then the set $\Gamma_n = \Gamma_n(c_1, \dots, c_{n-1})$ is a lattice group in $\mathrm{Tr}_1(n, \mathbb{Q})$ for every positive integer n.

Having constructed one lattice group with our bare hands, it is now quite easy to find lots of them. If G is any unipotent matrix group over $\mathbb{Q}$, it is clear that the intersection of all lattice groups containing G is again a lattice group containing G: we call it the *lattice hull* of G and denote it by G^{lat}. Now if $G \le \text{Tr}_1(n, \mathbb{Z})$, then Proposition 1 and Theorem 2 together show that

$$G \le G^{\text{lat}} \le \Gamma_n \cap \exp \mathscr{L}(G)$$

(Γ_n as defined in Proposition 1). Since Γ_n is polycyclic (why?) and every element of $\exp \mathscr{L}(G)$ has some positive power in G, we deduce that $|G^{\text{lat}}:G|$ is finite. More generally, we can state

Theorem 3 If G is a finitely generated unipotent matrix group over $\mathbb{Q}$, then G has finite index in its lattice hull G^{lat}.

Since for every $x \in GL_n(\mathbb{Q})$ we clearly have

$$(G^x)^{\text{lat}} = (G^{\text{lat}})^x,$$

Theorem 3 will follow from

Lemma 2 If $G \le GL_n(\mathbb{Q})$ is finitely generated and unipotent, then $G^x \le \text{Tr}_1(n, \mathbb{Z})$ for some $x \in GL_n(\mathbb{Q})$.

Proof We may assume that $G \le \text{Tr}_1(n, \mathbb{Q})$; this is something we didn't prove, but see section D of Chapter 5 for a reference. Say $G = \langle g_1, \ldots, g_s \rangle$. There exists $m \in \mathbb{N}$ such that

$$m(g_i - 1) \in \text{Tr}_0(n, \mathbb{Z}) \text{ for } i = 1, \ldots, s.$$

Let x be the diagonal matrix diag$(1, m, m^2, \ldots, m^{n-1})$. A trivial calculation shows that then $g_i^x \in \text{Tr}_1(n, \mathbb{Z})$ for each i, which gives the result.

Remark The construction of Γ_n, above, was very crude: thus it gives $\Gamma_3 = \Gamma_3(1, 8)$ whereas we have remarked above that $\Gamma_3(1, 2)$ is already a lattice group. It might be interesting to do

Exercise 5 Work out the lattice hull of $\text{Tr}_1(n, \mathbb{Z})$ for $n = 2, 3, 4, \ldots$ (For a general recipe see Section E.)

Instead of requiring $\log G$ to be a lattice in $\mathscr{L}(G)$, we could, more ambitiously, ask for $\log G$ to be a *Lie subring* of $\mathscr{L}(G)$: i.e. closed under the operation $(,)$ as well as under addition. Let us call a unipotent group G a *Lie-ring group* if $\log G$ is a Lie subring of the Lie algebra $\mathscr{L}(G)$ (the name 'Lie group' has unfortunately already been taken!); define the *Lie-ring hull* G^{Lie} of G in the obvious way. By copying the argument which led up to Proposition 1, the reader will have no difficulty in doing

Exercise 6 Define integers $c_1, c_2, \ldots$ recursively by

$$c_1 = 1, c_i = i!^{2i+1} c_{i-1}^{2i^2} \text{ for } i > 1.$$

Show that $\Gamma_n(c_1,\ldots,c_{n-1})$ is a Lie-ring group for every n. Deduce that for every finitely generated unipotent matrix group G over $\mathbb{Q}$, the index $|G^{\text{Lie}}:G|$ is finite.

For most purposes Theorem 3 gives all that is required. There is another approach to the construction of lattice hulls, however, which gives more precise information – in particular, an upper bound depending only on n for the index $|G^{\text{lat}}:G|$. It is based on the Baker–Campbell–Hausdorff formula, and is really a more delicate version of Lemma 1. In the proof of that lemma, we introduced a natural number r as a common denominator for the rational coefficients t_f occurring in (11). We now choose $r\in\mathbb{N}$ to satisfy the (possibly stronger) condition that for each e and every x, $y\in\mathrm{Tr}_1(n,\mathbb{Q})$,

$$q_e(\log x,\log y)_e\in r^{-1}\sum_f \mathbb{Z}\log[x,y]_f. \tag{18}$$

For $0\le s\le n$ denote by U_s the group of all matrices in $\mathrm{Tr}_1(n,\mathbb{Z})$ having $n-s-1$ lines of zeros above the main diagonal, as in (12) above.

Lemma 3   Let $K\lhd G\le\mathrm{Tr}_1(n,\mathbb{Q})$. There exists $t\in\mathbb{N}$, depending only on n, such that for all $g\in G$ and $k\in K$,

$$\log g + t\log k\in\log K\langle g\rangle \tag{19}$$
$$t\log g + \log k\in\log K\langle g^t\rangle. \tag{20}$$

Proof   We show by induction on i that if $K\le U_i$, then (19) and (20) hold with $t=r^{2^{i-1}}$. Since $K\le U_{n-1}=\mathrm{Tr}_1(n,\mathbb{Q})$, we can then take
$$t=r^{2^{n-2}}$$
to complete the proof. If $K\le U_1$ then (19) and (20) hold for every t, since U_1 is central in $\mathrm{Tr}_1(n,\mathbb{Q})$, and in particular for $t=r^{2^0}$. So the induction starts. Now fix $i>1$ and make the relevant inductive hypothesis. Suppose $K\le U_i$, and put $H=[K,G]$. Note that $H\le U_{i-1}$ and $H\lhd G$. Now let $g\in G$ and $k\in K$. For any $t\in\mathbb{Z}$ we have

$$\log g + t\log k - \log gk^t$$

$$= -\sum q_e(\log g, t\log k)_e$$

$$= -t\sum t^{n_e}\cdot q_e(\log g,\log k)_e$$

$$\in t\cdot r^{-1}\sum\mathbb{Z}\log[g,k]_f,$$

where n_e+1 is the number of occurrences of Y in $(X,Y)_e$. Taking $t=r^{2^{i-1}}$ and noting that $[g,k]_f\in H$ for each f, we have

$$\log g + t\cdot\log k - \log gk^t\in r^{2^{i-1}-1}\mathbb{Z}\log H.$$

Now the proof of Lemma 1 (see (14) above) shows that

$$r^{2^{i-2}-1}\mathbb{Z}\log H \subseteq \log H,$$

and so

$$\log g + t\log k - \log gk^t \in r^{2^{i-2}}\log H \qquad (21)$$

(because $2^{i-1} - 1 = 2^{i-2} - 1 + 2^{i-2}$). Now apply the inductive hypothesis (19) with gk^t for g and H for K, to deduce that

$$\log g + t\log k \in \log H\langle gk^t\rangle \subseteq \log K\langle g\rangle.$$

This gives the inductive step for (19). For (20), interchange g and k in the above argument: the analogue of (21) then gives

$$\log k + t\log g - \log kg^t \in r^{2^{i-2}}\log H;$$

the same inductive hypothesis then yields

$$\log k + t\log g \in \log H\langle kg^t\rangle \subseteq \log K\langle g^t\rangle$$

as required. This completes the proof.

Lemma 3 enables us to provide a simple sufficient condition for a lattice group. For $G \le \mathrm{Tr}_1(n, \mathbb{Q})$ and $s \in \mathbb{N}$, put

$$G^{1/s} = \langle g^{1/s} = \exp(s^{-1}\log g)|g \in G\rangle,$$

a subgroup of $\exp\mathscr{L}(G)$. We then have

Theorem 4 If $H \le \mathrm{Tr}_1(n, \mathbb{Q})$ and $H \lhd H^{1/t}$, where t is as in Lemma 3, then H is a lattice group.

Proof Let $x, y \in H$. Then $x = g^t$ for some $g \in H^{1/t}$. Apply Lemma 3 with $K = H$ and $G = H^{1/t}$ to get

$$\log x + \log y = t\log g + \log y \in \log H\langle g^t\rangle = \log H.$$

We shall apply this criterion with the help of some simple, purely group-theoretic, facts; the proofs are given below.

Proposition 2 If G is a nilpotent group of class at most c and $s \in \mathbb{N}$, then every element of the subgroup

$$G^{s^c} = \langle g^{s^c}|g \in G\rangle$$

is the sth power of an element of G.

Proposition 3 Let G be a nilpotent group of class at most c, and let $s \in \mathbb{N}$. Let X be a set of generators for G and put $H = \langle x^s|x \in X\rangle$. Then

$$G^{s^m} \le H$$

where $m = \frac{1}{2}c(c + 1)$.

Remark   This shows in particular that if $G = K^{1/s}$ for a subgroup K, then $G^{s^m} \le K$.

Exercise 7   Let G be a subgroup of $\mathrm{Tr}_1(n,\mathbb{Q})$ and let $s \in \mathbb{N}$. Show that the indices $|G:G^s|$ and $|G^{1/s}:G|$ are finite, and bounded by a function of n and s only (i.e. independent of G).
(*Hint*: Bound the orders of the factor groups

$$(G \cap U_i)U_{i-1}/(G^s \cap U_i)U_{i-1}$$

and

$$(G^{1/s} \cap U_i)U_{i-1}/(G \cap U_i)U_{i-1},$$

for $i = 1, \ldots, n-1$, using the above remark (of course G is nilpotent of class at most $n-1$ here).)

Suppose now that we have an arbitrary subgroup $G \le \mathrm{Tr}_1(n,\mathbb{Q})$. Let $t = t(n)$ be the number given in Lemma 3, and consider the subgroup

$$H = G^{t^{n-1}}.$$

By Proposition 2, $h^{1/t} \in G$ for every $h \in H$, so $H^{1/t} \le G$. But clearly $H \lhd G$, so $H \lhd H^{1/t}$ and Theorem 4 tells us that H is a lattice group. The same argument shows that the group

$$K = (G^{1/t^{n-1}})^{t^{n-1}}$$

is a lattice group. Evidently $K \ge G$, and the index $|K:G|$ is at most $|G^{1/t^{n-1}}:G|$. Together with the results of Exercise 7, this establishes

Theorem 5   There exist natural numbers α, β, γ, depending only on n, such that for every subgroup $G \le \mathrm{Tr}_1(n,\mathbb{Q})$,

G^α and $(G^{1/\alpha})^\alpha$ are lattice groups;

$|G:G^\alpha| \le \beta; \quad |(G^{1/\alpha})^\alpha : G| \le \gamma.$

Corollary 7   The index $|G^{\mathrm{lat}}:G|$ is finite and bounded for all unipotent subgroups G of $GL_n(\mathbb{Q})$.

There remains to prove Propositions 2 and 3. Both are easy consequences of

Lemma 4   Let G be a group with $\gamma_{c+1}(G) = 1$ and let $x_1, \ldots, x_r \in G$. Then for every positive integer k we have

$$(\gamma_c \langle x_1, \ldots, x_r \rangle)^{k^c} = \gamma_c \langle x_1^k, \ldots, x_r^k \rangle. \tag{22}$$

Proof   If $c = 1$ the result is clear. Suppose $c > 1$ and argue by induction on c. Without loss of generality we assume that $G = \langle x_1, \ldots, x_r \rangle$.

The inductive hypothesis applied to the group $G/\gamma_c(G)$ then gives

$$(\gamma_{c-1}(G))^{k^{c-1}}\cdot\gamma_c(G) = \gamma_{c-1}\langle x_1^k,\ldots,x_r^k\rangle\cdot\gamma_c(G). \tag{23}$$

Since $\gamma_c(G)$ is in the centre of G,

$$\begin{aligned}
(\gamma_c(G))^{k^c} &= \langle[\gamma_{c-1}(G),g]\,|\,g\in G\rangle^{k^c} \\
&= \langle[u,x_i]^{k^c}\,|\,u\in\gamma_{c-1}(G), i=1,\ldots,r\rangle \\
&= \langle[u^{k^{c-1}},x_i^k]\,|\,u\in\gamma_{c-1}(G), i=1,\ldots,r\rangle \\
&= \langle[\gamma_{c-1}(G)^{k^{c-1}},x_i^k]\,|\,i=1,\ldots,r\rangle \\
&= \langle[\gamma_{c-1}\langle x_1^k,\ldots,x_r^k\rangle,x_i^k]\,|\,i=1,\ldots,r\rangle \quad\text{by (23)} \\
&= \gamma_c\langle x_1^k,\ldots,x_r^k\rangle
\end{aligned}$$

as claimed.

Proof of Proposition 2 Given G with $\gamma_{c+1}(G)=1$ and $s\in\mathbb{N}$, we must show that every element of G^{s^c} is the sth power of some element of G. If $c=1$ this is clear, so suppose $c>1$ and argue by induction on c. Let $x_1,\ldots,x_r\in G$ and consider

$$g = x_1^{s^c}\ldots x_r^{s^c},$$

a typical element of G^{s^c}. Put

$$H = \langle x_1^s,\ldots,x_r^s\rangle, \quad K = \gamma_c(H).$$

By Lemma 4,

$$K = (\gamma_c\langle x_1,\ldots,x_r\rangle)^{s^c}. \tag{24}$$

Now $g\in H^{s^{c-1}}$ so the inductive hypothesis applied to H/K shows that there exist $y\in H$ and $z\in K$ with $g=y^s z$. Since $\gamma_c\langle x_1,\ldots,x_r\rangle$ is abelian, (24) shows that $z=u^{s^c}$ for some $u\in\gamma_c\langle x_1,\ldots,x_r\rangle$. Thus

$$g = (yu^{s^{c-1}})^s$$

is an sth power in G, as required.

Proof of Proposition 3 We have G and s as above, with $G=\langle X\rangle$ say. We must show that the subgroup $H=\langle x^s\,|\,x\in X\rangle$ contains $G^{s^{m(c)}}$ where $m(c)=\frac12 c(c+1)$. Again this is clear if $c=1$, so we take $c>1$ and argue by induction on c. A typical element of G has the form

$$g = x_1^{\pm1}\ldots x_r^{\pm1}$$

with $x_1,\ldots,x_r\in X$. Put $M=\langle x_1,\ldots,x_r\rangle$ and $N=\langle x_1^s,\ldots,x_r^s\rangle$. Inductive hypothesis applied to $M/\gamma_c(M)$ shows that

$$g^{s^{m(c-1)}} = uv$$

for some $u\in N$ and $v\in\gamma_c(M)$. But $s^{m(c)}=s^{m(c-1)}\cdot s^c$, so

$$g^{s^{m(c)}} = (uv)^{s^c} = u^{s^c}v^{s^c}$$

since v is central in M. By Lemma 4, $(\gamma_c(M))^{sc} = \gamma_c(N)$; therefore

$$g^{sm(c)} \in N^{sc}\gamma_c(N) \le N \le H.$$

The result follows.

C. Automorphisms

Let G be a subgroup of $\mathrm{Tr}_1(n,\mathbb{Q})$ and $\Lambda = \mathscr{L}(G) = \mathbb{Q}\log G$ the corresponding Lie algebra, as in Theorem 2. Then $\bar{G} = \exp\Lambda$ is the radicable hull of G. Theorem 1 showed that the group structure of $\bar{G}$ can be recovered from the Lie algebra structure of Λ, via the Baker–Campbell–Hausdorff formula. To get the Lie algebra structure of Λ, on the other hand, it would seem that we need to know not only the structure of $\bar{G}$ as a group, but also the way that $\bar{G}$ sits inside $GL_n(\mathbb{Q})$. However, it is a feature of the Baker–Campbell–Hausdorff formula that it can be 'disentangled', enabling us to reconstruct Λ directly from G. Precisely, we have

Proposition 4   Let $\Lambda \subseteq \mathrm{Tr}_0(n,\mathbb{Q})$ and $\Lambda' \subseteq \mathrm{Tr}_0(m,\mathbb{Q})$ be Lie subalgebras. Put $H = \exp\Lambda$ and $K = \exp\Lambda'$. For any mapping $\theta : \Lambda \to \Lambda'$, define $\hat\theta : H \to K$ by $(\exp a)\hat\theta = \exp(a\theta)$ for $a \in \Lambda$. Then θ is an isomorphism of Lie algebras if and only if $\hat\theta$ is an isomorphism of groups.

Proof   We may as well suppose that $m = n$, since the smaller set of triangular matrices can be embedded in the larger one. Theorem 1 shows that H and K are indeed groups, so the statement makes sense. Also it is clear that θ is bijective if and only if $\hat\theta$ is bijective; we assume henceforth that this is the case. Now if θ is a Lie algebra isomorphism, then $(a*b)\theta = a\theta * b\theta$ for all $a, b \in \Lambda$; so $(xy)\hat\theta = x\hat\theta \cdot y\hat\theta$ for all $x, y \in H$. Also if $x = \exp a \in H$, we have

$$x^{-1}\hat\theta = \exp(-a\theta) = (\exp(a\theta))^{-1} = (x\hat\theta)^{-1}.$$

Therefore $\hat\theta$ is a group isomorphism.

Suppose now that $\hat\theta$ is a group isomorphism. We must show that θ preserves scalar multiplication by $\mathbb{Q}$, addition, and the Lie bracket operation. Scalar multiplication is easy: if $a \in \Lambda$ and $r \in \mathbb{Z}$,

$$\exp((ra)\theta) = (\exp(ra))\hat\theta = (\exp a)^r \hat\theta = ((\exp a)\hat\theta)^r$$
$$= (\exp(a\theta))^r = \exp(r \cdot a\theta),$$

so $(ra)\theta = r \cdot a\theta$. Hence for non-zero integers p and q,

$$q \cdot (q^{-1}pa)\theta = (pa)\theta = p \cdot a\theta,$$

giving $(q^{-1}pa)\theta = q^{-1}p \cdot a\theta$ as required.

Now put $\Lambda_1 = \Lambda$ and for $i > 1$ put $\Lambda_i = (\Lambda_{i-1}, \Lambda)$. Then $\Lambda_n = 0$ since

$\Lambda \subseteq \mathrm{Tr}_0(n, \mathbb{Q})$. Let P_i be the statement

$$'(a + b)\theta = a\theta + b\theta \quad \text{for all } a \in \Lambda, \ b \in \Lambda_i',$$

and let Q_i be the statement

'if $r > 1, a_1, \ldots, a_r \in \Lambda$, and at least one of the

a_j is in Λ_i, then $(a_1, \ldots, a_r)\theta = (a_1\theta, \ldots, a_r\theta)'$.

What we have to do is to prove P_1 and Q_1, and we do this by reverse induction: it is easy to see that both P_n and Q_n are true.

Proof that $P_{i+1} \Rightarrow Q_i$ Fix $b \in \Lambda_i$, and consider the statement R_r:

'if $a_1, \ldots, a_r \in \Lambda$ and $b \in \{a_1, \ldots, a_r\}$, then $(a_1, \ldots, a_r)\theta = (a_1\theta, \ldots, a_r\theta)'$.

Certainly R_r holds for $r \geq n$, since in that case both sides of the equation are zero. Fix $r < n$ and suppose that R_j has been proved for all $j > r$. Write $x_j = \exp a_j$ for each j. From Theorem 1, Corollary 2 in section A, we have

$$(a_1, \ldots, a_r) = \log[x_1, \ldots, x_r] - \sum r_j c_j \tag{25}$$

where each c_j is of the form $(a_{j_1}, \ldots, a_{j_{s(j)}})$, with $s(j) > r$, and each a_k occurs among $a_{j_1}, \ldots, a_{j_{s(j)}}$. In particular, b must occur, so from $R_{s(j)}$ we have $c_j\theta = (a_{j_1}\theta, \ldots, a_{j_{s(j)}}\theta)$. Now since θ is a group isomorphism it follows that

$$(\log[x_1, \ldots, x_r])\theta = \log[x_1, \ldots, x_r]\theta = \log[x_1\theta, \ldots, x_r\theta].$$

Also each c_j is in Λ_{i+1} since $b \in \Lambda_i$. So by repeated applications of P_{i+1},

$$\begin{aligned}
(a_1, \ldots, a_r)\theta &= (\log[x_1, \ldots, x_r])\theta - \sum r_j c_j\theta \\
&= \log[x_1\theta, \ldots, x_r\theta] - \sum r_j(a_{j_1}\theta, \ldots, a_{j_{s(j)}}\theta) \\
&= (a_1\theta, \ldots, a_r\theta)
\end{aligned}$$

by the analogue of (25). Thus R_r holds. Hence by induction we see that R_r holds for every $r > 1$, and this establishes Q_i.

Proof that $P_{i+1} \Rightarrow P_i$ Let $a \in \Lambda$ and $b \in \Lambda_i$. Then from Theorem 1,

$$a + b = a * b - \sum q_e(a, b)_e.$$

Now $(a * b)\theta = a\theta * b\theta$ since $*$ corresponds to group multiplication in H and θ is an isomorphism. We have already shown that Q_i holds, and this gives $(a, b)_e\theta = (a\theta, b\theta)_e$ for each e. Since θ also preserves scalar multiplication, $(-q_e(a, b)_e)\theta = -q_e(a\theta, b\theta)_e$ for each e. But each of these terms lies in Λ_{i+1}; so repeated applications of P_{i+1} give

$$\begin{aligned}
(a + b)\theta &= (a * b)\theta + \sum(-q_e(a, b)_e)\theta = a\theta * b\theta - \sum q_e(a\theta, b\theta)_e \\
&= a\theta + b\theta.
\end{aligned}$$

Thus P_i holds; and this finishes the proof of Proposition 4.

Before discussing automorphisms, let us note an important consequence of Proposition 4: like the radicable hull, the Lie algebra associated to a $\mathfrak{X}$-group is essentially unique. More precisely, we have

Proposition 5   Let N be a $\mathfrak{X}$-group and $\alpha : N \to \mathrm{Tr}_1(m, \mathbb{Q})$ an injective homomorphism. Let $\beta = \beta_N : N \to \mathrm{Tr}_1(n, \mathbb{Z})$ be the canonical embedding (defined in Chapter 5). Then there is a Lie algebra isomorphism $\alpha^* : \mathscr{L}(N\beta) \to \mathscr{L}(N\alpha)$ such that the following diagram commutes:

$$
\begin{array}{ccc}
\mathscr{L}(N\beta) & \xrightarrow{\;\alpha^*\;} & \mathscr{L}(N\alpha) \\[4pt]
{\scriptstyle\log}\Big\uparrow & & \Big\uparrow{\scriptstyle\log} \\[4pt]
N\beta & & N\alpha \\[4pt]
& {\scriptstyle\beta}\nwarrow \quad \nearrow{\scriptstyle\alpha} & \\
& N &
\end{array}
$$

For by Exercise 1 in section A, the isomorphism $\beta^{-1}\alpha : N\beta \to N\alpha$ extends to an isomorphism $\varphi : \exp\mathscr{L}(N\beta) \to \exp\mathscr{L}(N\alpha)$; and Proposition 4 shows that $\alpha^* = \exp\circ\varphi\circ\log$ is an isomorphism of Lie algebras.

Exercise 8   Show that the assignment $N \to \mathscr{L}(N\beta_N)$ defines a functor Λ from the category of $\mathfrak{X}$-groups and injective homomorphisms to the category of (finite-dimensional nilpotent) Lie algebras over $\mathbb{Q}$ and injective homomorphisms.
(*Hint*: If $\theta : N \to M$ is injective, define $\Lambda(\theta)$ so as to make the following diagram commute:

$$
\begin{array}{ccc}
\mathscr{L}(N\beta_N) & \overset{\Lambda(\theta)}{\cdots\cdots\blacktriangleright} & \mathscr{L}(M\beta_M) \\[4pt]
{\scriptstyle(\theta\beta_M)^*}\nwarrow & & \nearrow \\[4pt]
& \mathscr{L}(N\theta\beta_M)\,. & \qquad)
\end{array}
$$

We turn now to the main topic of this section.

Proposition 6   Let G be a subgroup of $\mathrm{Tr}_1(n, \mathbb{Q})$ and put $\bar{G} = \exp\mathscr{L}(G)$. Then there is a canonical commutative diagram

$$
\begin{array}{ccc}
\mathrm{Aut}\,G & \xrightarrow{\;\sim\;} & \Gamma \\[4pt]
{\scriptstyle\varepsilon}\Big\downarrow & & \Big\uparrow \\[4pt]
\mathrm{Aut}\,\bar{G} & \xrightarrow{\;\sim\;} & \mathrm{Aut}\,\mathscr{L}(G)
\end{array}
\qquad\qquad (26)
$$

where Γ is the stabilizer in Aut $\mathscr{L}(G)$ of the subset $\log G$, and ε sends an automorphism of G to its unique extension to an automorphism of $\bar{G}$.

Proof The map ε is explained in Exercise 1, Corollary 2 in section A. The bottom isomorphism comes from Proposition 4, on taking $H = K = \bar{G}$ and $\Lambda = \Lambda' = \mathscr{L}(G)$; an automorphism γ of $\bar{G}$ corresponds to the automorphism $\exp\circ\,\gamma\circ\log$ of $\mathscr{L}(G)$. It is clear that the image of Aut G under this isomorphism is just the subgroup Γ of Aut $\mathscr{L}(G)$, so we can fill in the top row of the diagram in the obvious way.

To take full advantage of Proposition 6 we must explain the specially desirable features that the automorphism group of a Lie algebra possesses. This necessitates something of a digression into the realm of *algebraic groups*.

Definition A $\mathbb{Q}$-*group* of degree n is a subgroup

$$\mathscr{A} \leq GL_n(\mathbb{C})$$

which consists of all the matrices $g = (g_{ij})\in GL_n(\mathbb{C})$ which are common zeros of some set of polynomials in n^2 variables over $\mathbb{Q}$ (g is a *zero* of the polynomial P if $P(g_{11},\ldots,g_{ij},\ldots,g_{nn}) = 0$). For a subring R of $\mathbb{C}$, we write

$$\mathscr{A}(R) = \mathscr{A} \cap GL_n(R).$$

A subgroup Γ of $\mathscr{A}$ is called *arithmetic* in $\mathscr{A}$ if $\Gamma \leq \mathscr{A}(\mathbb{Q})$ and Γ is commensurable with $\mathscr{A}(\mathbb{Z})$, i.e. $\Gamma \cap \mathscr{A}(\mathbb{Z})$ has finite index in both Γ and $\mathscr{A}(\mathbb{Z})$.

The term '$\mathbb{Q}$-group' is short for 'algebraic matrix group defined over $\mathbb{Q}$'. A matrix group is called *arithmetic* if it is arithmetic in some $\mathbb{Q}$-group. We shall have more to say about arithmetic groups later on; but first, some examples.

Lemma 5 Let Λ be an n-dimensional $\mathbb{Q}$-algebra (not necessarily associative). Choose a basis for Λ and identify Aut Λ with a subgroup of $GL_n(\mathbb{Q})$. Then Aut $\Lambda = \mathscr{A}(\mathbb{Q})$ for some $\mathbb{Q}$-group $\mathscr{A}$ of degree n. If L is a lattice in Λ then the $N_{\mathrm{Aut}\Lambda}(L)$ is arithmetic in $\mathscr{A}$.

Proof A $\mathbb{Q}$-linear automorphism of Λ is an algebra automorphism if and only if it preserves the multiplication in Λ. So Aut Λ consists of those matrices $g\in GL_n(\mathbb{Q})$ such that

$$(\lambda g)\cdot(\mu g) - (\lambda\mu)g = 0 \quad \forall\lambda,\mu\in\Lambda. \tag{27}$$

For a fixed pair $\lambda, \mu\in\Lambda$, (27) is a vector equation, equivalent to n scalar equations of the form

$$P^{(i)}_{\lambda\mu}(g) = 0, \quad i = 1,\ldots,n \tag{28}$$

where each $P^{(i)}_{\lambda\mu}$ is a certain polynomial in the n^2 entries of the matrix g. It is tempting to assert now that Aut Λ is therefore equal to $\mathscr{A}(\mathbb{Q})$ where $\mathscr{A}$ is the subgroup of $GL_n(\mathbb{C})$ defined by the equations (28) for all λ and μ in Λ; this is correct, but one does have to check that the set of all common zeros of these equations in $GL_n(\mathbb{C})$ actually forms a group! There are general reasons why this must be so (the Zariski-closure of a group of matrices is again a group, for those in the know); but it is perhaps simpler to observe that the equations (28) do in fact define the subgroup Aut $(\Lambda \otimes \mathbb{C})$ of $GL_n(\mathbb{C})$. The final claim, regarding $N_{\mathrm{Aut}\,\Lambda}(L)$, is clear when L is the lattice spanned by our chosen basis of Λ, for in that case $N_{\mathrm{Aut}\,\Lambda}(L) = \mathscr{A}(\mathbb{Z})$ by definition. The general case follows from

Lemma 6 Let L be a lattice in $\mathbb{Q}^n$, and put $K = N_{GL_n(\mathbb{Q})}(L)$. Then $K \cap GL_n(\mathbb{Z})$ has finite index in both K and $GL_n(\mathbb{Z})$.

Proof Since L is finitely generated and $\mathbb{Q}L = \mathbb{Q}^n$, there exists a positive integer m such that

$$m\mathbb{Z}^n \subseteq L \subseteq m^{-1}\mathbb{Z}^n.$$

Since $GL_n(\mathbb{Z})$ fixes the two lattices $m\mathbb{Z}^n$ and $m^{-1}\mathbb{Z}^n$, it permutes all the intermediate lattices among themselves. But there are only finitely many such intermediate lattices, since the index $|m^{-1}\mathbb{Z}^n : m\mathbb{Z}^n|$ is finite; so there is a subgroup of finite index in $GL_n(\mathbb{Z})$ which leaves each one of them fixed. In particular, then,

$$K \cap GL_n(\mathbb{Z}) = N_{GL_n(\mathbb{Z})}(L)$$

has finite index in $GL_n(\mathbb{Z})$. The same argument interchanging the roles of $\mathbb{Z}^n$ and L shows that $K \cap GL_n(\mathbb{Z})$ has finite index in K also.

Further examples of algebraic and arithmetic groups are mentioned below. We have now collected enough information for

Theorem 6 Let G be a subgroup of $\mathrm{Tr}_1(n, \mathbb{Q})$ and put $\bar{G} = \exp \mathscr{L}(G)$. Identity Aut $\mathscr{L}(G) = A$, say, with a group of matrices over $\mathbb{Q}$ by fixing a $\mathbb{Q}$-basis for $\mathscr{L}(G)$, and put $\Gamma = N_A(\log G)$. Then

  (i) $A = \mathscr{A}(\mathbb{Q})$ for a certain $\mathbb{Q}$-group $\mathscr{A}$;
  (ii) Aut $\bar{G} \cong A$ via $\alpha \mapsto \exp \circ \alpha \circ \log$;
  (iii) Aut $G \cong \Gamma$ by the restriction of the map in (ii);
  (iv) if G is finitely generated, then Γ is arithmetic in $\mathscr{A}$.

Proof Part (i) is a special case of Lemma 5. Parts (ii) and (iii) come from Proposition 6. For Part (iv), we need some results from section B. We saw in Theorem 3 that G has finite index in its lattice hull G^{lat}; since G is now finitely generated, so also is G^{lat}, and therefore log $G^{\text{lat}} = L$, say, is actually a lattice in $\mathscr{L}(G)$ (see Exercise 3). Hence by Lemma 5, it will now suffice to show that Γ is a subgroup of finite index in $N_A(L)$. Suppose $\gamma \in \Gamma$. Then $\gamma = \exp \circ \alpha \circ \log$ for some $\alpha \in \text{Aut } \bar{G}$ with $G^\alpha = G$. Now $\log (G^{\text{lat}})^\alpha = L\gamma$ is again a lattice, so $(G^{\text{lat}})^\alpha$ is a lattice group containing $G^\alpha = G$, whence $(G^{\text{lat}})^\alpha \geq G^{\text{lat}}$. Replacing γ by γ^{-1} and α by α^{-1}, we get the reverse inclusion, and infer that $(G^{\text{lat}})^\alpha = G^{\text{lat}}$. Therefore $L\gamma = L$, and we conclude that $\Gamma \leq N_A(L)$. Now put $N_1 = N_{\text{Aut}\bar{G}}(G)$ and $N_2 = N_{\text{Aut}\bar{G}}(G^{\text{lat}})$. The isomorphism given in (ii) sends N_2 to $N_A(L)$ and N_1 to Γ. Therefore $N_1 \leq N_2$ and $|N_2:N_1| = |N_A(L):\Gamma|$. Since G^{lat} has only finitely many subgroups of index equal to $|G^{\text{lat}}:G|$ and N_2 permutes these among themselves, it follows that N_1 has finite index in N_2. Thus $|N_A(L):\Gamma|$ is finite, as required.

Corollary 8 The automorphism group of a $\mathfrak{T}$-group is isomorphic to an arithmetic group.

Exercise 9 Suppose $G = N\beta_N$ for some $\mathfrak{T}$-group N. Prove parts (i), (iii) and (iv) of Theorem 6 without using Proposition 4.
(*Hint*: Use the fact that β_N extends to a map of $N]$ Aut N into $GL_n(\mathbb{Z})$, so that every automorphism of G can be obtained by conjugation with some matrix in $GL_n(\mathbb{Z})$; deduce the existence of diagram (26), apart from the surjectivity of the bottom row.)

Remark For a different derivation of Corollary 8, see Exercise 10 of Chapter 8.

There is a large body of theory pertaining to arithmetic groups, the results of which can be applied via Theorem 6 to questions about $\mathfrak{T}$-groups. Before mentioning an example, let us extend this theorem a little. We need some elementary results about arithmetic groups:

Lemma 7 (i) Every finite group is isomorphic to an arithmetic group.

(ii) A subgroup of finite index in an arithmetic group is arithmetic.

(iii) The direct product of two arithmetic groups is arithmetic.

Proof (i) Let G be a group of order $n < \infty$. Embed G in $GL_n(\mathbb{Z})$ by its regular representation, so G becomes a group of (permutation) matrices

$$G = \{g^{(1)}, \ldots, g^{(n)}\} \subseteq GL_n(\mathbb{Z})$$

say. Then for a matrix $g \in GL_n(\mathbb{C})$,

$$g \in G \Leftrightarrow \prod_{q=1}^{n} (g_{i_q j_q} - g_{i_q j_q}^{(q)}) = 0$$

$\qquad$ for all choices of n-tuples $(i_1 j_1, \ldots, i_n j_n)$. $\qquad\qquad$ (29)

So these equations (29) actually define a $\mathbb{Q}$-group $\mathscr{G}$ such that

$$\mathscr{G} = \mathscr{G}(\mathbb{Q}) = \mathscr{G}(\mathbb{Z}) = G.$$

(ii) is virtually immediate from the definition of an arithmetic group.

(iii) Suppose G and H are arithmetic in $\mathscr{G}$ and $\mathscr{H}$ respectively.

Think of $G \times H$ as a subgroup of $\mathscr{G} \times \mathscr{H}$ realized as

$$\left\{ \begin{pmatrix} g & 0 \\ 0 & h \end{pmatrix} \middle| g \in \mathscr{G}, h \in \mathscr{H} \right\};$$

it is easy to see that in this form, $\mathscr{G} \times \mathscr{H}$ is a $\mathbb{Q}$-group, and that $G \times H$ is arithmetic as a subgroup of it.

Now given an arbitrary finitely generated nilpotent group, we can embed it in the direct product of a $\mathfrak{T}$-group and a finite group, with a corresponding embedding of its automorphism group. You can work out the details in

Exercise 10 Let G be a finitely generated nilpotent group of class c, and let e be the exponent of its torsion subgroup T. Put $m = e^c$. Show (with the help of Proposition 2) that

$$G^m \cap T = 1.$$

Write $\bar{G} = G/T$ and $\tilde{G} = G/G^m$, and let $\iota: G \to \bar{G} \times \tilde{G}$ be the natural embedding. Show that $G^m \iota = \bar{G}^m \times 1$, so that $G\iota$ has finite index in $\bar{G} \times \tilde{G}$. Show that there is an induced embedding $\iota^*: \operatorname{Aut} G \to \operatorname{Aut} \bar{G} \times \operatorname{Aut} \tilde{G}$, which maps $C_{\operatorname{Aut}G}(G/G^m)$ onto $C_{\operatorname{Aut}\bar{G}}(\bar{G}/\bar{G}^m) \times 1$ (this requires a moment's thought); deduce that the image $(\operatorname{Aut} G)\iota^*$ has finite index in $\operatorname{Aut} \bar{G} \times \operatorname{Aut} \tilde{G}$.

Theorem 6, Lemma 7 and Exercise 10 now immediately give

Corollary 9 The automorphism group of a finitely generated nilpotent group is isomorphic to an arithmetic group.

Among the interesting group-theoretic properties of arithmetic groups let us quote the following, a deep result due to Borel and Harish-Chandra:

Theorem 7 An arithmetic group is finitely presented, and it has only finitely many conjugacy classes of finite subgroups.

Exercise 11 Can an arithmetic group have an infinite periodic subgroup?

Corollary 10 The automorphism group of a finitely generated nilpotent group is finitely presented, and has only finitely many conjugacy classes of finite subgroups.

We shall see in Chapter 8 that this result can be extended to polycyclic groups. Let us conclude this section with some more examples of $\mathbb{Q}$-groups and arithmetic groups.

Exercise 12 Show that if $\mathscr{A}$ is a $\mathbb{Q}$-group of degree n and $x \in GL_n(\mathbb{Q})$ then $\mathscr{A}^x$ is a $\mathbb{Q}$-group, and if Γ is an arithmetic subgroup of $\mathscr{A}$ then Γ^x is arithmetic in $\mathscr{A}^x$.

Exercise 13 Let G be a finitely generated unipotent group of matrices over $\mathbb{Q}$. Show that $\exp \mathscr{L}(G) = \mathscr{G}(\mathbb{Q})$ for a certain $\mathbb{Q}$-group $\mathscr{G}$ and that G is arithmetic in $\mathscr{G}$.
(*Hint*: In view of Exercise 12 and Lemma 2, we may assume that $G \leq \mathrm{Tr}_1(n, \mathbb{Z})$. The vector space $\mathscr{L}(G) = \mathbb{Q} \log G$ is defined as a subspace of $\mathrm{Tr}_0(n, \mathbb{Q})$ by a family of linear equations $L_i(X) = 0$ say (X an indeterminate $n \times n$ matrix). So $\exp \mathscr{L}(G)$ is the set of common zeros in $\mathrm{Tr}_1(n, \mathbb{Q})$ of the equations $L_i(\log g) = 0$. You must check that these equations define a subgroup $\mathscr{G}$ of $\mathrm{Tr}_1(n, \mathbb{C})$, and that G has finite index in $\mathscr{G}(\mathbb{Z})$.)

Exercise 14 Show that every finitely generated nilpotent group is isomorphic to an arithmetic group.

Although polycyclic groups share many of the nice properties that arithmetic groups have, we shall see in Chapter 11 that the result of Exercise 14 cannot be extended to polycyclic groups in general.

D. Isomorphisms

An important application of the ideas we have been discussing is the solution of the *isomorphism problem* for nilpotent groups. The problem is to construct

Algorithm 1 This decides, given finite presentations for two nilpotent groups G and H, whether $G \cong H$.

This problem has a long but rather meagre history, see section E. To get a feeling for it, do

Exercise 15 Construct an algorithm to decide whether two abelian groups, specified by finitely many generators and relations, are isomorphic.
(*Hint*: look up an old-fashioned proof of the structure theorem for finitely generated abelian groups.)

What the methods of this chapter provide, strictly speaking, is not a procedure for Algorithm 1, but rather a reduction of this algorithm to two further algorithms of a more evidently Diophantine nature. I shall state what these do, and refer the reader to the literature for further information (see section E); the methods required for these algorithms are deep and lie beyond the scope of this book.

Algorithm A   This finds a finite set of generators for any given arithmetic group.

Algorithm B   Given a $\mathbb{Q}$-rational representation $\rho : GL_n(\mathbb{C}) \to GL_m(\mathbb{C})$ and given two vectors $a, b \in \mathbb{Q}^m$, this decides whether there exists $g \in GL_n(\mathbb{Z})$ such that $a\rho(g) = b$.

(*A* homomorphism ρ as above is called a $\mathbb{Q}$-*rational representation* if there exist m^2 polynomials ρ_{ij} over $\mathbb{Q}$ $(i, j = 1, \ldots, m)$ in $n^2 + 1$ variables such that

$$\rho(g)_{ij} = \rho_{ij}(g_{11}, \ldots, g_{nn}, (\det g)^{-1})$$
$$\text{for } 1 \leq i, j \leq m \text{ and all } g \in GL_n(\mathbb{C}).)$$

We saw in the previous section (Corollary 9) that if G is a finitely generated nilpotent group then Aut G is isomorphic to an arithmetic group. This observation leads directly from Algorithm A to

Algorithm 2   which finds a finite set of generators for Aut G where G is a given finitely generated nilpotent group.

Of course there are numerous details which have to be checked: it is not completely obvious that, given a finite presentation of G, one can effectively construct the relevant arithmetic group Γ (whatever this might mean), and effectively specify the isomorphism $\Gamma \cong \text{Aut } G$. However, I think it is fairly plausible that this can indeed be done; the reader might profitably try it before looking up the literature. Similar comments apply to the other algorithms about to be described: all that we can do here is give the underlying ideas.

Three steps reduce Algorithm 1 to Algorithms A and B.

First reduction to the case of $\mathfrak{T}$-groups. Suppose G and H are finitely generated nilpotent groups, given by finite presentations. Let c be an upper bound for their nilpotency classes, and e a common multiple for the exponents of $T = \tau(G)$ and $S = \tau(H)$. Put $m = e^c$. Just are in Exercise 10, we have embeddings

$$\iota_1 : G \to G/T \times G/G^m, \quad \iota_2 : H \to H/S \times H/H^m$$

with $G^m \iota_1 = G^m T/T \times 1$ and $H^m \iota_2 = H^m S/S \times 1$. If $H \cong G$ then $H/S \cong G/T$ and $H/H^m \cong G/G^m$. Supposing we can solve the isomorphism problem for $\mathfrak{T}$-groups, we can decide whether $H/S \cong G/T$. If the answer is 'yes', we pick a specific isomorphism $\varphi : G/T \to H/S$. It is then easy to see that $G \cong H$ if and only if there exist $\gamma \in \operatorname{Aut} G/T$ and an isomorphism $\alpha : G/G^m \to H/H^m$ such that

$$(G\iota_1)(\gamma\phi, \alpha) = H\iota_2. \tag{30}$$

This leaves us with infinitely many things to test, since in general $\operatorname{Aut} G/T$ will be infinite. The next two lemmas reduce the problem to a finite amount of checking.

Lemma 8 Let $\varphi : G/T \to H/S$ and $\alpha : G/G^m \to H/H^m$ be isomorphisms, and let $\gamma \in \operatorname{Aut} G/T$. Let $\pi : H/S \times H/H^m \to H/H^m S \times H/H^m$ denote the natural map. Then (30) holds if and only if

$$(G\iota_1)(\gamma\varphi, \alpha)\pi = H\iota_2\pi. \tag{31}$$

Proof Certainly $(30) \Rightarrow (31)$. Suppose (31) holds. Then

$$(G\iota_1)(\gamma\varphi, \alpha) \cdot \ker \pi = (H\iota_2) \cdot \ker \pi.$$

But

$$\ker \pi = H^m S/S \times 1 = H^m \iota_2 \le H\iota_2$$

and, since $(\gamma\varphi, \alpha)$ is an isomorphism,

$$H^m S/S \times 1 = (G^m T/T \times 1)(\gamma\varphi, \alpha) = (G^m \iota_1)(\gamma\varphi, \alpha) \le (G\iota_1)(\gamma\varphi, \alpha);$$

so we recover (30).

Lemma 9 Let F be a finite group of order d and let X be a set of generators for F. Then every element of F is a product of at most $d - 1$ elements of X.

Proof Since for $x \in X$, $x^{-1} = x^{d-1}$, every element of F is of the form

$$f = x_1 x_2 \ldots x_l$$

with $x_1, \ldots, x_l \in X$. If $l \ge d$ then at least two of the $l + 1$ 'initial segments' $x_1 x_2 \ldots x_i$ $(i = 0, 1, \ldots)$ must be equal; say the ith and the jth are equal, where

$i < j$. Then

$$f = x_1 \ldots x_i x_{j+1} \ldots x_l$$

is a shorter expression for f; and we can go on shortening the expression for f until we get $l < d$.

Let us return to the question of whether $\gamma \in \operatorname{Aut} G/T$ and $\alpha : G/G^m \overset{\sim}{\to} H/H^m$ exist so that (30) holds. Using Algorithm 2, find a finite generating set Y for $\operatorname{Aut} G/T$. Say $\operatorname{Aut} G/G^m T$ has order d. If γ_1 and γ_2 in $\operatorname{Aut} G/T$ have the same effect on $G/G^m T$, then $(\gamma_1 \varphi, \alpha)\pi = (\gamma_2 \varphi, \alpha)\pi$ in (31), since $(G^m T/T)\varphi = H^m S/S$. So if we set

$$W = \{ y_1 y_2 \ldots y_t \mid 0 \le t < d, \; y_1, \ldots, y_t \in Y \},$$

then Lemma 9 shows that

$$\{ (\gamma\varphi, \alpha)\pi \mid \gamma \in \operatorname{Aut} G/T, \alpha : G/G^m \cong H/H^m \}$$
$$= \{ (\gamma\varphi, \alpha)\pi \mid \gamma \in W, \alpha : G/G^m \cong H/H^m \}. \tag{32}$$

Thus to see whether (30) holds for some γ and α, it suffices to test for each $\gamma \in W$ and for each α whether (31) holds; and this can be done in finitely many steps, by inspection inside the finite group $(H/S \times H/H^m)\pi$. (The essential point of this trick is that we have no *a priori* way to identify the left-hand side of (32), or equivalently the image in $\operatorname{Aut} G/G^m T$ of $\operatorname{Aut} G/T$.)

Exercise 16 How can one determine the nilpotency class of a finitely generated nilpotent group which is given by (finitely many) generators and relations?

Second reduction to Lie rings. Let G and H be $\mathfrak{X}$-groups. Exercise 3 of Chapter 5 shows that $G \cong H$ if and only if $n(G) = n(H) = n$, say, and $G\beta_G$ is conjugate in $GL_n(\mathbb{Z})$ to $H\beta_H$. If this is so, then $G\beta_G^{\mathrm{Lie}}$ and $H\beta_H^{\mathrm{Lie}}$ must be conjugate (see section B), which implies that the Lie rings $L_1 = \log G\beta_G^{\mathrm{Lie}}$ and $L_2 = \log H\beta_H^{\mathrm{Lie}}$ are conjugate under the action of $GL_n(\mathbb{Z})$, hence that L_1 and L_2 are isomorphic as Lie rings. Since the group-theoretic structure of $G\beta_G^{\mathrm{Lie}} = \exp L_1$ is determined by the Lie ring structure of L_1 via the Baker–Campbell–Hausdorff formula (Theorem 1), and similarly for $H\beta_H^{\mathrm{Lie}}$, we see, conversely, that $L_1 \cong L_2 \Rightarrow G\beta_G^{\mathrm{Lie}} \cong H\beta_H^{\mathrm{Lie}}$. The upshot is that $G \cong H$ if and only if there is an isomorphism $\theta : L_1 \cong L_2$ such that the induced isomorphism $\theta^+ : G\beta_G^{\mathrm{Lie}} \to H\beta_H^{\mathrm{Lie}}$ maps $G\beta_G$ onto $H\beta_H$. This question is similar in form to what we discussed in the first reduction, above. To decide it what is needed is

Algorithm 3 If L is a Lie ring, free of finite rank as $\mathbb{Z}$-module, this finds a finite set of generators for $\operatorname{Aut} L$; and

Algorithm 4   If L_1 and L_2 are Lie rings, both free of finite rank as $\mathbb{Z}$-modules, this decides whether $L_1 \cong L_2$ as Lie rings.

(We shall see in the 'third reduction' below that these algorithms do not depend on the Lie-ness of the rings; they work just as well if the word 'Lie' is replaced by the phrase 'not necessarily associative'.)

Using Algorithm 4, decide whether $L_1 \cong L_2$. If not, then $G \not\cong H$. If 'yes', choose an isomorphism $\psi : L_1 \overset{\sim}{\to} L_2$. Then $G \cong H$ if and only if there exists $\gamma \in \operatorname{Aut} L_1$ such that

$$(G\beta_G)(\gamma\psi)^+ = H\beta_H \tag{33}$$

where $(\gamma\psi)^+ = \log \circ \gamma\psi \circ \exp$. We can now reduce the question of whether γ exists so that (33) holds to a finite amount of checking, just as we did in the first reduction:

Exercise 17   Given a finitely generated subgroup $U \le \operatorname{Tr}_1(n, \mathbb{Z})$, show how one can determine a positive integer m such that $(U^{\operatorname{Lie}})^m \le U$. (*Hint*: Either find an m depending only on n, by imitating the proof of Theorem 5 (section B) – this is quite tricky; or find an m, depending on U now, by combining Exercise 6 with an estimate of the index $|\operatorname{Tr}_1(n, \mathbb{Z}) \cap \exp \mathscr{L}(U) : U|$.)

Returning to our problem, we find a finite generating set Y for $\operatorname{Aut} L_1$, by Algorithm 3, and find an m as in Exercise 17 which works for both $U = G\beta_G$ and $U = H\beta_H$. Put

$$d = |\operatorname{Aut}(H\beta_H^{\operatorname{Lie}}/(H\beta_H^{\operatorname{Lie}})^m)|$$

and let W be the set of all products of at most $d-1$ elements of Y. Then Lemma 9 shows that (33) holds for some $\gamma \in \operatorname{Aut} L_1$ if and only if

$$(G\beta_G)(\gamma\psi)^+\pi = (H\beta_H)\pi \text{ for some } \gamma \in W,$$

where $\pi : H\beta_H^{\operatorname{Lie}} \to H\beta_H^{\operatorname{Lie}}/(H\beta_H^{\operatorname{Lie}})^m$ is the natural map. This can be decided in finitely many steps by looking inside the finite group $(H\beta_H^{\operatorname{Lie}})\pi$. Thus the original problem is reduced to Algorithms 3 and 4.

Third reduction   Given two rings L_1 and L_2, not necessarily associative, whose additive groups L_1^+, L_2^+ are free of finite rank over $\mathbb{Z}$, we want to decide whether $L_1 \cong L_2$. Assume that $L_1^+ \cong L_2^+ \cong \mathbb{Z}^n$; if the ranks were different there would be no question of isomorphism. Fixing $\mathbb{Z}$-bases $(e_1, \ldots, e_n), (f_1, \ldots, f_n)$ for L_1, L_2 respectively, we can identify the set of all $\mathbb{Z}$-module isomorphisms $L_1 \overset{\sim}{\to} L_2$ with $GL_n(\mathbb{Z})$. A $\mathbb{Z}$-module isomorphism $\gamma : L_1 \overset{\sim}{\to} L_2$ is a ring isomorphism if and only if it preserves the multiplication, and this holds if and only if

$$e_i\gamma \cdot e_j\gamma = (e_i \cdot e_j)\gamma \quad (i, j = 1, \ldots, n), \tag{34}$$

the '·' on the left representing multiplication in L_2, that on the right multiplication on L_1. These operations are defined by the *structure constants* of L_1 and L_2:

$$e_i \cdot e_j = \sum d_{ij}^k e_k$$

$$f_i \cdot f_j = \sum b_{ij}^k f_k$$

say; and (34) is equivalent to

$$\left(\sum_p \gamma_{ip} f_p \right) \cdot \left(\sum_q \gamma_{jq} f_q \right) = \sum_r \sum_k a_{ij}^k \gamma_{kr} f_r \qquad \text{(all } i,j)$$

$$\|$$

$$\sum_r \sum_{p,q} \gamma_{ip} b_{pq}^r \gamma_{jq} f_r.$$

Equating coefficients of f_r gives

$$\sum_{p,q} \gamma_{ip} b_{pq}^r \gamma_{jq} - \sum_k a_{ij}^k \gamma_{kr} \qquad \text{(all } i, j, r). \tag{35}$$

Writing $\bar\gamma$ for the matrix γ^{-1}, (35) is equivalent to a system of equations

$$\sum_{p,q,r} \gamma_{ip} b_{pq}^r \gamma_{jq} \bar\gamma_{rs} = a_{ij}^s \qquad \text{(all } i,j,s). \tag{36}$$

(36) expresses the hypothesis that the operation of γ transforms the multiplication defined by the system of structure constants $\mathbf{a} = (a_{ij}^s)$ into that defined by the system $\mathbf{b} = (b_{pq}^r)$. Thinking of these systems of constants as vectors in n^3-dimensional space, we can consider (36) as defining an action ρ of $GL_n(\mathbb{C})$ on $\mathbb{C}^{n^3}$: formally, for $g \in GL_n(\mathbb{C})$ define $\rho(g) \in GL_{n^3}(\mathbb{C})$ by

$$\rho(g)_{uv} = \bar g_{i\lambda} \bar g_{j\mu} g_{vs} \tag{37}$$

where $u = (\lambda - 1)n^2 + (\mu - 1)n + v$,

$$v = (s - 1)n^2 + (i - 1)n + j$$

(where $\bar g = g^{-1}$). Then (34) is equivalent to

$$\mathbf{b}\rho(\gamma^{-1}) = \mathbf{a}. \tag{38}$$

We must check that ρ is a $\mathbb{Q}$-rational representation of $GL_n(\mathbb{C})$. It is visibly defined by polynomials over $\mathbb{Q}$ in the entries of g and in $(\det g)^{-1}$; I leave it to the reader to do

Exercise 18   Show that $\rho : GL_n(\mathbb{C}) \to GL_{n^3}(\mathbb{C})$ defined by (37) is a homomorphism.

Algorithm B will therefore tell us whether there exists $\gamma \in GL_n(\mathbb{Z})$ such that $\mathbf{a}\rho(\gamma) = \mathbf{b}$, and this finishes Algorithm 4.

Algorithm 3 is merely a special case of Algorithm A; for taking $L_1 = L_2 = L$ in the above discussion we see that $\operatorname{Aut} L$ can be identified with the stabilizer in $GL_n(\mathbb{Z})$, acting via ρ, of the 'structure constants' vector **a** corresponding to L. Thus

Aut $L = \mathscr{G}(\mathbb{Z})$

is an arithmetic group, where

$$\mathscr{G} = \{g \in GL_n(\mathbb{C}) | \mathbf{a}\rho(g) = \mathbf{a}\}$$

is evidently a $\mathbb{Q}$-group. (Another look at Lemma 5, in the previous section, shows that we are really just repeating its proof here, in a more explicit form; thus $\mathscr{G}$ is the automorphism group of the algebra $\mathbb{C} \otimes L$.)

This is as far as we shall go with the isomorphism problem for nilpotent groups; the possibility of extending its solution to polycyclic groups will be discussed in the next chapter. It is worth remarking, however, that the preceding discussion has a significance which goes beyond its application to a decision problem. What we have established is that, roughly speaking, the classification of $\mathfrak{T}$-groups up to isomorphism is equivalent to the classification of the orbits of $GL_n(\mathbb{Z})$ acting rationally on a certain vector space. This does not necessarily make it easier, but it does show that the problem of classifying these groups fits into a general pattern familiar in other parts of mathematics. A paradigm for such problems is the classification theory of integral quadratic forms, and this theory has played a pivotal role in the genesis of Algorithms A and B: they are based on a modern generalization, due to Borel and Harish-Chandra, of the classical reduction theory of quadratic forms. In Chapter 11 we shall see a much more direct link between $\mathfrak{T}$-groups and quadratic forms.

The ideas of this section also form the basis for the deep finiteness theorems discussed in Chapters 9 and 10.

E. Notes

The proper setting for the Baker–Campbell–Hausdorff formula is the theory of formal power series in non-commuting variables; the version quoted in Theorem 1 is obtained by specialising the variables. For the formula itself see Jacobson (1962), Chapter V section 5, or Amayo & Stewart (1974), Chapter 5; the latter contains also an account of the application to nilpotent groups.

The construction which I have given for the Lie algebra associated to a $\mathfrak{T}$-group is not the only possible one, nor even the most elegant. It is perhaps more down-to-earth than some more obviously functorial versions. For these, see Baumslag (1971), Amayo & Stewart (1974), and Warfield (1976), Chapter 12. There are also various ways to construct the radicable hull of a

$\mathfrak{T}$-group; see each of the books just mentioned, and also Hall (1969), Chapter 6. The correspondence between nilpotent groups and nilpotent Lie algebras was originally set up by Mal'cev (1949).

The subject of torsion-free nilpotent groups and their various completions, Lie algebras etc. is a large one, and I have only touched the surface in this chapter. To get better acquainted with it one should really have a look at all the books mentioned above, and further references therein.

The ubiquity of lattice groups was established in Moore (1965). Most of the detailed results in section B, above, I worked out anew; but they may well be known to the experts. Proposition 2 is due to Mal'cev (1958). Regarding Exercise 5, we have

$$\mathrm{Tr}_1(n,\mathbb{Z})^{\mathrm{lat}} = \mathrm{Tr}_1(n,\mathbb{Z})^{\mathrm{Lie}} = \Gamma_n(e_1,\ldots,e_{n-1})$$

where

$$e_i = \prod_{p\,prime} p^{[(i-1)/(p-1)]}$$

(and $[x]$ denotes the integer part of a real number n). This result is due to Geoff Smith.

The connection between nilpotent groups and arithmetic groups, discussed in section C, was I believe first exploited by Auslander & Baumslag (1967); see Chapter 4 of Baumslag (1971) and references given there.

The construction of the radicable hull $\bar{G}$, say, of a $\mathfrak{T}$-group G described by Hall (*loc. cit.*) actually exhibits $\bar{G}$ as the group of $\mathbb{Q}$-points of an 'affine algebraic group' (though not in that language), the $\mathbb{Z}$-points of which comprise the group G. The general theory of these groups shows that they can be realised as algebraic matrix groups, and this gives a (superficially) different approach to the embedding results of Chapter 5 and the arithmeticity result of this chapter.

The solution of the isomorphism problem of nilpotent groups, modulo Algorithms A and B above, appears in Grunewald & Segal (1980, II); the approach there is slightly different in that instead of using the Lie-ring hull of $G\beta_G$, we describe an algorithm which decides whether or not two given unipotent subgroups of $GL_n(\mathbb{Z})$ are conjugate (see also Chapter 8, section D, below). The effective procedures given in that paper will do just as well to fill in most of the gaps in the account presented above. Though quite long, the paper is basically elementary and would make a reasonable supplement to section D for the reader interested in constructive procedures in group theory. Algorithms A and B are given in Grunewald & Segal (1980, I); this is not an elementary paper. See also Grunewald & Segal (1979 *b*) for a brief outline of the ideas involved, and some other applications.

One might remark that very little seems to be known about the isomorphism problem for finitely presented groups in general. It was formulated by Max Dehn in 1911, along with the word problem and the conjugacy problem. It is known that all three problems are in general effectively insoluble – thus for example the question of whether an arbitrary finitely presented group is the trivial group or not is effectively undecidable; see Lyndon & Schupp (1977), Chapter IV section 4. The only positive result of a general nature regarding the isomorphism problem seems to be our Algorithm 1. There is no doubt that this can be extended to polycyclic groups; this is explained in section D of Chapter 7 below. My feeling is that the isomorphism problem for the class of finitely presented metabelian groups should be effectively undecidable, but as far as I know nothing has been proved in this direction (even for soluble groups of any derived length).

7
Semi-simple splitting

The prototype for semi-simple splitting is the decomposition of the full triangular group of matrices $Tr_n(k)$, over a field k, as the semi-direct product $Tr_1(n,k)]D_n(k)$. If G is a polycyclic group, we know from Chapter 5 that G can be embedded in $GL_n(\mathbb{Z})$, for some n; and Mal'cev's theorem (Chapter 2) then shows that there is a subgroup G_1 of finite index in G and a matrix $x \in GL_n(k)$, where k is a suitable algebraic number field, such that $G_1^x \leq Tr_n(k)$. Let $\mathfrak{o}$ be the ring of integers in k and $\mathfrak{o}^*$ the group of units of $\mathfrak{o}$. The eigenvalues of a matrix in $GL_n(\mathbb{Z})$ are algebraic integers, so if we identify G_1 with its image in $Tr_n(k)$ we find that the diagonal entries of all the matrices in G_1 (which are just the eigenvalues) all belong to $\mathfrak{o}^*$. Taking a positive integer e such that ex and ex^{-1} lie in $M_n(\mathfrak{o})$, and conjugating G_1 with the matrix diag $(1, e^2, \ldots, e^{2(n-1)})$, we 'clear denominators' in the above-diagonal entries of all the matrices in G_1^x. We end up having embedded G_1 into the group $Tr_n(\mathfrak{o})$.

Now $Tr_n(\mathfrak{o}) = Tr_1(n,\mathfrak{o})]D_n(\mathfrak{o}) = U]D$ say. Here U is a $\mathfrak{T}$-group, it is the Fitting subgroup of $Tr_n(\mathfrak{o})$, and D is a finitely generated abelian group. Of course, the connection between G_1 and $U]D$ is rather tenuous here – $U]D$ could be far larger than necessary. What we shall do in this chapter is describe a canonical construction: given a polycyclic group G, we find a subgroup G_1 of finite index and embed G_1 into a split extension $\bar{G}_1 = U]D$, with $U = \text{Fitt}\,(\bar{G}_1) \in \mathfrak{T}$ and D finitely generated abelian. $\bar{G}_1$ will be, in a special sense, the minimal over-group of G_1 having this form, and is called a *semi-simple splitting* of G_1, for reasons which will become apparent later.

As well as results from Chapter 3 and Chapter 6, the construction depends on properties of the *Jordan decomposition* of matrices; the relevant facts are established in section A. The construction itself is explained in section B. The rest of the chapter is devoted to applications: the canonical embedding of a polycyclic group into $GL_n(\mathbb{Z})$, in section C; and an outline solution of the *isomorphism problem* for polycyclic groups, in section D.

The origin of semi-simple splittings is in the theory of soluble Lie groups, and there are important applications to the theory of *solvmanifolds*. These applications will not be touched on here; but in a series of exercises at the end of section B, I outline a version of the semi-simple splitting construction which is relevant for that theory (no knowledge of Lie groups is required for these exercises.).

A. Jordan decomposition

Fix a positive integer n and a perfect field k. Let $g \in GL_n(k)$. There exists $x \in GL_n(k_1)$, for some finite extension field k_1 of k, such that g^x is a matrix in Jordan canonical form: thus g^x is block-diagonal,

$$g^x = \operatorname{diag}(J_1, \ldots, J_r),$$

say, where each block has the form

$$J_i = \begin{bmatrix} \lambda_i & * & & \\ & \lambda_i & * & 0 \\ & & \ddots & \ddots \\ 0 & & & * \\ & & & \lambda_i \end{bmatrix} = \lambda_i U_i \in GL_{n_i}(k_1),$$

say. Here $\lambda_1, \ldots, \lambda_r$ are the distinct eigenvalues of g and each $*$ is either 0 or 1; thus $U_i \in \operatorname{Tr}_1(n_i, k_1)$ and $n_1 + \cdots + n_r = n$. Putting

$$s = (\operatorname{diag}(\lambda_1 1_{n_1}, \ldots, \lambda_r 1_{n_r}))^{x^{-1}}$$
$$u = (\operatorname{diag}(U_1, \ldots, U_r))^{x^{-1}},$$

we see that s is diagonalizable, u is unipotent, and $g = su = us$ (recall that a matrix in $GL_n(k)$ is called *diagonalizable* if it is diagonalizable over some extension field of k).

Lemma 1 If $h \in GL_n(k)$ and h commutes with g, then h commutes with s and with u.

Proof Since h^x commutes with g^x, h^x respects the generalized eigenspaces of g^x (i.e. the subspaces of k_1^n annihilated by $(g^x - \lambda_i)^n$ for some i). But s^x acts as the scalar λ_i on the ith generalized eigenspace, for each i, and k_1^n is the direct sum of these subspaces; so h^x commutes with s^x. Therefore h commutes with s; since $u = s^{-1}g$, h commutes with u also.

Lemma 2 Suppose $g = vt = tv$ with v unipotent and t diagonalizable. Then $v = u$ and $t = s$.

Proof t commutes with $g = v^{-1}t$, so t commutes with s, by Lemma 1. Therefore st^{-1} is diagonalizable. Similarly $u^{-1}v$ is unipotent. But

$u^{-1}v = st^{-1}$; since this matrix is both unipotent and diagonalizable, it is the identity. The result follows.

Proposition 1   Let $g \in GL_n(k)$. Then there exist unique matrices g_u, g_s in $GL_n(\bar{k})$ (where $\bar{k}$ = algebraic closure of k) such that g_u is unipotent, g_s is diagonalizable, and $g = g_u g_s = g_s g_u$. Moreover, g_u and g_s both lie in $GL_n(k)$.

Proof   We established the existence at the beginning of this section. Taking $k = \bar{k}$ in Lemma 2 we get the uniqueness. For the final claim, suppose g_u and g_s belong to $GL_n(k_1)$ where k_1 is a finite extension of k; we may take k_1 to be normal over k. If $\sigma \in \mathrm{Gal}(k_1/k)$, then $g = g^\sigma = g_u^\sigma g_s^\sigma = g_s^\sigma g_u^\sigma$, g_s^σ is diagonalizable, and g_u^σ is unipotent. By the uniqueness it follows that $g_u^\sigma = g_u$ and $g_s^\sigma = g_s$. Thus the entries of g_u and of g_s are fixed by $\mathrm{Gal}(k_1/k)$; since k is a perfect field, these matrix entries must lie in k. Thus g_u and g_s are in $GL_n(k)$.

The decomposition $g = g_u g_s$ is called the (multiplicative) *Jordan decomposition* of the matrix g.

There is an analogous decomposition for the elements of an (abstract) finite group. Suppose G is a finite group, p is a prime, and $g \in G$. Then g has finite order $p^e q$, say, with

$$\text{h.c.f.} (p^e, q) = 1 = ap^e + bq$$

for suitable $a, b \in \mathbb{Z}$. Putting $u = g^{bq}$ and $s = g^{ap^e}$ we get $g = us = su$, $u^{p^e} = 1$ and $s^q = 1$ (note that u and s are generators for, respectively, the p-component and the p'-component of the finite cyclic group $\langle g \rangle$; here p' means 'coprime to p'). The connection with our previous topic comes via

Lemma 3   Suppose k is a field of prime characteristic p. Let x be an element of finite order m in $GL_n(k)$. Then x is unipotent if and only if m is a power of p; x is diagonalizable if and only if h.c.f. $(p, m) = 1$.

Proof   Since char $k = p$, we have $(x - 1)^{p^l} = x^{p^l} - 1$ for every l. Taking $p^l \geq \max\{m, n\}$ we easily deduce the first claim. Suppose x is diagonalizable, and $m = p^e q$. Then x^q is diagonalizable, but also unipotent by the first part since $(x^q)^{p^e} = 1$. Therefore $x^q = 1$ and $p^e = 1$. Finally, suppose h.c.f. $(p, m) = 1$. Then $x_u^m x_s^m = x^m = 1$ so $x_u^m = x_s^{-m}$ is both unipotent and diagonalizable; therefore $x_u^m = 1$. But x_u has p-power order by the first part; since h.c.f. $(p, m) = 1$ it follows that $x_u = 1$. Thus $x = x_s$ is diagonalizable.

From this lemma and Proposition 1 we infer

Corollary 1 If char $k = p$ is prime and $g \in GL_n(k)$ has finite order $p^e q$, with h.c.f. $(p, q) = 1 = ap^e + bq$, then

$$g_u = g^{bq} \text{ and } g_s = g^{ap^e}.$$

We return now to our arbitrary perfect field k, and consider a generalisation of Lemma 1:

Proposition 2 Let $G \leq GL_n(k)$, and suppose P is a polynomial in n^2 variables over k such that $P(g) = 0$ for every $g \in G$. Then $P(g_u) = P(g_s) = 0$ for all $g \in G$.

As usual, $P(x)$ means $P(x_{11}, x_{12}, \dots, x_{nn})$ for a matrix $x = (x_{ij})$.

Proof If k is a finite field, Corollary 1 shows that for each $g \in G$ we have $g_u, g_s \in \langle g \rangle \leq G$; so in this case there is nothing to prove. To deal with the general case, let $g \in G$ and assume without loss of generality that there exists $x \in GL_n(k)$ such that g_s^x is diagonal. Let R be the subring of k generated by the coefficients of P and the entries of the matrices g_u, g_s, x, x^{-1} and g^{-1}. Then R is a finitely generated integral domain; in such a ring, the Jacobson radical is always zero (see e.g., Atiyah & Macdonald: *Introduction to Commutative Algebra*, page 70, Exercise 22). So if we suppose that $P(g_u) \neq 0$, then there exists a maximal ideal M of R with $P(g_u) \notin M$. The field $F = R/M$ is finite (see Atiyah & Macdonald, page 84, Exercise 6). Let $\pi: GL_n(R) \to GL_n(F)$ be the natural map ('reduction modulo M'). Then $g\pi = g_u\pi \cdot g_s\pi = g_s\pi \cdot g_u\pi$; $g_s\pi$ is diagonalizable because $(g_s\pi)^{x\pi}$ is actually diagonal, and $g_u\pi$ is unipotent because $(g_u\pi - 1)^n$ is the image of $(g_u - 1)^n$ which is the zero matrix. Therefore $g_u\pi = (g\pi)_u$ and $g_s\pi = (g\pi)_s$. Also g has finite order because $GL_n(F)$ is a finite group. It follows by Corollary 1 that

$$g_u\pi = (g\pi)^\alpha, \quad g_s\pi = (g\pi)^\beta$$

for some α and $\beta \in \mathbb{Z}$. Therefore modulo M, $g_u \equiv g^\alpha$ and $P(g_u) \equiv P(g^\alpha) = 0$ since $g^\alpha \in G$. This contradicts our choice of M, and shows that $P(g_u) = 0$. The same argument shows that $P(g_s) = 0$.

Recalling the definition of a $\mathbb{Q}$-group (Chapter 6, section C), we can state, as a consequence of Propositions 1 and 2,

Corollary 2 If $\mathscr{G}$ is a $\mathbb{Q}$-group then g_u and g_s belong to $\mathscr{G}$ for every $g \in \mathscr{G}$. If $g \in \mathscr{G}(\mathbb{Q})$ then $g_u \in \mathscr{G}(\mathbb{Q})$ and $g_s \in \mathscr{G}(\mathbb{Q})$.

Exercise 1 If $g \in GL_n(\mathbb{Z})$ does it follow that g_u and g_s lie in $GL_n(\mathbb{Z})$?

I shall use ( )$_u$ and ( )$_s$ as symbols denoting mappings $GL_n(k) \rightarrow GL_n(k)$, so that for $G \le GL_n(k)$,

$$G_u = \{g_u | g \in G\}, \quad G_s = \{g_s | g \in G\}.$$

Exercise 2 Show that ( )$_u$ and ( )$_s$ are not in general homomorphisms.

$$\left(\text{Hint: try } \begin{pmatrix} 1 & 1 \\ 0 & 1 \end{pmatrix} \text{ and } \begin{pmatrix} 1 & 0 \\ 1 & 1 \end{pmatrix}. \right)$$

Exercise 3 Suppose $G \le GL_n(k)$ is abelian. Show that G_u and G_s are subgroups of $GL_n(k)$ and that the maps $g \rightarrow g_u, g \rightarrow g_s$ from G to G_u and G_s respectively are homomorphisms. What are their kernels?

Let us try to improve on the last exercise. Consider the 'p-decomposition' in a finite group G, discussed above: if $g \in G$ has order $p^e q$ with h.c.f. $(p, q) = 1 = ap^e + bq$, write $g_p = g^{bq}$ and $g_{p'} = g^{ap^e}$. Suppose that the two maps $g \mapsto g_p, g \mapsto g_{p'}$ are homomorphisms, and denote their images by P, Q respectively. Then P and Q are subgroups of G and $G = PQ$. It is easy to see that P consists of all the p-elements of G and Q of all the p'-elements of G ('p-element' = element of p-power order; 'p'-element' = element of order coprime to p); therefore $P \cap Q = 1, P \triangleleft G$ and $Q \triangleleft G$. Thus $G = P \times Q$ and P is the unique sylow p-subgroup of G. Conversely, if we assume that G has this form, then it is easy to see that the maps ( )$_p$ and ( )$_{p'}$ on G are exactly the projections onto P and Q respectively, so they are homomorphisms.

Now a finite group which is nilpotent is the direct product of its Sylow subgroups; so whatever the prime p, such a group splits up in the way we have been considering. Recalling Corollary 1 above, we may therefore interpret the outcome of the present discussion in the form of

Lemma 4 Let F be a finite field and H a nilpotent subgroup of $GL_n(F)$. Then $H = H_u \times H_s$ and the maps ( )$_u$, ( )$_s$ restricted to H are homomorphisms.

It will come as no surprise, after the proof of Proposition 2, that this lemma can be generalised:

Proposition 3 Let G be a nilpotent subgroup of $GL_n(k)$. Then G_u and G_s are subgroups of $GL_n(k)$, $G \le G_u \times G_s$, and the maps ( )$_u : G \rightarrow G_u$ and ( )$_s : G \rightarrow G_s$ are homomorphisms.

Proof Let g, $h \in G$. We want to show that $(gh)_u = g_u h_u$; suppose this is false. Let $x, y \in GL_n(k_1)$ be such that g_s^x and h_s^y are diagonal, where k_1 is some extension field of k; let R be the subring of k_1 generated by the entries of the matrices g_u, g_s, h_u, h_s, x, y and of their inverses. As in the proof of Proposition 2, we find a maximal ideal M of R such that $(gh)_u \not\equiv g_u h_u$ mod M. Then $F = R/M$ is a finite field. Let $\pi: GL_n(R) \to GL_n(F)$ be the natural map and put $H = \langle g, h \rangle \pi$. Lemma 4 shows that

$$((gh)\pi)_u = (g\pi \cdot h\pi)_u = (g\pi)_u \cdot (h\pi)_u.$$

But just as in the proof of Proposition 2, we have

$$((gh)\pi)_u = (gh)_u \pi, \quad (g\pi)_u = g_u \pi, \quad (h\pi)_u = h_u \pi.$$

The conclusion is that

$$(gh)_u \equiv g_u h_u \text{ mod } M.$$

This contradicts our choice of M, and shows that $(gh)_u = g_u h_u$. Since g and h were arbitrary elements of G, it follows that G_u is a subgroup of $GL_n(k)$ and that $(\ )_u: G \to G_u$ is a homomorphism (you can check directly that $(g^{-1})_u = g_u^{-1}$). Similarly, G_s is a subgroup and $(\ )_s: G \to G_s$ is a homomorphism. Certainly $G_u \cap G_s = 1$ and $G \subseteq G_u G_s$; so it remains to show that G_u and G_s commute elementwise, and this will follow if we can show that G_u and G_s normalise each other. Now if $g, h \in G$ then $(g_u)^h$ is unipotent, $(g_s)^h$ is diagonalizable, and $g^h = (g_u)^h (g_s)^h = (g_s)^h (g_u)^h$. So $(g_u)^h = (g^h)_u \in G_u$. Thus

$$G_u^{h_s} = G_u^{h_u^{-1} h} = G_u^h \le G_u.$$

It follows that G_s normalizes G_u, and the converse is proved similarly. (Alternatively, we can lift the result back from the case of a finite field, as we did with the other properties.)

For the final topic, let us salvage something from the problem posed by Exercise 1.

Lemma 5 Let G be a nilpotent subgroup of $GL_n(\mathbb{Z})$. Then G has a subgroup H of finite index such that $H_u H_s \le GL_n(\mathbb{Z})$.

Proof Put $K = G_u \cap GL_n(\mathbb{Z})$ and $H = \{g \in G \mid g_u \in K\}$. Then H is a subgroup of G, since $g \mapsto g_u$ is a homomorphism, and $H_u = K \le GL_n(\mathbb{Z})$. Also $H_s \le H \cdot H_u \le GL_n(\mathbb{Z})$; so it remains to show that $|G:H|$ is finite. But $|G:H| = |G_u:K|$, and G_u is a finitely generated unipotent subgroup of $GL_n(\mathbb{Q})$ (it is finitely generated because it is a homomorphic image of G, and G is polycyclic by Corollary 1 of Chapter 2); so the result will follow from the next lemma:

Lemma 6   If G is a finitely generated unipotent subgroup of $GL_n(\mathbb{Q})$ then $G \cap GL_n(\mathbb{Z})$ has finite index in G.

Proof   By Lemma 2 of Chapter 6, there exists $x \in GL_n(\mathbb{Q})$ such that $G^x \le \mathrm{Tr}_1(n, \mathbb{Z})$. Let L be the lattice $\mathbb{Z}^n \cdot x^{-1}$ in $\mathbb{Q}^n$ (i.e. the $\mathbb{Z}$-module spanned by the rows of the matrix x^{-1}). Then

$$L \cdot G = \mathbb{Z}^n x^{-1} G = \mathbb{Z}^n \cdot G^x \cdot x^{-1} = \mathbb{Z}^n \cdot x^{-1} = L.$$

Now there exists a positive integer m such that $mL \subseteq \mathbb{Z}^n \subseteq m^{-1}L$. Since G fixes L, it fixes mL and $m^{-1}L$, therefore G permutes the finitely many lattices lying between mL and $m^{-1}L$ among themselves. It follows that $G \cap GL_n(\mathbb{Z})$, which is just the stabilizer in G of the lattice $\mathbb{Z}^n$, has finite index in G, as claimed.

Remark   Here we have actually used the fact that a group consisting of unipotent matrices can be put into unitriangular form. The full proof, though not hard, would constitute a digression; but it may be worthwhile to prove here a very easy special case which suffices for present purposes.

Fact   If G is a nilpotent subgroup of $GL_n(k)$ and every element of G is unipotent, then G acts nilpotently on the vector space k^n (and can therefore be put into unitriangular form over k).

Proof   Write $V = k^n$. Let $1 \ne z \in \zeta_1(G)$ and let U be the set of fixed points of z in V. Then U is a k-subspace of V and $0 < U < V$, since z acts nilpotently on V and $z \ne 1$. Since z is central in G, U is actually a kG-submodule of V, so the action of G on V gives rise to representations of G on U and on V/U. Arguing by induction on the dimension, we may suppose that G acts nilpotently on U and on V/U; the result follows.

Exercise 4   (i) Suppose $x \in GL_n(k)$ is diagonalizable, and let U be a $k\langle x \rangle$-submodule of k^n. Show that the restriction of x to U is diagonalizable. (ii) Let $x \in GL_n(k)$ and suppose $k^n = U + V$ with U and V both $k\langle x \rangle$-submodules of k^n. Show that x is diagonalizable if and only if its restrictions to U and to V are both diagonalizable.
(*Hint*: Enlarge k, so that 'diagonalizable' can be changed to 'diagonalizable over k'. Then use Lemma 3 of Chapter 2.)

B. The construction

To get some idea how a semi-simple splitting should look, let us consider – as we did in the introduction – a subgroup G of $\mathrm{Tr}_n(\mathfrak{o})$, where $\mathfrak{o}$ is some ring of algebraic integers. Write $U = \text{Tr}_1(n, \mathfrak{o})$ and $D = D_n(\mathfrak{o})$. Thus $G \le U]D$. Now we only want to keep enough of U and of D to 'cover' G adequately, so let us put

$$T = D \cap UG, \quad \bar{G} = \langle G, T \rangle, \quad \text{and} \quad M = U \cap \bar{G}.$$

The following are easy to verify:

Exercise 5 (i) $\bar{G} = MG = M]T$. (ii) If $N = U \cap G$, then $M \cap G = N$ and $T \cong G/N$; (iii) M is a $\mathfrak{X}$-group and T is finitely generated abelian.

Now what we want to do is to achieve a similar situation starting with an abstract polycyclic group G, without an ambient matrix group to work in. For this, we need some intrinsic property of the action of T on M which corresponds to the fact that $M \le \text{Tr}_1(n, \mathfrak{o})$ and $T \le D_n(\mathfrak{o})$; the relevant concepts were developed in section C of the previous chapter. In that chapter, we associated to an arbitrary $\mathfrak{X}$-group N, say, a Lie algebra $\Lambda = \mathscr{L}(N\beta_N) = \mathbb{Q} \log(N\beta_N)$ over $\mathbb{Q}$, in such a way that automorphisms of N could be identified with automorphisms of Λ. Moreover, we showed that

$$\text{Aut}\, \Lambda = \mathscr{G}(\mathbb{Q}) \le GL_m(\mathbb{Q})$$

for a certain $\mathbb{Q}$-group $\mathscr{G}$, where $m = \dim_{\mathbb{Q}} \Lambda$; and identifying Aut N with the normaliser in Aut Λ of the subset $\log(N\beta_N)$, Aut N becomes an arithmetic subgroup of G.

Now let us call an automorphism of a $\mathfrak{X}$-group N *semi-simple* if the corresponding automorphism of $\Lambda = \mathscr{L}(N\beta_N)$ is diagonalizable, and *unipotent* if the corresponding automorphism of Λ is unipotent. Returning to the situation $G \le M]T$ discussed above, we can now state

Exercise 6 Show that T acts by semi-simple automorphisms on M.

(*Hint*: (a) If x is a diagonal matrix in $GL_n(k)$, show that conjugation by x is a diagonalizable automorphism of $M_n(k)$ as k-vector space, for any field k (in fact the 'matrix units' e_{ij} are eigenvectors). Deduce by Exercise 4 that T acts diagonalizably on $\mathbb{Q} \log M \subseteq \text{Tr}_0(n, k)$, where k is the field of fractions of $\mathfrak{o}$. (b) Embed $M_n(k)$ in $M_{nd}(\mathbb{Q})$, where $d = (k : \mathbb{Q})$, thereby embedding M into $\text{Tr}_1(nd, \mathbb{Q})$ by a map α, say. Show that $\mathbb{Q} \log M$ can be identified with $\mathbb{Q} \log(M\alpha) \subseteq \text{Tr}_0(nd, \mathbb{Q})$. Then use Proposition 5 of Chapter 6 to show that $\mathbb{Q} \log M$ is isomorphic as a $\mathbb{Q} T$-module to the Lie algebra $\mathbb{Q} \log(M\beta_M)$.)

Exercises 5 and 6 have been included in order to motivate the definition of semi-simple splittings; we shall not be needing the results for what follows. Before giving that definition, it will be useful to discuss some further properties of the automorphisms of $\mathfrak{X}$-groups.

Let N be a $\mathfrak{X}$-group, $\Lambda = \mathscr{L}(N\beta_N)$ its Lie algebra and $\mathscr{G}$ the $\mathbb{Q}$-group such that Aut $\Lambda = \mathcal{G}(\mathbb{Q})$. Identify Aut N with a subgroup of $\mathcal{G}(\mathbb{Q})$. Suppose now that H is a nilpotent subgroup of Aut N. Then $H \cap \mathcal{G}(\mathbb{Z}) = H_1$, say, has finite index in H, because Aut N, being arithmetic in $\mathcal{G}$, is commensurable with $\mathcal{G}(\mathbb{Z})$. By Lemma 5, in section A, H_1 has a subgroup K of finite index such that $K_u K_s \leq GL_m(\mathbb{Z})$ (where $\mathcal{G} \leq GL_m$), and Proposition 3 then shows that $K_u K_s \leq \mathcal{G}(\mathbb{Z})$. Since Aut N is commensurable with $\mathcal{G}(\mathbb{Z})$ and $(\ )_u, (\ )_s$ are homomorphisms on K, it follows that K has a subgroup L of finite index such that $L_u L_s \leq$ Aut N. Thus we have established

Lemma 7   Let N be a $\mathfrak{T}$-group and H a nilpotent subgroup of Aut N. Identify Aut N with a $\mathbb{Q}$-linear group by considering Aut $N \leq$ Aut $\mathscr{L}(N\beta_N)$. Then H has a subgroup L of finite index such that

$$L \leq L_u \times L_s \leq \text{Aut } N.$$

We also need a group-theoretic interpretation for an automorphism of N to be unipotent or semi-simple.

Lemma 8   Let N be a $\mathfrak{T}$-group. Then an element or subgroup acts unipotently on N if and only if it acts nilpotently on N.

Proof   Let us consider a subgroup H of Aut N (the argument is the same if we are interested in an element of Aut N). The map β_N embeds $N]H$ into $GL_r(\mathbb{Z})$ for some r; write $G = N\beta_N$, $X = H\beta_N$, and $\Lambda = \mathbb{Q} \log G$. We must show that X acts nilpotently on G if and only if X acts unipotently on Λ, the action both times being conjugation. We argue by induction on the dimension of Λ, assuming merely that G is a subgroup of $\text{Tr}_1(r, \mathbb{Q})$ and $X \leq GL_r(\mathbb{Q})$ normalizes G. Put $\Lambda_2 = \mathbb{Q} \log G'$. The results of Chapter 6, section A, show that log induces a homomorphism of G into the additive group of Λ/Λ_2. Since Λ/Λ_2 is torsion-free abelian (being a $\mathbb{Q}$-vector space), log induces a homomorphism

$$\lambda : G/T \to \Lambda/\Lambda_2$$

where $T/G' = \tau(G/G')$. It is easy to verify that in fact λ is injective, and it is clear that the image of λ spans Λ/Λ_2 over $\mathbb{Q}$. Since conjugation by elements of X commutes with the operation log, we see that Λ/Λ_2 is isomorphic as a $\mathbb{Q}X$-module with $\mathbb{Q} \otimes (G/T)$. Hence X acts nilpotently on G/T if and only if it acts nilpotently, or equivalently unipotently, on Λ/Λ_2.

Since G is torsion-free and nilpotent, $T \neq G$ (see Chapter 1); therefore $\Lambda/\Lambda_2 \neq 0$ and $\dim_\mathbb{Q} \Lambda_2 < \dim_\mathbb{Q} \Lambda$. Also it follows from Theorem 1 of Chapter 6 that $\mathbb{Q} \log T = \mathbb{Q} \log G' = \Lambda_2$. So we may apply the inductive hypothesis and suppose that X acts nilpotently on T if and only if X acts unipotently on Λ_2. Together with the result of the previous paragraph this shows that X acts nilpotently on G if and only if X acts unipotently on Λ.

It does not seem so easy to characterise semi-simple automorphisms of a $\mathfrak{T}$-group; the following result will suffice for our purposes:

Lemma 9 Let N be a $\mathfrak{T}$-group, x an automorphism of N, and M an x-invariant subgroup of N. If x acts nilpotently on M, then x_s acts as the identity on M. (Here we think of x_s as an automorphism of the radicable hull $\bar{N} = \exp \mathscr{L}(N\beta_N)$; though we shall only apply the lemma when x_s is actually in $\operatorname{Aut} N$.)

Proof Proposition 5 of Chapter 6 shows that $\mathscr{L}(M\beta_M) \cong \mathscr{L}(M\beta_N)$, and this is an isomorphism of $\mathbb{Q}\langle x \rangle$-modules; so by Lemma 8, x acts as a unipotent automorphism on $\mathscr{L}(M\beta_N)$. But certainly x_u acts unipotently on $\mathscr{L}(M\beta_N)$, and x_u commutes with x. Since a product of commuting unipotent matrices is again unipotent it follows that $x_s = x_u^{-1}x$ acts unipotently on $\mathscr{L}(M\beta_N)$. But Exercise 4 shows that x_s acts diagonalisably on $\mathscr{L}(M\beta_N)$. Therefore x_s acts as the identity on $\mathscr{L}(M\beta_N)$, and the result follows.

We are now ready to introduce semi-simple splittings.

Definition Let G be a polycyclic group. A polycyclic group $\bar{G}$ containing G is a *semi-simple splitting* for G if the following hold:
  (i) $\bar{G} = M]T$ where $M = \operatorname{Fitt}(\bar{G}) \in \mathfrak{T}$ and T is free abelian;
  (ii) T acts by semi-simple automorphisms on M;
  (iii) $G \triangleleft \bar{G}$, $G \cap T = 1$, and $\bar{G} = MG = GT$;
  (iv) $M = (M \cap G) \cdot C_M(T)$.

Note that we have slightly strengthened the conditions suggested by Exercises 5 and 6, above: this will make it harder to establish the *existence* of a suitable $\bar{G}$, but easier to establish something like *uniqueness*. The point of the precise conditions chosen will become apparent as we proceed.

A polycyclic group G is called *splittable* if G has a semi-simple splitting. Let us look at some consequences of splittability:

Lemma 10 Suppose $G \leq \bar{G} = M]T$ as above, and put $N = \operatorname{Fitt}(G)$. Then $N \triangleleft \bar{G}$ and the following hold:
  (i) $N = M \cap G \in \mathfrak{T}$ and $G/N \cong T$ is free abelian;
  (ii) $C =_{\text{def}} C_G(T)$ is a $\mathfrak{T}$-group, and $G = NC$;

(iii) C is a maximal nilpotent subgroup of G;
(iv) for $g \in \bar{G}$ let $g^* \in \operatorname{Aut} N$ denote conjugation by g. Then $C_s^* = T^*$;
(v) $C_T(G) = 1$.

Proof Since N is characteristic in G and $G \lhd \bar{G}$, $N \lhd \bar{G}$. Since N is also nilpotent, $N \leq M \cap G$. But $M \cap G$ is nilpotent and normal in G, so $M \cap G = N$. This also implies that

$$G/N = G/(M \cap G) \cong MG/M = \bar{G}/M \cong T,$$

and establishes (i). Since T is abelian, $C_{\bar{G}}(T) = C_M(T) \times T$, so $C \leq C_{\bar{G}}(T)$ is a $\mathfrak{T}$-group. From (iv) in the definition above, we have

$$G = G \cap MT = G \cap ((G \cap M)C_M(T)T) = (G \cap M)C_{\bar{G}}(T) = NC.$$

Thus (ii) holds. To prove (iii), suppose $C \leq E \leq G$ and E is nilpotent. To show that $E = C$, we must check that T centralises E, and since $E = (E \cap N)C$, from (ii), it is enough to show that T centralises $E \cap N$. Take $t \in T$. Since $\bar{G} = MG = MC$, by (ii) again, we have $t = mx$, say, with $m \in M$ and $x \in C$. Put

$$K = \langle E \cap N, m \rangle \leq M.$$

Then $K^x = K$, because $m^x = (tx^{-1})^x = tx^{-1} = m$ and $(E \cap N)^x = E^x \cap N = E \cap N$ since $x \in C \leq E$; therefore $K^t = K$ since $m \in K$. Now

$$K/K' = (E \cap N)K'/K'.\langle m \rangle K'/K';$$

x acts nilpotently on $E \cap N$ and centralises m, so x acts nilpotently on K/K'. But $t = mx$ acts like x on K/K' so t acts nilpotently on K/K'. By Proposition 14 of Chapter 1, t acts nilpotently on K. But t acts semi-simply on M, so Lemma 9 shows that t centralises K. Thus t centralises $E \cap N$, which is what we had to show.

Let us prove (iv). Let $x \in C$. Then $x = mt$ with $m \in M$ and $t \in T$, and $mt = tm$ since x centralises T. Therefore $x^* = m^*t^* = t^*m^*$. Now m^* is a unipotent automorphism of N, by Lemma 8, and t^* is a semi-simple automorphism of N, by Exercise 4 (recall that $\mathscr{L}(N\beta_N)$ is isomorphic to $\mathscr{L}(N\beta_M)$ as $\mathbb{Q}\langle t^* \rangle$-module, see Proposition 5 of Chapter 6). It follows that $t^* = x_s^*$. Thus we see that $C_s^* \subseteq T^*$. Moreover, every $t \in T$ arises in this way, since $T \leq \bar{G} = MC$ by (ii); so in fact $C_s^* = T^*$ as required.

Finally, since $\bar{G} = GT$ and T is abelian we have

$$C_T(G) = T \cap \zeta_1(\bar{G}) \leq T \cap M = 1.$$

Thus (v) holds, and the proof is complete.

We have now assembled some necessary conditions for a polycyclic group to be splittable. But the very statement of Lemma 10 makes it more or less clear that these conditions are also *sufficient*: for they actually give us a recipe to construct $\bar{G}$.

Theorem 1 Let G be a polycyclic group, with Fitting subgroup N. Then G is splittable if and only if the following hold:

(i) $N \in \mathfrak{T}$ and G/N is free abelian;

(ii) $G = NC$ for some nilpotent subgroup C of G, which also satisfies

(iii) $C_s^* \leq \operatorname{Aut} N$, where $* : C \to \operatorname{Aut} N$ denotes conjugation (i.e., C_s^* fixes the subset $\log(N\beta_N)$ inside $\mathscr{L}(N\beta_N)$).

Proof If G is splittable, let $\bar{G} = M \,]\, T$ be a semi-simple splitting. Taking $C = C_G(T)$, we get (i), (ii) and (iii) by Lemma 10. Suppose conversely that conditions (i), (ii) and (iii) are satisfied. What we have to do is cook up a subgroup T of $\operatorname{Aut} G$ in such a way that

$$\bar{G} = G \,]\, T$$

becomes a semi-simple splitting for G. We know that T should centralise C and act like C_s^* on N; since $G = NC$ this will specify T completely. There are thus two tasks: one, to show that there is such a subgroup T of $\operatorname{Aut} G$; and two, to show that for such a T, $G \,]\, T$ is indeed a semi-simple splitting for G.

Let us construct the group T. For each $c \in C$, define a map $t_c : G \to G$ by putting

$$(ax)^{t_c} = a^{c_s^*} \cdot x \text{ for } a \in N \text{ and } x \in C.$$

We have to check that this is a legitimate definition. So suppose $ax = by$ with $a, b \in N$ and $x, y \in C$. Then $xy^{-1} = a^{-1}b \in N \cap C$; since C is nilpotent, c^* acts nilpotently on $N \cap C$, therefore c_s^* centralises $N \cap C$, by Lemma 9. So

$$(a^{-1}b)^{c_s^*} = a^{-1}b = xy^{-1},$$

giving

$$b^{c_s^*} y = a^{c_s^*} x$$

as required.

Now that we know that each t_c is a well defined map of $G \to G$, we can easily show that in fact $c \mapsto t_c$ gives a homomorphism $\tau : C \to \operatorname{Aut} G$. Note first that $C' \leq N$, since G/N is abelian, so $(C^*)'$ is unipotent (by Lemma 8), whence

$$[C^*, C_s^*] \leq (C^*)_s' = 1 \tag{1}$$

(recall that $C^* \leq C_u^* \times C_s^*$, by Proposition 3). Now let $a, b \in N$ and $x, y, c \in C$.

Then

$$((ax)(by))^{t_c} = (ab^{x^{-1}} \cdot xy)^{t_c}$$

$$= a^{c_s^*} b^{x^{*-1} c_s^*} \cdot xy$$

$$= a^{c_s^*} b^{c_s^* x^{*-1}} \cdot xy \qquad\qquad \text{by (1)}$$

$$= a^{c_s^*} x \cdot b^{c_s^*} y = (ax)^{t_c} \cdot (by)^{t_c}.$$

Thus t_c is an endomorphism of the group G. If $c, d \in C$ then

$$(ax)^{t_c t_d} = (a^{c_s^*} x)^{t_d} = a^{c_s^* c_d^*} x = a^{(cd)_s^*} x = (ax)^{t_{cd}},$$

since $*$ and $(\ )_s$ are homomorphisms on C, C^* respectively. It follows that $t_c t_{c^{-1}} = t_{c^{-1}} t_c = t_1$, the identity automorphism of G, so $t_c \in \mathrm{Aut}\, G$; and it follows also that $\tau: C \to \mathrm{Aut}\, G$, where $c\tau = t_c$, is a homomorphism.

Now put

$$T - C\tau \leq \mathrm{Aut}\, G;$$

this completes the first of our two tasks.

Taking $\bar{G} = G]\, T$ and $M = \mathrm{Fitt}(\bar{G})$, we must now check conditions (i)–(iv) in the definition of a semi-simple splitting.

T *is free abelian.* For $c \in C$, $t_c = 1$ if and only if $c_s^* = 1$, i.e. if and only if c acts nilpotently on N (by Lemma 8). Since $G = NC$, $N = \mathrm{Fitt}(G)$, and G/N is abelian, c acts nilpotently on N if and only if $c \in N \cap C$. Thus $\ker \tau = N \cap C$ and so

$$T \cong C/(N \cap C) \cong NC/N = G/N;$$

this is free abelian by hypothesis.

$\bar{G} = M]\, T$. Let us identify M explicitly. Define a map $u: C \to \bar{G}$ by $u(c) = c \cdot t_c^{-1}$. Since T centralises C, u is a homomorphism, so $u(C)$ is a subgroup of G. Now for $c \in C$ and $a \in N$,

$$a^{u(c)} = (a^{c^*})^{t_c^{-1}} = a^{c^* c_s^{*-1}} = a^{c_u^*};$$

thus $u(C)$ acts on N like the unipotent group C_u^*, and Lemma 8 shows that $u(C)$ acts nilpotently on N. Since $u(C)$ (like C) is nilpotent it follows that the group $N \cdot u(C)$ is nilpotent. Since $\bar{G} = GT$, $C \leq u(C) \cdot T$, and $G = NC$, we have

$$\bar{G} = N \cdot u(C) \cdot T. \qquad\qquad (2)$$

But T centralises $u(C)$ and normalises N, so $N \cdot u(C) \lhd \bar{G}$. Thus certainly $N \cdot u(C) \leq M = \mathrm{Fitt}(\bar{G})$. In fact $M = N \cdot u(C)$. For we have $M = Nu(C) \cdot (M \cap T)$; but $M \cap T$ acts nilpotently on N, since M is nilpotent, and semi-simply on N, from the definition of T. So by Exercise 4, $M \cap T$ centralises N. Therefore $M \cap T$ centralises G and so $M \cap T = 1$ (since $T \leq \mathrm{Aut}\, G$).

Thus from (2), $\bar{G} = MT$. We have just seen that $M \cap T = 1$, so $\bar{G} = M]\, T$ as claimed.

$G \triangleleft \bar{G}, G \cap T = 1,$ *and* $\bar{G} = MG = GT$. These all follow from the definition of T, except for $\bar{G} = MG$: this follows from (2) above. That $M = (M \cap G)C_M(T)$ follows from the fact that $M = N \cdot u(C)$.

T acts by semi-simple automorphisms on M. Certainly T acts semi-simply on N, by its definition; and T acts trivially on $u(C)$. By Proposition 5 of Chapter 6, we can identify the Lie algebra $\mathscr{L}(N\beta_N)$ with $\mathscr{L}(N\beta_M)$; so we know that T acts by diagonalisable automorphisms on $\mathscr{L}(N\beta_M)$, and by the identity on $\mathscr{L}(u(C)\beta_M)$. By Exercise 4(ii), we may infer that T acts semi-simply on M provided we have

$$\mathscr{L}(M\beta_M) = \mathscr{L}(N\beta_M) + \mathscr{L}(u(C)\beta_M). \tag{3}$$

To prove (3), consider an element $\lambda \in \log(M\beta_M)$. Then $\lambda = \log a = \log(xy)$ with $a \in M$, $x \in N$ and $y \in u(C)$. Corollaries 1 and 3 of Chapter 6 show that

$$\log(xy) \equiv \log x + \log y \mod \mathbb{Q} \log(M\beta_M)'. \tag{4}$$

But $(M\beta_M)' = M'\beta_M \leq N\beta_M$, since, as is easily seen, $u(C)' \leq N$. So the right-hand side of (4) lies in $\mathscr{L}(N\beta_M) + \mathscr{L}(u(C)\beta_M)$, and this establishes (3).

This concludes the proof of Theorem 1. Of course, this whole discussion will only have some point if there are splittable polycyclic groups. The results of Chapter 3 will show that in fact splittable groups are not far to seek. First, a simple observation:

Lemma 11 If G is a splittable polycyclic group and $\mathrm{Fitt}(G) \leq H \leq G$, then H is splittable.

Proof. By Theorem 1, $G/\mathrm{Fitt}(G)$ is free abelian and $\mathrm{Fitt}(G)$ is a $\mathfrak{T}$-group. It follows that $H \triangleleft G$, that $\mathrm{Fitt}(H) = \mathrm{Fitt}(G)$, and that $H/\mathrm{Fitt}(H)$ is free abelian. We also know that $G = \mathrm{Fitt}(G)C$ for some $\mathfrak{T}$-group C, such that C_s^* acts on $\mathrm{Fitt}(G)$. Then $H = \mathrm{Fitt}(H) \cdot (H \cap C)$, and it is clear that conditions (i)–(iii) of Theorem 1 hold for H.

Theorem 2 Let G be a polycyclic-by-finite group. Then G has a splittable characteristic subgroup of finite index.

Proof $\mathrm{Fitt}(G)$ has a torsion-free subgroup N of finite index, which we may assume characteristic in G. Since G/N is residually finite, G has a normal subgroup G_1 of finite index with $G_1 \cap \mathrm{Fitt}(G) = N$; it is easy to see that then $\mathrm{Fitt}(G_1) = N$. By Mal'cev's theorem (Theorem 4 of Chapter 2), G_1/N is abelian-by-finite, so G_1/N has a free abelian normal subgroup G_2/N of finite index, and again $\mathrm{Fitt}(G_2) = N$. Now recall Theorem 3 of Chapter 3:

this tells us that G_2 has a nilpotent subgroup, D say, with $|G_2:ND|$ finite. For $x \in D$ let $x^* \in \operatorname{Aut} N$ denote conjugation by x. By Lemma 7, there is a subgroup C of finite index in D such that $C_s^* \leq \operatorname{Aut} N$. Put $G_3 = NC$. Theorem 1 shows that G_3 is splittable (note that $\operatorname{Fitt}(G_3) = N$, because $G_3 \lhd G_2$ since G_2/N is abelian); and G_3 has finite index in G, by construction. Finally, let

$$H = \bigcap_{\alpha \in \operatorname{Aut} G} G_3^\alpha.$$

Then $N \leq H \leq G_3$, so H is splittable by Lemma 11. Evidently H is characteristic in G; and H has finite index in G because G is finitely generated and $|G:G_3|$ is finite (compare Chapter 1, Lemma 2).

Now that we have found plenty of splittable groups, let us investigate 'how unique' their semi-simple splittings are going to be. Once again, the answer is based on a result from Chapter 3, and the clue is provided by Lemma 10, above. The main step is accomplished in

Lemma 12 Let G be a polycyclic group with semi-simple splittings $\bar{G}_1 = G]T_1$ and $\bar{G}_2 = G]T_2$, with T_1, T_2 playing the usual role. Put $C_i = C_G(T_i)$ for $i = 1, 2$. If C_1 and C_2 are conjugate in G then $\bar{G}_1$ and $\bar{G}_2$ are isomorphic, by a map which is the identity on G.

Proof By Lemma 10, $G = NC_1 = NC_2$ where $N = \operatorname{Fitt}(G)$. So there exists $x \in N$ such that $C_2^x = C_1$. Let $\bar{x}$ denote the inner automorphism of G given by x. Now T_1 and T_2 act faithfully on G, by Lemma 10(v), so we can identify T_1 and T_2 with subgroups of $\operatorname{Aut} G$; then $\bar{G}_1$ and $\bar{G}_2$ become subgroups of $\operatorname{Hol} G = G] \operatorname{Aut} G$.

I shall show that

$$\bar{G}_2^{\bar{x}} = \bar{G}_1. \tag{5}$$

Given (5), the lemma follows easily. For then conjugation by the element $x \cdot \bar{x}^{-1}$ in $\operatorname{Hol} G$ maps $\bar{G}_1$ isomorphically onto $\bar{G}_2$, and it certainly acts as the identity on G. To establish (5), recall from Lemma 10 that

$$T_i^* = (C_i^*)_s \leq \operatorname{Aut} N, \qquad i = 1, 2,$$

(where for $h \in \bar{G}_i$, $h^* \in \operatorname{Aut} N$ denotes conjugation by h). Now for $c \in C_2$,

$$(c^x)^* = (c^{*x^*})_s = (c_s^*)^{x^*};$$

since $C_2^x = C_1$, it follows that

$$(T_2^{\bar{x}})^* = T_2^{*x^*} = (C_2^*)_s^{x^*} = (C_2^x)_s^* = (C_1^*)_s = T_1^*.$$

Thus $T_2^{\bar{x}}$ acts like T_1 on N. But $T_2^{\bar{x}}$ and T_1 both centralise $C_1 = C_2^x$, and

$G = NC_1$; therefore $T_2^{\bar{x}} = T_1$ as groups of automorphisms of G. Thus

$$\bar{G}_2^{\bar{x}} = G^{\bar{x}}] T_2^{\bar{x}} = G] T_1 = \bar{G}_1,$$

which is (5).

For use in the next section, we note a consequence of the above proof: taking $\theta : \bar{G}_2 \to \bar{G}_1$ to be conjugation by $\bar{x}$ in Hol G, we have

Corollary 3   Under the hypotheses of Lemma 12, there is an isomorphism $\theta : \bar{G}_2 \to \bar{G}_1$ and an element $x \in \text{Fitt}(G)$ such that $T_2 \theta = T_1$ and $g\theta = g^x$ for all $g \in G$.

Let us call two semi-simple splittings $\bar{G}_1, \bar{G}_2$ of a group G *equivalent* if there is an isomorphism $\theta : \bar{G}_1 \to \bar{G}_2$ with $\theta|_G = $ identity. Lemma 10(ii) and (iii) and Lemma 12 show that inequivalent semi-simple splittings for G give rise to non-conjugate maximal nilpotent supplements for $\text{Fitt}(G)$ in G. According to Theorem 4 of Chapter 3, however, the maximal nilpotent supplements for $\text{Fitt}(G)$ lie in finitely many conjugacy classes in G. Hence we conclude:

Theorem 3   A splittable polycyclic group has up to equivalence only finitely many semi-simple splittings.

Exercise 7   In the proof of Theorem 1, we constructed a semi-simple splitting for G using a nilpotent supplement C for $\text{Fitt}(G)$, without assuming that C was maximal. Would we get a different semi-simple splitting if we replaced C by a larger subgroup?

This concludes the development of our theory for polycyclic groups. As I mentioned in the introduction, however, this theory was originally created in the context of soluble Lie groups. It applied not to polycyclic groups, but rather to free abelian extensions of nilpotent Lie groups. As this is of interest to topologists, and as the theory is actually simpler, it seems worth while to sketch here how it goes. This is done in the following sequence of exercises.

As the theory is purely algebraic, we can use any field k of characteristic zero in place of the real numbers; but we need the definition and some elementary properties of 'nilpotent groups with exponents in k', *k-powered* groups for short. Rather than develop these here, I refer the reader to Chapter 6 of P. Hall's notes *Nilpotent Groups*, where a full account is given from an axiomatic point of view. In fact we are only concerned with k-powered groups of finite dimension: one can think of such a group as a subgroup N of $\text{Tr}_1(n, k)$, for some n, with the property that $\log N$ is actually a k-subspace of $\text{Tr}_0(n, k)$. It follows from (our) Chapter 6, section A, that

$\log N$ is then a Lie subalgebra over k of $\mathrm{Tr}_0(n,k)$. An example to bear in mind is the radicable hull of a $\mathfrak{X}$-group, k being here the field $\mathbb{Q}$. In general, the operation of k on N is given by the formula

$$g^\lambda = \exp(\lambda \log g) \quad \text{for} \quad g \in N, \lambda \in k.$$

k-subgroups and k-automorphisms of N are defined in the obvious way. We shall need the following fact (see Hall, *op. cit*):

Lemma 13   If N is a k-powered group and K is a normal k-subgroup of N, then N/K inherits the structure of a k-powered group from N.

Now the soluble groups we consider belong to the following class:

Definition $\mathfrak{X}(k)$ is the class of groups G such that:

  (i) G has a normal nilpotent k-powered subgroup N of finite dimension;
 (ii) the inner automorphisms of G restrict to k-automorphisms of N;
(iii) G/N is free abelian of finite rank.

Exercise 8   Generalise Theorem 2** and Theorem 1* of Chapter 3, replacing $\mathbb{Q}$ by k throughout.

Exercise 9   Let G be an $\mathfrak{X}(k)$ group, with normal subgroup N as in the above definition. (i) Show that G has a nilpotent subgroup C such that $G = NC$. (ii) Suppose $G = NC_1 = NC_2$ with C_1 and C_2 nilpotent. Show that there exists $x \in N$ such that $\langle C_1, C_2^x \rangle$ is nilpotent. Deduce that maximal nilpotent supplements for N in G are conjugate (if they exist!).
(*Hint*: Copy the proofs of Theorems 3 and 4 of Chapter 3. Remember to check that everything in sight is k-powered! (But note that apart from this, the argument is actually simpler, since we don't have to keep dropping to subgroups of finite index.))

Definition   Let G be an $\mathfrak{X}(k)$ group with normal k-powered subgroup N as in the definition above. A *semi-simple splitting* for G is a group $\bar{G}$ containing G such that
   (i) $\bar{G} = M]T$ where M is a nilpotent $\mathfrak{X}(k)$ group and T is free abelian;
  (ii) $G \triangleleft \bar{G} = MG = G \cdot T$, $G \cap T = 1$, and $M \cap G = N$;
 (iii) T acts by semi-simple k-automorphisms on N (i.e. T acts by diagonalizable automorphisms on the k-space $\log N$);
  (iv) $M = N \cdot C_M(T)$ and $C_T(G) = 1$.

Exercise 10   Let G be an $\mathfrak{X}(k)$ group. Show that G has a semi-simple splitting, and that this is unique up to equivalence.

(*Hint*: Let C be a nilpotent supplement for N in G, and let C^* denote the image of C, acting by conjugation, in Aut N. Define $T \leq$ Aut G by setting $T|_N = C_s^*$, and making T act as the identity on C (think about why $C_s^* \leq$ Aut N); copy the proof of Theorem 1 to show that this works and that $\bar{G} = G]\,T$ has the required properties. For the uniqueness, copy the proof of Theorem 3 and use Exercise 9(ii). (Curiously, this roundabout argument shows that there exist *maximal* nilpotent supplements for N in G, a fact which doesn't seem to be obvious otherwise.))

C. Canonical embeddings

In Chapter 5, section B, we associated to a $\mathfrak{X}$-group N a positive integer $n = n(N)$ and an injective homomorphism $\beta_N : N]\,\text{Aut}\,N \to GL_n(\mathbb{Z})$, with $N\beta_N \leq \text{Tr}_1(n, \mathbb{Z})$. The embedding β_N was uniquely determined up to equivalence, in the sense of representation theory, i.e. up to composition with some inner automorphism of $GL_n(\mathbb{Z})$. Now suppose we have a splittable polycyclic group G, with semi-simple splitting $\bar{G} = M]\,T$: here $M = \text{Fitt}(\bar{G})$ is a $\mathfrak{X}$-group and $T \leq \text{Aut}\,M$ (we may think of T as contained in Aut M, because $C_T(M) = 1$). Thus

$$G \leq \bar{G} = M]\,T \leq M]\,\text{Aut}\,M,$$

and we get an embedding

$$\beta_{M|G} : G \to GL_n(\mathbb{Z}) \tag{6}$$

where $n = n(M)$. With Theorem 2 above and Lemma 8 of Chapter 5, this provides an alternative proof of the embedding theorem: every polycyclic-by-finite group has a faithful linear representation over $\mathbb{Z}$.

Of course, the proof given in Chapter 5, section C, was much simpler. The point of the present approach is that we can make our embedding *canonical*. The embedding (6) is determined up to equivalence once we have chosen the semi-simple splitting $\bar{G}$ for the given group G: but what if choose a different $\bar{G}$? We have seen in Theorem 3 that there are only finitely many possibilities for $\bar{G}$ up to equivalence; so let us see what happens to (6) when $\bar{G}$ is replaced by an equivalent semi-simple splitting. It is easier in fact to consider the (possibly) stronger hypothesis described in Corollary 3:

Lemma 14   Let G be a polycyclic group and $\bar{G}_i = G]\,T_i$, for $i = 1, 2$, be semi-simple splittings for G. Put $M_i = \text{Fitt}(\bar{G}_i)$ and write

$\beta_i : G \to GL_{n_i}(\mathbb{Z})$ for $\beta_{M_i|G}(i = 1, 2)$. If there is an isomorphism $\theta : \bar{G}_2 \to \bar{G}_1$ and an element $x \in G$ such that $T_2\theta = T_1$ and $g\theta = g^x$ for all $g \in G$, then $n_1 = n_2$ and β_1 and β_2 are equivalent representations of G.

> *Proof* Certainly $M_2\theta = M_1$, so $M_2 \cong M_1$ and
> $$n_2 = n(M_2) = n(M_1) = n_1.$$

Let $\theta^* : M_2] \operatorname{Aut} M_2 \to M_1] \operatorname{Aut} M_1$ be the isomorphism induced by $\theta|_{M_2}$. Since $T_i \leq \operatorname{Aut} M_i$ for each i, we can compare $T_2\theta^*$ with $T_2\theta$: let $t \in T_2$, $a \in M_2$, $b = a\theta \in M_1$.
Then
$$b^{t\theta^*} = (b\theta^{-1})^t\theta = (a^t)\theta$$
by the definition of θ^*; while
$$b^{t\theta} = (a\theta)^{t\theta} = (a^t)\theta$$
since θ is an isomorphism. Thus $t\theta^*$ and $t\theta$ act in the same way on M_1, therefore $t\theta^* = t\theta$. Since $\theta^*|_{M_2} = \theta|_{M_2}$ and $\bar{G}_2 = M_2 \cdot T_2$, it follows that $\theta^*|_{G_2} = \theta$. So far, we have established the commutativity of the solid-arrow part of the following diagram, where x^* denotes conjugation by the element x:

$$(7)$$

Now Exercise 3 of Chapter 5 shows that there is an inner automorphism γ of $GL_n(\mathbb{Z})$ making the right-hand square commute as well. Put $\xi = (x\beta_1)^{-1} \in GL_n(\mathbb{Z})$. Then for each $g \in G$ we have
$$g\beta_1 = (g^{x^*})^{x^{-1}}\beta_1 = (g^{x^*}\beta_1)^{(x\beta_1)^{-1}}$$
$$= (g\beta_2)^{\gamma\xi}.$$

Thus β_1 and β_2 are equivalent representations of G, as claimed.

Let us call the semi-simple splittings $\bar{G}_1, \bar{G}_2$ of G *strongly equivalent* if they are related as in the last lemma. Strictly speaking this is a relation between the pairs $(\bar{G}_i, T_i)$: it depends on the chosen subgroups T_i of $\operatorname{Aut} G$. But then so do the embeddings $\beta_{M_i|G}$ where $M_i = \operatorname{Fitt}(\bar{G}_i)$. Now just as we deduced Theorem 3 from Theorem 4 of Chapter 3, using Lemma 12, we can use Corollary 3 to infer that a splittable polycyclic group has only finitely many semi-simple splittings up to strong equivalence. Let G be such a group and let $\bar{G}_1, \ldots, \bar{G}_k$ be representatives for the distinct strong equivalence classes of semi-simple splittings of G. Let

$$\beta_i: G \to GL_{n_i}(\mathbb{Z})$$

be the restriction of $\beta_{M_i}: \text{Hol } M_i \to GL_{n_i}(\mathbb{Z})$, where $M_i = \text{Fitt}(\bar{G}_i)$, $i = 1, \ldots, k$. We then obtain a canonical embedding of G by defining

$$n_G = n_1 + \ldots + n_k,$$

$$\beta_G: G \to GL_{n_G}(\mathbb{Z}); \quad g \mapsto \text{diag}(g\beta_1, \ldots, g\beta_k). \tag{8}$$

The representation β_G is uniquely determined up to equivalence; for if we replace some $\bar{G}_i$ by a strongly equivalent semi-simple splitting, then Lemma 14 shows that β_i only gets replaced by an equivalent representation; while a change in the order of our representatives simply has the effect of conjugating each $g\beta_G$ by some (fixed) permutation matrix. To sum up, we have established most of

Theorem 4 There is a canonical way to embed a splittable polycyclic group G into a suitable $GL_n(\mathbb{Z})$, given by (8) above. This embedding has the following properties:

(i) β_G maps $\text{Fitt}(G)$ into $\text{Tr}_1(n_G, \mathbb{Z})$;

(ii) If G and H are splittable polycyclic groups and $\alpha: G \to H$ is an isomorphism, then $n_H = n_G = n$, say, and there is an inner automorphism γ of $GL_n(\mathbb{Z})$ making the following diagram commute:

$$\begin{array}{ccc} G & \xrightarrow{\ \beta_G\ } & GL_n(\mathbb{Z}) \\ \alpha \downarrow & & \downarrow \gamma \\ H & \xrightarrow[\ \beta_H\ ]{} & GL_n(\mathbb{Z}). \end{array}$$

Statement (i) is clear, since $\text{Fitt}(G) \le M_i$ for each i, and β_{M_i} sends M_i into $\text{Tr}_1(n_i, \mathbb{Z})$. To prove (ii), note that if $\bar{G}_i = G]T_i$, $i = 1, \ldots, k$, are as above, then we can take $\bar{H}_i = H](\alpha^{-1} T_i \alpha)$, for $i = 1, \ldots, k$, as representatives for the strong equivalence classes of semi-simple splittings of H: I leave it to the reader to check the details (draw a diagram like (7), with $G \hookrightarrow \bar{G} = G]T$ in the top row and $H \hookrightarrow \bar{H} = H](\alpha^{-1} T \alpha)$ in the bottom row).

If we want to embed an arbitrary polycyclic-by-finite group canonically, we must specify a canonical way to choose a splittable subgroup of finite index. One way to do this is as follows: given a polycyclic-by-finite group G, put $F = \mathrm{Fitt}(G)$ and let s be the smallest positive integer for which F^s is torsion-free. Taking $N = F^s$ and looking at the proof of Theorem 2, above, we see that G has a splittable subgroup G_3 of finite index, with $N = \mathrm{Fitt}(G_3)$. Lemma 11 shows that every subgroup H of G with $N \le H \le G_3$ is splittable; so in particular the group $NG^{t!}$ is splittable for all sufficiently large positive integers t, namely all $t \ge t_G$ where

$$t_G = \min\{t \in \mathbb{N} \mid NG^{t!} \text{ is splittable}\}$$
$$\le |G:G_3|.$$

We now define $s_G = \min\{s \in \mathbb{N} \mid \mathrm{Fitt}(G)^s \in \mathfrak{X}\}$, and put

$$G_{\mathrm{split}} = \mathrm{Fitt}(G)^{s_G} \cdot G^{t_G!}.$$

Then G_{split} is a splittable characteristic subgroup of finite index in G, and it is clear that under any isomorphism $\varphi : G \to H$, we have $s_H = s_G$, $t_H = t_G$, and $G_{\mathrm{split}}\varphi = H_{\mathrm{split}}$. Note that if G is already splittable, then $G_{\mathrm{split}} = G$. So we can extend the definition of β_G to arbitrary polycyclic-by-finite groups by letting β_G be the representation of G induced from $\beta_{G_{\mathrm{split}}}$ (see Lemma 8 of Chapter 4 for this).

It is then not difficult to verify.

Corollary 4   For polycyclic-by-finite G, put $n_G = |G : G_{\mathrm{split}}| \cdot n_{G_{\mathrm{split}}}$ and let $\beta_G : G \to GL_{n_G}(\mathbb{Z})$ be induced from $\beta_{G_{\mathrm{split}}}$. Then the representation β_G is uniquely determined up to equivalence; if $\varphi : G \to H$ is an isomorphism then $n_H = n_G = n$ say, and there exists an inner automorphism γ of $GL_n(\mathbb{Z})$ making the following diagram commute:

$$
\begin{array}{ccc}
G & \xrightarrow{\;\beta_G\;} & GL_n(\mathbb{Z}) \\[2pt]
{\scriptstyle\phi}\big\downarrow & & \big\downarrow{\scriptstyle\gamma} \\[2pt]
H & \xrightarrow[\;\beta_H\;]{} & GL_n(\mathbb{Z})
\end{array}
\tag{9}
$$

Exercise 11   Prove Corollary 4 in detail.

For some purposes, as we shall see in the next section, it is convenient to be a little more general. For G as above and any $t \ge t_G$, put $G(t) = \mathrm{Fitt}(G)^{s_G} \cdot G^{t!}$. Then $G(t)$ is splittable; we define $n_G(t) = |G : G(t)| \cdot n_{G(t)}$ and let $\beta_G(t)$ be the representation of G induced from $\beta_{G(t)}$ (thus β_G was just $\beta_G(t_G)$). It is clear that the last Corollary can be generalised: if $\varphi : G \to H$ is an isomorphism, then $t_H = t_G$, and for each $t \geq t_G$, there is a commutative diagram like (9) with $n = n_G(t) = n_H(t)$ and $\beta_G(t)$, $\beta_H(t)$ in place of β_G, β_H respectively.

D. The isomorphism problem revisited

At the time of writing, the solution of the isomorphism problem for polycyclic groups has not been completed in all its details. However, work is in progress, and there is little doubt that a positive result is imminent. Here, I shall outline briefly how the results of the previous section can be used to attack this problem.

To solve the isomorphism problem for polycyclic groups, I claim that the following three algorithms will suffice.

Algorithm 1 Given a polycyclic-by-finite group G, by finitely many generators and relations, the algorithm finds an upper bound for the number t_G (defined in section C).

Algorithm 2 Given a group G as above, and given a positive integer $t \geq t_G$, the algorithm computes the number $n_G(t)$, and computes the representation $\beta_G(t): G \rightarrow GL_{n_G(t)}(\mathbb{Z})$ (i.e. gives the image under $\beta_G(t)$ of each given generator of G).

Algorithm 3 Given a positive integer n, and given finite sets of generators for two soluble-by-finite subgroups P and Q of $GL_n(\mathbb{Z})$, the algorithm decides whether P and Q are conjugate in $GL_n(\mathbb{Z})$.

Algorithm 3 requires some new ideas, and will be described in the next chapter. Algorithms 1 and 2 are discussed below, but first let us justify the claim that these algorithms together solve the isomorphism problem. So suppose G and H are two polycyclic-by-finite groups, each given by finitely many generators and relations. Using Algorithm 1, we find a positive integer t such that $t \geq t_G$ and $t \geq t_H$. Algorithm 2 then computes $n_G(t)$ and $n_H(t)$. If these are unequal, we conclude that $G \not\cong H$. Suppose then that $n_G(t) = n_H(t) = n$, say. We saw at the end of section C that

$$G \cong H \Rightarrow G\beta_G(t) \text{ conjugate to } H\beta_H(t) \text{ in } GL_n(\mathbb{Z}).$$

But $\beta_G(t)$ and $\beta_H(t)$ are injective homomorphisms, so we have, conversely, that if $G\beta_G(t)$ and $H\beta_H(t)$ are conjugate then

$$G \cong G\beta_G(t) \cong H\beta_H(t) \cong H.$$

So to decide whether or not $G \cong H$, we find finite generating sets for $P = G\beta_G(t)$ and for $Q = H\beta_H(t)$, by Algorithm 2, and then decide, using

Algorithm 3, whether or not P and Q are conjugate in $GL_n(\mathbb{Z})$.

Let us discuss the algorithms. Algorithm 1 is in principle fairly straightforward. We have to compute s_G, the least positive integer s for which $\mathrm{Fitt}(G)^s$ is torsion-free. Having done this, we start listing nilpotent subgroups of G, and keep going until we find one, D say, such that ND has finite index in G, where $N = \mathrm{Fitt}(G)^{s_G}$. We then find a subgroup C of finite index in D such that CN/N is free abelian and $C_s^* \leq \mathrm{Aut}\, N$ (see the proof of Theorem 2, in section B). The group NC is then splittable and has finite index in G; so taking $t = |G:NC|$ we have $t \geq t_G$ as required.

Algorithm 2 is rather more problematic. Given $t \geq t_G$, we must first construct the group $G(t) = \mathrm{Fitt}(G)^{s_G} \cdot G^{t!}$. If we can compute $n_{G(t)}$ and $\beta_{G(t)}:G(t) \to GL_{n_{G(t)}}(\mathbb{Z})$, then there is in principle no problem in computing $n_G(t) = |G:G(t)| \cdot n_{G(t)}$ and in constructing the induced representation $\beta_G(t):G \to GL_{n_G(t)}(\mathbb{Z})$. So the main problem is the canonical embedding of a given splittable group. Assume then that G is splittable. We must find a complete set of representatives for the strong equivalence classes of semi-simple splittings of G. The proof of Lemma 12 showed that the equivalence classes correspond bijectively with conjugacy classes of maximal nilpotent supplements C for $N = \mathrm{Fitt}(G)$ in G which satisfy the condition $C_s^* \leq \mathrm{Aut}\, N$.

Our first task is to construct a complete list of representatives for the conjugacy classes of maximal nilpotent supplements for N in G; this is quite a complicated matter, and it depends on having 'constructive' versions of the existence and finiteness theorems (Theorems 3 and 4) of Chapter 3. Having obtained the required list of nilpotent subgroups of G, we then delete those subgroups C in the list for which $C_s^* \nleq \mathrm{Aut}\, N$.

Call the remaining groups in the list $C_1, \ldots, C_k$. The proof of Theorem 1 shows how to construct $\bar{G}_i = G]\, T_i$ such that $C_G(T_i) = C_i$, for each i; one then has to find $M_i = \mathrm{Fitt}(\bar{G}_i)$, and compute the map β_{M_i}, as defined in section B of Chapter 5, for each i. The canonical embedding of G can then finally be read off as the direct sum of the $\beta_{M_i|G}$ for $1 \leq i \leq k$.

E. Notes

An account of the Jordan decomposition is given in Chapter 7 of Wehrfritz (1973b).

Semi-simple splittings were first constructed in Wang (1956), but for soluble Lie groups. A 'polycyclic' version was developed by Tolimieri (1971). Wang's construction already showed that a polycyclic group can be embedded into $GL_n(\mathbb{C})$ for a suitable n; using a similar approach, Auslander (1967) was able to obtain an embedding into $GL_n(\mathbb{Z})$. A connected account of this topic, with its topological applications, is given in the survey article Auslander (1973).

As far as I know, the approach to semi-simple splittings which I have adopted in this chapter is original; it was motivated by work on the paper Grunewald, Pickel & Segal (1980) (though the version in that paper is slightly different: what we show, roughly speaking, is that every torsion-free nilpotent-by-free abelian polycyclic group can be embedded as a subgroup of finite index in a splittable group – to get $C_s^* \leq \operatorname{Aut} N$, we enlarge N instead of shrinking C).

As well as being relevant to the isomorphism problem, the semisimple splitting will play an essential role in the argument of Chapter 10.

For more on nilpotent groups with exponents in a ring ('k-powered groups'), see Hall (1969), Chapter 6, or Warfield (1976), Chapter 10.

A different canonical embedding of a polycyclic group into an affine algebraic group over $\mathbb{Q}$ – and hence into $GL_n(\mathbb{Q})$ – is described in Donkin (1982). His construction is more 'natural' than mine, and I think that this is an exciting development which deserves attention. In the long run, the ideas of this paper may well provide a new and smoother approach to the results of this chapter and of Chapter 10.

At the time of writing, Algorithms 1 and 2 (section D) are not fully complete; but most of the serious difficulties seem to have been overcome. This work is due to Neil Maxwell. A sketch of Algorithm 3 is given in section D of Chapter 8.

(*Added in proof*: all the algorithms have now been constructed; the isomorphism problem for polycyclic-by-finite groups is thus completely solved.)

8
Soluble ℤ-linear groups

In this chapter we are going to take a closer look at the way in which a polyclic group can sit inside $GL_n(\mathbb{Z})$. As we saw in Chapter 7, a good understanding of isomorphisms between polycyclic groups would seem to depend on knowing when soluble subgroups of $GL_n(\mathbb{Z})$ are conjugate, and an algorithm for deciding this question is sketched in section D, below. Sections B and C investigate the normalizer in $GL_n(\mathbb{Z})$ of an arbitrary soluble-by-finite subgroup. Using the fact (proved in Chapter 5) that the holomorph of a polycyclic-by-finite group G can be embedded in some $GL_n(\mathbb{Z})$, we shall derive some rather striking consequences: (*a*) Aut G has a normal subgroup K, isomorphic to an arithmetic group, with (Aut G)/K finitely generated and abelian-by-finite; (*b*) Aut G is a finitely presented group; (*c*) the finite subgroups of Aut G lie in finitely many conjugacy classes; and (*d*) for each natural number m, there exist only finitely many non-isomorphic extensions of G by a group of order m.

An important tool in these investigations is the *Zariski topology* in a linear group. This has been lurking in the background to some of the previous chapters; but the time has come for a fuller discussion of the topic, and this is given in section A.

A. The Zariski topology

Let k be a field and m a natural number. An *algebraic set* in the vector space k^m is the set of common zeros of a family of polynomials in m variables over k; that is, a set of the form

$$\mathscr{V}(S) = \{(x_1, \ldots, x_m) \in k^m \mid f(x_1, \ldots, x_m) = 0 \ \forall f \in S\}$$

where S is any subset of the polynomial ring $k[x_1, \ldots, x_m]$.

Exercise 1 (i) $\varnothing$ and k^m are algebraic sets. (ii) The intersection of a family of algebraic sets is an algebraic set. (iii) The union of two algebraic sets is an algebraic set.
(*Hint*: for (iii), show that

$$\mathscr{V}(S_1) \cup \mathscr{V}(S_2) = \mathscr{V}(S_1 S_2)$$

where $S_1 S_2 = \{fg \mid f \in S_1, g \in S_2\}$.)

156

The exercise shows that the algebraic sets form the closed sets of a topology on k^m: this topology is the *Zariski topology*.

Exercise 2 Show that points are closed (i.e. the Zariski topology is T_1); is this topology Hausdorff?

Definition Let W be a subset of k^m and let $\theta: W \to k^n$ be a map. The map θ is *rational* if there exist n pairs of polynomials (P_i, Q_i), in m variables over k, with $Q_i(w) \neq 0$ for all $w \in W$ and each i, such that

$$w\theta = (P_1(w)/Q_1(w), \ldots, P_n(w)/Q_n(w)) \tag{1}$$

for all $w \in W$.

Lemma 1 Give $W \subseteq k^m$ the subspace topology induced by the Zariski topology on k^m, and give k^n its own Zariski topology. Then every rational map from W into k^n is continuous.

Proof Suppose $\theta: W \to k^n$ is given by rational functions as in (1). Let U be an open set in k^n. We must show that $U\theta^{-1}$ is open in W. Now U is the complement of some algebraic set $\mathscr{V}(S)$ in k^n, where $S \subseteq k[X_1, \ldots X_n]$. for each $f \in S$, put

$$f^*(X_1, \ldots, X_m) = Q_1(\mathbf{X})^{e_1} \ldots Q_n(\mathbf{X})^{e_n} \cdot f(P_1(\mathbf{X})/Q_1(\mathbf{X}), \ldots, P_n(\mathbf{X})/Q_n(\mathbf{X}))$$

where e_i is the degree to which the ith variable occurs in the polynomial f, for $i = 1, \ldots, n$. Thus f^* is the polynomial obtained by 'clearing denominators' in the rational expression $f(P_1/Q_1, \ldots, P_n/Q_n)$. Since none of the polynomials Q_i vanishes anywhere on W, we see that for $w \in W$,

$$w \in U\theta^{-1} \Leftrightarrow w\theta \notin \mathscr{V}(S)$$
$$\Leftrightarrow f(P_1(w)/Q_1(w), \ldots, P_n(w)/Q_n(w)) \neq 0 \quad \text{for some } f \in S$$
$$\Leftrightarrow f^*(w) \neq 0 \quad \text{for some } f \in S.$$

Thus

$$U\theta^{-1} = W \backslash \mathscr{V}(S^*)$$

where $S^* = \{f^* | f \in S\}$; so $U\theta^{-1}$ is an open subset of W as required.

Let us consider more closely the relation between algebraic sets and sets of polynomials. For a subset $W \subseteq k^m$, put

$$\mathscr{I}(W) = \{f \in k[X_1, \ldots, X_m] | f(w) = 0 \quad \forall w \in W\}.$$

It is easy to see that $\mathscr{I}(W)$ is an *ideal* in the ring $k[X_1, \ldots, X_m]$. Moreover, W is closed in k^m if and only if $W = \mathscr{V}(\mathscr{I}(W))$. The 'if' statement is clear;

while if we suppose W closed, then $W = \mathscr{V}(S)$ for some $S \subseteq k[X_1,...,X_m]$; then $\mathscr{I}(W) \supseteq S$, so

$$\mathscr{V}(\mathscr{I}(W)) \subseteq \mathscr{V}(S) = W,$$

and as $W \subseteq \mathscr{V}(\mathscr{I}(W))$ it follows that $W = \mathscr{V}(\mathscr{I}(W))$. This observation allows us to translate ideal-theoretic results into topological ones. For example, Hilbert's basis theorem shows that $k[X_1,...,X_m]$ is a Noetherian ring: so every ascending chain of ideals becomes stationary after finitely many steps, and we deduce

Lemma 2   Every descending chain of closed sets in k^m becomes stationary after finitely many steps.

Proof   Let $W_1 \supseteq W_2 \supseteq \ldots$ be a chain of closed sets in k^m. Put $I_j = \mathscr{I}(W_j)$ for each j. Then $I_1 \subseteq I_2 \subseteq \ldots$ is an ascending chain of ideals in $k[X_1,...,X_m]$, so for some n we have $I_n = I_{n+1} = \ldots$. Consequently

$$W_j = \mathscr{V}(I_j) = \mathscr{V}(I_n) = W_n \quad \text{for all } j \geq n,$$

since each W_j is closed.

Now our concern in this chapter is with groups of matrices. So let us give $M_n(k)$ a Zariski topology by thinking of each $n \times n$ matrix as a vector in k^{n^2}; $GL_n(k)$ gets the subspace topology, and this is what we shall call the *Zariski topology* on $GL_n(k)$. All topological terms will henceforth (in this chapter) refer to this topology. Note that the $\mathbb{Q}$-groups discussed in Chapter 6 are closed subgroups of $GL_n(\mathbb{C})$ (though not every such closed subgroup is a $\mathbb{Q}$-group; I return to this point later).

Lemma 3   Let H be a non-empty closed subset of $GL_n(k)$. If $xy \in H$ for every x and y in H, then H is a subgroup of $GL_n(k)$.

Proof   Let $h \in H$. Since the entries of the matrix xh are polynomials in the entries of x, Lemma 1 shows that the map $x \mapsto xh$ of $GL_n(k)$ onto itself is continuous. The same holds for the inverse map, namely $x \mapsto xh^{-1}$; so $x \mapsto xh$ gives a homeomorphism of $GL_n(k)$ onto itself. As H is closed in $GL_n(k)$, it follows that each of the sets Hh^i, $i = 0, 1, 2, \ldots$ is closed. By hypothesis, $H \supseteq Hh$, which implies that

$$H \supseteq Hh \supseteq Hh^2 \supseteq \ldots \supseteq Hh^j \supseteq \ldots$$

is a descending chain of closed sets. Thus Lemma 2 shows that

$$Hh^{r+1} = Hh^r$$

for some r. Multiply by h^{-r} to deduce that $Hh = H$. From this we find, first, that $eh = h$ for some $e \in H$; this e must be the identity matrix, and so $1 \in H$. We next find that $gh = 1$ for some $g \in H$, whence $h^{-1} = g \in H$. It follows that H is indeed a subgroup of $GL_n(k)$.

Proposition 1   Let G be a subgroup of $GL_n(k)$ and $\bar{G}$ its closure in $GL_n(k)$. Then $\bar{G}$ is also a subgroup.

Proof   By Lemma 3, it will suffice to show that $\bar{G}$ is closed w.r.t. multiplication. Put

$$H = \{g \in GL_n(k) \mid \bar{G}g \subseteq \bar{G}\}. \tag{2}$$

I claim (i) H is a closed subset of $GL_n(k)$, and (ii) $G \subseteq H$. Accepting (i) and (ii), we infer that $\bar{G} \subseteq H$; but this just says that $\bar{G}$ is closed w.r.t. multiplication, which is what we had to show.

Proof of (i)   For each $y \in \bar{G}$, left multiplication by y^{-1} gives a homeomorphism of $GL_n(k)$ onto itself (compare the proof of Lemma 3). Therefore $y^{-1}\bar{G}$ is closed in $GL_n(k)$ for each such y, and consequently

$$H = \bigcap_{y \in \bar{G}} y^{-1}\bar{G}$$

is closed also.

Proof of (ii)   Let $g \in G$. Since $\bar{G}g^{-1}$ is closed and contains $Gg^{-1} = G$, it follows that $\bar{G}g^{-1} \supseteq \bar{G}$. Therefore $\bar{G} \supseteq \bar{G}g$ and so $g \in H$ as required.

Proposition 2   If $G \le GL_n(k)$ is soluble, then $\bar{G}$ is also soluble, with the same derived length.

Proof   Let us prove that

$$(\bar{G})' \le \overline{G'} \tag{3}$$

(note that both $\bar{G}$ and $\overline{G'}$ are groups, by Proposition 1); the result then follows easily by induction on derived length (the case $G = 1$ is covered by Exercise 2). To prove (3), observe that for each $g \in GL_n(k)$, the map

$$\theta_g : GL_n(k) \to GL_n(k) ; \; x \mapsto [x, g]$$

is continuous, by Lemma 1. It follows that

$$[\bar{G}, g] = \bar{G}\theta_g \subseteq \overline{G\theta_g} = \overline{[G, g]}. \tag{4}$$

Letting g run over G we obtain

$$[G, \bar{G}] = [\bar{G}, G] \subseteq \left\langle \bigcup_{g \in G} [G, g] \right\rangle \subseteq \bar{G}'. \tag{5}$$

Hence for each $h \in \bar{G}$, $G\theta_h \subseteq \bar{G}'$, and the continuity of θ_h gives

$$[\bar{G}, h] \subseteq \bar{G}'.$$

This establishes (3).

Proposition 3   Let H be a subset of $GL_n(k)$, with closure $\bar{H}$ in $GL_n(k)$. Put $G = GL_n(k)$. Then

(i) $C_G(H)$ is a closed subgroup of G;
(ii) $N_G(H) \subseteq N_G(\bar{H})$;
(iii) $N_G(\bar{H})$ is a closed subgroup of G.

Proof   (i) For $h \in H$, the map $g \mapsto gh - hg$ is a rational map of G into $M_n(k)$. By Lemma 1 and Exercise 2, the inverse image of 0 under this map is closed in G. Thus $C_G(h)$ is closed, and therefore $C_G(H) = \bigcap_{h \in H} C_G(h)$ is closed also.

(ii) For $g \in G$, the inner automorphism induced by g is a homeomorphism of G onto itself, by Lemma 1. Therefore $\overline{H^g} = \bar{H}^g$. Hence if $g \in N_G(H)$, then

$$\bar{H}^g = \overline{H^g} = \bar{H}$$

and so $g \in N_G(\bar{H})$.

(iii) For $h \in \bar{H}$, the map $g \mapsto g^{-1}hg$ is a rational map of G into G. Since $\bar{H}$ is closed in G, Lemma 1 shows that the set

$$\{g \in G | g^{-1}hg \in \bar{H}\} = P_h, \text{ say,}$$

is closed in G. Similarly the set

$$Q_h = \{g \in G | ghg^{-1} \in \bar{H}\}$$

is closed in G. Therefore

$$N_G(\bar{H}) = \bigcap_{h \in \bar{H}} (P_h \cap Q_h)$$

is closed in G.

A homomorphism between linear groups which is also a rational map is called a *rational homomorphism*.

Proposition 4   Assume that the field k is perfect, and let G be a closed subgroup of $GL_n(k)$. If $\theta : G \to GL_m(k)$ is a rational homomorphism,

then

$$g_u \theta = (g\theta)_u, \quad g_s \theta = (g\theta)_s$$

for every $g \in G$.

(Here, g_u and g_s are the Jordan components of g; see Chapter 7, section A. We saw there that if $g \in G$ and G is closed, then $g_u \in G$ and $g_s \in G$.)

Proof Since θ is a homomorphism, for each $g \in G$ we have

$$g\theta = (g_u g_s)\theta = (g_s g_u)\theta = (g_u \theta)(g_s \theta) = (g_s \theta)(g_u \theta).$$

So it will suffice to show that $g_u \theta$ is unipotent and that $g_s \theta$ is diagonalizable.

There exist polynomials Q_{ij}, P_{ij} over k, in the n^2 matrix entries, such that

$$(x\theta)_{ij} = P_{ij}(x)/Q_{ij}(x) \quad \forall x \in G, \quad 1 \le i, j \le m. \tag{6}$$

Let R_1 be the subring of k generated by the coefficients of all these polynomials. Put $u = g_u$ and $y = u\theta$. We have to show that $y_s = 1$, so let us suppose that $y_s \ne 1$ and seek a contradiction. There exists a matrix $\alpha \in GL_m(k_1)$, for some extension field k_1 of k, such that y_s^α is diagonal; let R be the subring of k_1 generated by R_1 together with the entries of the following matrices:

$$u, u^{-1}, y_s, y_s^{-1}, \alpha, \text{ and } \alpha^{-1}.$$

Put

$$\sigma = \prod_{i,j} Q_{ij}(1_n)$$

and note that $0 \ne \sigma \in R_1 \subseteq R$. Since $y_s \ne 1$, there exist i and j with $(y_s)_{ij} - \delta_{ij} \ne 0$, and then $\sigma((y_s)_{ij} - \delta_{ij}) \ne 0$. So R has a maximal ideal M with $\sigma \ne 0$ mod M and $(y_s)_{ij} \not\equiv \delta_{ij}$ mod M (the maximal ideals of R intersect in zero; see Atiyah & Macdonald: *Introduction to Commutative Algebra*, Exercise 22 on page 70). Let F be the finite field R/M (Atiyah & Macdonald, page 84, Exercise 6), and let

$$\pi : GL_n(R) \to GL_n(F), \quad \pi' : GL_m(R) \to GL_m(F)$$

be the natural maps. Now I leave it as an exercise to show that

$$x \in G, x\pi = 1 \Rightarrow x\theta\pi' = 1; \tag{7}$$

this uses (6) and the fact that $\sigma \not\equiv 0$ mod M. (7) shows that we can define a homomorphism $\bar{\theta} : G\pi \to G\theta\pi'$ making the following diagram commute:

$$
\begin{array}{ccc}
G & \xrightarrow{\ \theta\ } & G\theta \\
{\scriptstyle \pi}\downarrow & & \downarrow{\scriptstyle \pi'} \\
G\pi & \xrightarrow[\ \bar{\theta}\ ]{} & G\theta\pi'
\end{array}
$$

Since u is unipotent, $u\pi$ is unipotent and therefore a p-element of $G\pi$, where $p = \text{char } F$. Hence $yx' = u\theta\pi' = u\pi\bar{\theta}$ is a p-element in $GL_m(F)$. But $y_u\pi'$ is also a p-element, and it commutes with $y\pi'$; consequently $y_s\pi' = (y_u\pi')^{-1}(y\pi')$ has p-power order. Since $y_s\pi'$ is also diagonalizable over F (by the choice of R), it follows that $y_s\pi' = 1$. Thus $(y_s)_{ij} \equiv \delta_{ij} \mod M$, and this contradicts the choice of M.

The proof that $g_s\theta$ is diagonalizable is entirely analogous, and is left to the reader.

For the next result, which is very important, we need a lemma from algebraic geometry; for the proof, see Shafarevich (1974) Chapter 1, Theorem 6, or Northcott (1980) Chapter 3, Theorem 33.

Lemma 4 Let V be a non-empty algebraic set in k^n and $\theta: V \to k^m$ a rational map. Let W be the closure of $V\theta$ in k^m. If k is algebraically closed, then $V\theta$ contains a non-empty open subset of W.

Proposition 5 Assume k algebraically closed. Let G be a closed subgroup of $GL_n(k)$ and let $\theta: G \to GL_m(k)$ be a rational homomorphism. Then $G\theta$ is a closed subgroup of $GL_m(k)$.

Proof First of all, we must embed $GL_n(k)$ in $SL_{n+1}(k)$, by identifying each matrix $g \in GL_n(k)$ with the matrix

$$\begin{bmatrix} & & 0 \\ g & \vdots & \\ & & 0 \\ 0\ldots 0 & (\det g)^{-1} \end{bmatrix} \in SL_{n+1}(k).$$

In this way, $GL_n(k)$ may be identified with a certain algebraic set in $M_{n+1}(k)$. As G is closed in $GL_n(k)$, we may consider G an algebraic set in $M_{n+1}(k)$. So Lemma 4 can be applied, to show that $G\theta$ contains a non-empty open subset U, say, of $\overline{G\theta}$. Now let $x \in \overline{G\theta}$. Then right multiplication by x effects a homeomorphism of $\overline{G\theta}$ onto itself (by Proposition 1), so Ux is an open subset of $\overline{G\theta}$. Therefore $Ux \cap G\theta \neq \emptyset$; thus $ux \in G\theta$ for some $u \in U \subseteq G\theta$, and since $G\theta$ is a group it follows that $x \in G\theta$. Thus $\overline{G\theta} \subseteq G\theta$ and the proposition follows.

Proposition 6 Let G be a subgroup of $GL_n(k)$ and N a normal subgroup of G which is closed in G. Then there exists a rational homomorphism $\theta: G \to GL_m(k)$, for some m, with $\ker \theta = N$.

This result is also basic, but I shall omit the proof. See Wehrfritz's book *Infinite Linear Groups*, Theorem 6.4 on page 86.

Next, let us consider what happens when the field k is allowed to vary. If K is an extension field of k, then k^n has two Zariski topologies: its own, and the subspace topology induced by the Zariski topology on K^n. Fortunately, however, confusion is averted by

Lemma 5 If K is an extension field of k, then the Zariski topology on K^n induces the Zariski topology on k^n, considered as a subset of K^n.

Proof If S is a subset of $k[X_1, \ldots, X_n]$, then, using an obvious notation, we plainly have

$$\mathscr{V}_k(S) = k^n \cap \mathscr{V}_K(S).$$

Thus every Zariski-closed set in k^n is also closed in the subspace topology on k^n. Conversely, suppose S is a subset of $K[X_1, \ldots, X_n]$; we must show that

$$\mathscr{V}_K(S) \cap k^n = \mathscr{V}_k(T) \tag{8}$$

for some subset T of $k[X_1, \ldots, X_n]$. To define T, let $(\alpha_i)_{i \in I}$ be a k-basis for K. Then each $f \in S$ can be expressed in the form

$$f = \sum \alpha_i g_i$$

with each $g_i \in k[X_1, \ldots, X_n]$. Let T be the set of all such polynomials g_i which occur as f runs through S. It is clear, with a moment's thought, that we then indeed obtain (8); this proves Lemma 5.

On the other hand, if $k \subseteq K$ as above, we obtain a topology on K^n, weaker in general than the Zariski topology, by letting the closed sets be those of the form

$$\mathscr{V}(S), \quad S \subseteq k[X_1, \ldots, X_n]. \tag{9}$$

This topology is called the *k-topology* on K^n, and the subsets $\mathscr{V}(S)$ as in (9) are the *k-closed* subsets of K^n. The main result regarding these is

Proposition 7 Assume that K is a normal and separable extension field of k. Let W be a closed set in K^n. Then W is k-closed if and only if $W^\sigma = W$ for every k-automorphism σ of K.

Here σ is supposed to act co-ordinatewise on K^n:

$$(x_1, \ldots, x_n)^\sigma = (x_1^\sigma, \ldots, x_n^\sigma).$$

Proof Denote the group of all k-automorphisms of K by Γ. This group acts on the ring $K[X_1, \ldots, X_n]$ by acting on the coefficients of each polynomial. Now if W is k-closed in K^n then $W = \mathcal{V}(S)$ for some subset S of $k[X_1,\ldots,X_n]$; and it is clear that for $\sigma \in \Gamma$,

$$W^\sigma = \mathcal{V}(S^\sigma) = \mathcal{V}(S) = W.$$

The non-trivial assertion of the proposition is the converse. So suppose W is closed in K^n and $W = W^\sigma$ for every $\sigma \in \Gamma$. Then $I = \mathcal{I}(W)$ satisfies $I = I^\sigma$ for all $\sigma \in \Gamma$. Now if $I = J \cdot K[X_1,\ldots,X_n]$ for some $J \subseteq k[X_1,\ldots,X_n]$, then

$$W = \mathcal{V}(I) = \mathcal{V}(J)$$

will be k-closed as required; so the proposition will follow from

Lemma 6   Let K be a separable and normal field extension of k, with Galois group Γ. Put $R_K = K[X_1,\ldots,X_n]$ and $R_k = k[X_1,\ldots,X_n]$. If I is an ideal of R_K with $I = I^\sigma$ for all $\sigma \in \Gamma$, then $I = (I \cap R_k)R_K$.

Proof   Put $J = I \cap R_k$, and suppose that $I > JR_K$. Since $R_K = KR_k$, we can write

$$R_K = JR_K \oplus \bigoplus_i Kr_i$$

with each $r_i \in R_k$ (thus the r_i form a k-basis for R_K modulo JR_K). For each $x \in R_K$ we then have a unique expression

$$x = c + \sum_{j=1}^{t} \lambda_j r_{i(j)}$$

with $c \in JR_K$ and each $\lambda_j \in K \backslash \{0\}$. Define $t(x) = t$. Now choose $x \in I \backslash JR_K$ with $t(x)$ as small as possible. Then $t(x) \geq 1$, and replacing x by $\lambda_t^{-1}x$ we may suppose $\lambda_t = 1$. Then for each $\sigma \in \Gamma$,

$$x - x^\sigma = (c - c^\sigma) + \sum_{j=1}^{t} (\lambda_j - \lambda_j^\sigma)r_{i(j)}$$

$$= (c - c^\sigma) + \sum_{j=1}^{t-1} (\lambda_j - \lambda_j^\sigma)r_{i(j)}.$$

Since $x - x^\sigma \in I$ and $t(x - x^\sigma) < t = t(x)$, we must have $x - x^\sigma \in JR_K$. It follows that $\lambda_j - \lambda_j^\sigma = 0$ for each j. As this holds for every $\sigma \in \Gamma$, Galois theory shows that $\lambda_j \in k$ for each j. Therefore, since $c \in JR_K \subseteq I$,

$$x - c \in I \cap R_k = J$$

and so $x \in JR_K$, a contradiction. Thus $I = JR_K$ as claimed. (If you are unfamiliar with the Galois theory of infinite field extensions, note that an element λ of K which is transcendental over k will always be moved by some k-automorphism of K: extend $\{\lambda\}$ to a transcendence basis for K over k and let L be the corresponding pure transcendental extension of k. Then

$\lambda \mapsto \lambda + 1$ extends to a k-automorphism of L, and this extends to a k-automorphism of K since K is normal and algebraic over L. Thus we are left with λ algebraic over k. Let Λ be the normal closure of $k(\lambda)$ over k; then $[\Lambda:k]$ is finite, and we can apply ordinary Galois theory (having made sure that every k-automorphism of Λ extends to an automorphism of K, which we do in two steps: transcendental, algebraic, as above).)

In fact we have proved a little more than stated in Proposition 7: a subset W of K^n is said to be *defined over k* if the ideal $\mathcal{J}(W) \lhd K[X_1,\ldots,X_n]$ can be generated by polynomials with coefficients in k. Then Lemma 6 gives

Corollary 1 If K is separable and normal over k then every k-closed set in K^n is defined over k.

If $W \subseteq K^n$ and $\theta: W \to k^m$ is a rational map, we call θ k-*rational* provided the polynomials P_i and Q_i which give θ, as in (1) above, can be chosen to have coefficients in k. In this case, if $\sigma \in \Gamma$ we have

$$W^\sigma = W \Rightarrow (W\theta)^\sigma = W\theta$$
$$\Rightarrow (\mathcal{J}(W\theta))^\sigma = \mathcal{J}(W\theta)$$
$$\Rightarrow \overline{W\theta}^\sigma = \overline{W\theta}$$

for each $\sigma \in \Gamma$, where $\overline{W\theta} = \mathcal{V}(\mathcal{J}(W\theta))$ is the closure of $W\theta$ in K^m. So Proposition 7 gives

Corollary 2 Let $k \subseteq K$ be as above, let $W \subseteq K^n$ and let $\theta: W \to K^m$ be a k-rational map. If $W = W^\sigma$ for every k-automorphism σ of K, then the closure of $W\theta$ in K^m is defined over k.

In this connection, one should note that Proposition 6 can be improved slightly: if $G \le GL_n(k)$ and $N \lhd G$ is k-closed in G, then the map θ given in Proposition 6 may be taken k-rational; this follows from the proof quoted above.

Exercise 3 Show that Proposition 5 may fail if k is not algebraically closed.
(*Hint:* Try the map $x \mapsto x^2$ on $GL_1(\mathbb{Q})$.)

Now a $\mathbb{Q}$-group is nothing other than a $\mathbb{Q}$-closed subgroup of $GL_n(\mathbb{C})$, for some n. So we may re-state some of the results in different language:

Corollary 3 If $\mathcal{G}$ is a $\mathbb{Q}$-group and $\mathcal{H}$ a normal subgroup of G which is also a $\mathbb{Q}$-group, then there exists a $\mathbb{Q}$-rational homomorphism θ of $\mathcal{G}$ onto a $\mathbb{Q}$-group $\mathcal{K}$ such that $\ker\theta = \mathcal{H}$.

We must also discuss arithmetic groups. To clarify the situation, note

Lemma 7   Let H be a subgroup of $GL_n(\mathbb{Z})$.

(i) H is arithmetic (in some $\mathbb{Q}$-group of degree n) if and only if H has finite index in its own closure in $GL_n(\mathbb{Z})$.

(ii) If H has a subgroup of finite index which is arithmetic, then H is arithmetic.

(iii) If H is arithmetic and $H_1 \leq GL_n(\mathbb{Z})$ is also arithmetic, then $H \cap H_1$ is arithmetic.

Proof   (i) Let $\bar{H}$ be the closure of H in $GL_n(\mathbb{C})$. Propositions 1 and 7 show that $\bar{H}$ is a $\mathbb{Q}$-group, and the closure of H in $GL_n(\mathbb{Z})$ is just $\bar{H} \cap GL_n(\mathbb{Z}) = \bar{H}(\mathbb{Z})$. The 'if' statement follows at once. Conversely, if H is arithmetic in some $\mathbb{Q}$-group $\mathscr{G}$, then $H \leq \mathscr{G}(\mathbb{Q})$ implies $\bar{H} \leq \overline{\mathscr{G}(\mathbb{Q})} \leq \mathscr{G}$, consequently

$$|\bar{H}(\mathbb{Z}):H| \leq |\mathscr{G}(\mathbb{Z}):H| < \infty.$$

(ii) Suppose $K \leq_f H$ and K is arithmetic. Then there exists $L \lhd_f H$ with $L \leq K$, and L is also arithmetic since $L \leq_f K$. By Proposition 3, H normalizes $\bar{L}$, and so H normalizes $\bar{L}(\mathbb{Z})$. Hence $H \cdot \bar{L}(\mathbb{Z})$ is a group, and it is the union of finitely many cosets of $\bar{L}(\mathbb{Z})$. Therefore $H \cdot \bar{L}(\mathbb{Z})$ is closed in $GL_n(\mathbb{Z})$ and so $\bar{H}(\mathbb{Z}) \leq H \cdot \bar{L}(\mathbb{Z})$. Thus $|\bar{H}(\mathbb{Z}):H|$ is finite and H is arithmetic by (i).

(iii) Certainly $\bar{H}(\mathbb{Z}) \cap \bar{H}_1(\mathbb{Z})$ is closed in $GL_n(\mathbb{Z})$. But

$$|\bar{H}(\mathbb{Z}) \cap \bar{H}_1(\mathbb{Z}):H \cap H_1| = |\bar{H}(\mathbb{Z}) \cap \bar{H}_1(\mathbb{Z}):\bar{H}(\mathbb{Z}) \cap H_1| \cdot |\bar{H}(\mathbb{Z}) \cap H_1:H \cap H_1|$$

$$\leq |\bar{H}_1(\mathbb{Z}):H_1| \cdot |\bar{H}(\mathbb{Z}):H| < \infty,$$

so $H \cap H_1$ is arithmetic by (i).

Lemma 8   If $H \leq GL_n(\mathbb{Z})$ is arithmetic, then so is $N_{GL_n(\mathbb{Z})}(H)$.

Proof   Put $\mathscr{N} = N_{GL_n(\mathbb{C})}(\bar{H})$. Propositions 3 and 7 show that $\mathscr{N}$ is a $\mathbb{Q}$-group, so it will be enough if we show that $N_{GL_n(\mathbb{Z})}(H) = N$, say, is a subgroup of finite index in $\mathscr{N}(\mathbb{Z})$. Certainly $N \leq \mathscr{N}(\mathbb{Z})$, by Proposition 3. Also H has finite index in $\bar{H}(\mathbb{Z})$, by Lemma 7. A theorem of Borel and Harish-Chandra (Theorem 7 of Chapter 6; see section E for references) shows that $\bar{H}(\mathbb{Z})$ is finitely generated, so $\bar{H}(\mathbb{Z})$ has only finitely many subgroups of index equal to $|\bar{H}(\mathbb{Z}):H|$. These subgroups are permuted by $\mathscr{N}(\mathbb{Z})$, acting by conjugation; as N is just the stabilizer of H under this action, it follows that $|\mathscr{N}(\mathbb{Z}):N|$ is finite.

Corollary 3 showed that a quotient of $\mathbb{Q}$-groups is isomorphic to a $\mathbb{Q}$-group. The analogous result for arithmetic groups is a theorem of Borel; it lies much deeper than the other results of this section:

Theorem 1 Let $\mathscr{G}$ be a $\mathbb{Q}$-group and $\theta: \mathscr{G} \to GL_m(\mathbb{C})$ a $\mathbb{Q}$-rational homomorphism. If Γ is an arithmetic subgroup of G then $\Gamma\theta$ is arithmetic in the $\mathbb{Q}$-group $\mathscr{G}\theta$.

For the proof, see the references given in section E. To see what is the hard part of this theorem, do

Exercise 4 In the situation of Theorem 1, show that $\Gamma\theta \cap GL_m(\mathbb{Z})$ has finite index in $\Gamma\theta$. Show also that $\Gamma\theta\psi \le GL_m(\mathbb{Z})$ for a suitable inner automorphism ψ of $GL_m(\mathbb{Q})$.
(*Hint*: Show that the entries of all matrices in Γ have bounded denominators; deduce that the same holds for $\Delta = \Gamma\theta \le GL_m(\mathbb{Q})$. Now consider what Δ does to the $\mathbb{Z}$-lattices lying between $t\mathbb{Z}^m\Delta$ and $\mathbb{Z}^m\Delta$ where t is a common multiple for those denominators. For the second part, try conjugating by a matrix $\mu \in GL_m(\mathbb{Q})$ such that $\mathbb{Z}^m\Delta = \mathbb{Z}^m\mu$.)

The Zariski topology is particularly straightforward in unipotent groups: the following facts are essentially a repetition, in different language, of results from Chapter 6.

Proposition 8

(i) If $G \le \mathrm{Tr}_1(n, \mathbb{Q})$ then the closure of G in $GL_n(\mathbb{Q})$ is equal to the isolator of G in $\mathrm{Tr}_1(n, \mathbb{Q})$, i.e. the set of $x \in \mathrm{Tr}_1(n, \mathbb{Q})$ with $x^e \in G$ for some $e \ne 0$.

(ii) If $H \le \mathrm{Tr}_1(n, \mathbb{Z})$ then H has finite index in its closure in $GL_n(\mathbb{Z})$.

Proof (ii) follows from (i) since $\mathrm{Tr}_1(n, \mathbb{Z})$ is a finitely generated nilpotent group. Now $\mathrm{Tr}_1(n, \mathbb{Q})$ is evidently closed in $GL_n(\mathbb{Q})$, so if $\bar{G}$ is the closure of G in $GL_n(\mathbb{Q})$ than $\bar{G}$ is also the closure of G in $\mathrm{Tr}_1(n, \mathbb{Q})$. Now log and exp are mutually inverse rational maps between $\mathrm{Tr}_1(n, \mathbb{Q})$ and $\mathrm{Tr}_0(n, \mathbb{Q})$, hence homeomorphisms. It follows that

$$\bar{G} = \exp((\log G)^-)$$

where $(\log G)^-$ is the closure of the set $\log G$ w.r.t. the Zariski topology in $\mathrm{Tr}_0(n, \mathbb{Q})$. From Theorem 2 of Chapter 6, we know that the isolator of G in $\mathrm{Tr}_1(n, \mathbb{Q})$ is exactly $\exp(\mathbb{Q} \log G)$. So to prove (i), we must show that

$$(\log G)^- = \mathbb{Q} \log G. \tag{10}$$

Now $\mathbb{Q} \log G$ is a linear subspace of $\mathrm{Tr}_0(n, \mathbb{Q})$, so it is the solution-set of a family of linear equations in the matrix entries. Therefore $\mathbb{Q} \log G$ is closed and we have $(\log G)^- \subseteq \mathbb{Q} \log G$. To obtain the reverse inclusion, let

$\lambda \in \mathbb{Q} \log G$ and suppose $f \in \mathcal{I}(\log G)$; thus $f \in \mathbb{Q}[X_{11}, \ldots, X_{nn}]$ and $f(x) = 0$ for all $x \in \log G$. Define a polynomial φ in one variable over $\mathbb{Q}$ by

$$\varphi(Y) = f(Y\lambda).$$

Since $\lambda \in \mathbb{Q} \log G$ we have $t\lambda \in \log G$ for some positive integer t, and then $st\lambda \in \log G$ for every $s \in \mathbb{Z}$. Thus $\varphi(st) = 0$ for all $s \in \mathbb{Z}$. Hence φ is the zero polynomial, and so

$$f(\lambda) = \varphi(1) = 0.$$

This shows that $\mathbb{Q} \log G \subseteq \mathcal{V}(\mathcal{I}(\log G)) = (\log G)^-$, and establishes (10).

To conclude this section, here are some exercises relating the Zariski topology to the *congruence topology*, discussed in Chapter 4.

Let p be a prime number. The *p-adic topology* on $M_n(\mathbb{Z})$ is defined by taking the additive subgroups

$$p^m M_n(\mathbb{Z}), \quad m \in \mathbb{N}$$

as a base for the neighbourhoods of zero. The *p-congruence topology* on $GL_n(\mathbb{Z})$ is the subspace topology induced from the p-adic topology on $M_n(\mathbb{Z})$. It is easy to see that a base for the neighbourhoods of 1 is then given by the subgroups

$$K_{p^m} = (1 + p^m M_n(\mathbb{Z})) \cap GL_n(\mathbb{Z}), m \in \mathbb{N}.$$

Exercise 5   Show that a Zariski-closed subset of $M_n(\mathbb{Z})$ is closed in the p-adic topology, for every prime p.
(*Hint*: Observe that if $f \in \mathbb{Z}[X_{11}, \ldots, X_{nn}]$ and $x \in M_n(\mathbb{Z})$ satisfy $f(x) \equiv 0$ mod p^m for all m, then $f(x) = 0$.)

Exercise 6   Let G be a subgroup of $GL_n(\mathbb{Z})$ which is closed in the p-congruence topology. Let h be an integer coprime to p. Show that if $x \in K_p$ satisfies $x^{p^r h} \in G$ then $x^{p^r} \in G$.
(*Hint*: Exercise 8 of Chapter 5, section C, shows that K_p/K_{p^m} is a p-group for each $m \geq 1$. Given $m \geq 1$, say $x^{p^e} \in K_{p^m}$. Then $ap^e + bh = 1$ with $a, b \in \mathbb{Z}$, and x^{p^r} is congruent modulo p^m to $x^{bhp^r} \in G$.)

Exercise 7   Put $K_6 = (1 + 6M_n(\mathbb{Z})) \cap GL_n(\mathbb{Z})$. Let G be a Zariski-closed subgroup of $GL_n(\mathbb{Z})$. Show that $x \in K_6$ and $x^h \in G$ with $h \neq 0$ imply $x \in G$. Deduce that if $y \in GL_n(\mathbb{Z})$ and $y^h \in G$ with $h \neq 0$, then $y^e \in G$ where e is the exponent of the finite group $GL_n(\mathbb{Z}/6\mathbb{Z})$.

(For this, use the results of Exercises 5 and 6.) For any group H and subgroup C of H, the *isolator* of C in H is the set

$$i_H(C) = \{x \in H \mid x^m \in C \quad \text{for some } m \neq 0\}.$$

C is *isolated* if $i_H(C) = C$; thus Exercise 7 shows that closed subgroups of K_6 are isolated. In general, we say that H has the *isolator property* if for every subgroup C of H, $i_H(C)$ is a subgroup of H and $|i_H(C):C|$ is finite. Proposition 8 and Proposition 1 together show that all subgroups of $\text{Tr}_1(n, \mathbb{Z})$ have the isolator property. It follows that $\mathfrak{T}$-groups have this property, and hence (easy exercise!) so do all finitely generated nilpotent groups. As a nice illustration of the present technique, let us extend this result to polycyclic groups:

Exercise 8 Show that in every polycyclic group there is a subgroup of finite index which has the isolator property.
(*Hint*: By results from Chapter 5 and Chapter 2, it suffices to show that if $H \leq K_6$ is triangularizable, then H has the isolator property. Let N be the set of all unipotent matrices in H, and assume without loss of generality that $H' \leq N \leq \text{Tr}_1(n, \mathbb{Z})$. Use $^-$ to denote Zariski-closure in H, and let C be any subgroup of H. Show that (i) $i_H(C) \subseteq \bar{C}$; (ii) $i_H(C) \cap N = i_N(N \cap C) = \overline{N \cap C}$ $\leq N$; (iii) C normalizes $N \cap C$ and $|(\overline{N \cap C}) \, C : C|$ is finite; (iv) $\overline{N \cap C} \triangleleft \bar{C}$ and $\bar{C}/\overline{N \cap C}$ is a finitely generated abelian group; (v) $i_H(C)/(\overline{N \cap C})C$ is the torsion subgroup of $\bar{C}/(\overline{N \cap C})C$.)

B. Automorphisms of polycyclic groups

We showed in Chapter 6 that the automorphism group of a $\mathfrak{T}$-group is isomorphic to an arithmetic group. In this section we extend that result. One preliminary lemma is required.

Lemma 9 Let k be a field and $M \leq \text{Tr}_n(k)$. Put $X = N_{GL_n(k)}(M)$. Then for each $x \in X$,

$$[M, x^{(n!)!}] \subseteq \text{Tr}_1(n, k).$$

Proof For $a = (a_{ij}) \in M$ put

$$d(a) = \text{diag}(a_{11}, \ldots, a_{nn}).$$

Then X acts on the set $d(M)$ by

$$d(a) \cdot x = d(a^x).$$

Since the eigenvalues of a^x are the same as those of a, each orbit of X in its · action on $d(M)$ has length at most n! Therefore putting $r = (n!)$! we have

$$d(a) \cdot x^r = d(a) \quad \forall a \in M, x \in X.$$

As $a \mapsto d(a)$ is a homomorphism of M onto $d(M)$, it follows that

$$d([a, x^r]) = d(a^{-1} a^{x^r}) = d(a)^{-1}(d(a) \cdot x^r) = 1$$

for all $a \in M$ and $x \in X$. Thus $[a, x^r] \in \mathrm{Tr}_1(n, k)$ for all such a and x.

For any matrix group M, we write

$$u(M) = \langle N \lhd M \mid N \text{ is unipotent} \rangle.$$

Exercise 9 $u(M)$ is a unipotent normal subgroup of M, for every $M \leq GL_n(k)$. If M is triangularizable then $u(M)$ contains all the unipotent elements of M and $M' \leq u(M)$.

(*Hint*: M has a maximal unipotent normal subgroup, by Zorn's Lemma; call it U. If $N \lhd M$ and N is unipotent, show that UN is unipotent (see Exercise 2 of Chapter 5); deduce that $U = u(M)$. For the second part, observe that if $M \leq \mathrm{Tr}_n(k)$ then $u(M) = M \cap \mathrm{Tr}_1(n, k)$.)

Now the key result of the section is

Proposition 9   Let G be a soluble-by-finite subgroup of $GL_n(\mathbb{Z})$, and M a triangularizable characteristic subgroup of finite index in G. Put

$$U = u(M), \quad S = C_{GL_n(\mathbb{Z})}(G/U).$$

Then

(i) $N_{GL_n(\mathbb{Z})}(G)/S$ is finitely generated and abelian-by-finite;
(ii) S is an arithmetic group;
(iii) $S/C_{GL_n(\mathbb{Z})}(G)$ is isomorphic to an arithmetic group.

Proof (i) Write $N = N_{GL_n(\mathbb{Z})}(G)$. By hypothesis, there exist a field k and a matrix $\xi \in GL_n(k)$ such that $M^\xi \leq \mathrm{Tr}_n(k)$. Exercise 9 shows that $M^\xi \cap \mathrm{Tr}_1(n, k) = U^\xi$; since N^ξ normalizes M^ξ, we may apply Lemma 9 and then conjugate with ξ^{-1} to deduce that $N/C_N(M/U)$ is periodic. Now M/U is a finitely generated abelian group, so every periodic group of automorphisms of M/U is in fact finite; see Chapter 5, Exercise 8 and Corollary 3 (but the abelian case used here is much easier!). Therefore

$$N/C_N(M/U) \text{ is finite.}$$

Since G/M is finite, $N/C_N(G/M)$ is also finite, consequently

$$N/D \text{ is finite, where } D = C_N(G/M) \cap C_N(M/U).$$

Proposition 11 of Chapter 1 shows that $D/C_N(G/U)$ is a finitely generated abelian group. Since $C_N(G/U)$ is plainly equal to S, (i) follows.

(ii) Without loss of generality, we may suppose that

$$U = M \cap \mathrm{Tr}_1(n, \mathbb{Z})$$

(conjugate by a matrix in $GL_n(\mathbb{Z})$ to put U into unitriangular form). Write $\bar{U}$ for the closure of U in $GL_n(\mathbb{Q})$ and put $K = N_{GL_n(\mathbb{Q})}(\bar{U})$. Since $U \lhd N$, Proposition 3 shows that $N \le K$. Now by Proposition 6 there is a $\mathbb{Q}$-rational homomorphism $\theta : K \to GL_m(\mathbb{Q})$, for some m, with $\ker \theta = \bar{U}$. Putting

$$H = (C_{GL_m(\mathbb{Q})}(G\theta))\theta^{-1},$$

we see that H is a closed subgroup of K (Lemma 1 and Proposition 3). As K is closed in $GL_n(\mathbb{Q})$, by Proposition 3, it follows that H is closed in $GL_n(\mathbb{Q})$. Therefore

$$H_1 = H \cap GL_n(\mathbb{Z})$$

is closed in $GL_n(\mathbb{Z})$. Now in fact

$$H_1 = C_K(G\bar{U}/\bar{U}) \cap GL_n(\mathbb{Z}),$$

since $\ker \theta = \bar{U}$; since $S \le N \subseteq K \cap GL_n(\mathbb{Z})$ and S centralizes G/U, it follows that $S \le H_1$. Hence, by Lemma 7, to establish (ii) it will suffice to show that $|H_1 : S|$ is finite.

Put $V = \bar{U} \cap GL_n(\mathbb{Z})$. Then H_1 centralizes GV/V; and $|V:U|$ is finite by Proposition 8. Let

$$H_2 = N_{H_1}(G), H_3 = C_{H_2}((G \cap V)/U).$$

Since $|(G \cap V):U| \le |V:U|$ is finite, $|H_2:H_3|$ is finite. Since $|GV:G| = |V:G \cap V| \le |V:U|$ is finite, $|H_1:H_2|$ is finite (as GV is finitely generated, it has only finitely many subgroups of index equal to $|GV:G|$, and these are permuted by H_1). Finally, H_3 centralizes the factors $G/(G \cap V) \cong GV/V$ and $(G \cap V)/U$, consequently $H_3/C_{H_3}(G/U)$ is finite, by Proposition 11 of Chapter 1. As $C_{H_3}(G/U) \le S$, it follows that $|H_1:S|$ is finite as required.

(iii) Let $\mathscr{S}$ be the closure of S in $GL_n(\mathbb{C})$ and put $\mathscr{C} = C_{GL_n(\mathbb{C})}(G)$. Propositions 1, 3 and 7 show that $\mathscr{S}$ and $\mathscr{C}$ are $\mathbb{Q}$-groups. Also $\mathscr{C} \lhd \mathscr{S}$; for S normalizes G, so S normalizes $\mathscr{C}$ and the claim follows by Proposition 3. Proposition 6 therefore gives a $\mathbb{Q}$-rational homomorphism $\rho : \mathscr{S} \to GL_r(\mathbb{C})$, for some r, with $\ker \rho = \mathscr{C}$; by part (ii) and Theorem 1, $S\rho$ is an arithmetic group. But

$$S\rho \cong S/(\ker \rho \cap S) = S/C_S(G),$$

so we have (iii).

Later on we shall need a corollary of the above proof:

Corollary 4   If M is a triangularizable subgroup of $GL_n(\mathbb{Z})$, then $N_{GL_n(\mathbb{Z})}(M)$ is an arithmetic group.

For taking $G = M$ in the argument above, we find that N/S is finite; thus part (ii) and Lemma 8 show that N is arithmetic.

We are ready now for the main result:

Theorem 2   Let G be a polycyclic-by-finite group and $A = $ Aut G. Then there exists $K \lhd A$ with A/K finitely generated and abelian-by-finite and K isomorphic to some arithmetic group.

Proof By Theorem 7 of Chapter 5, the semi-direct product $G]A$ may be embedded in $GL_n(\mathbb{Z})$, for some n. Consider this done, put $N = N_{GL_n(\mathbb{Z})}(G)$ and $C = C_{GL_n(\mathbb{Z})}(G)$. By Mal'cev's theorem (Chapter 2), G has a triangularizable subgroup M of finite index, which we may take to be characteristic in G. Put $U = u(M)$ and $S = C_{GL_n(\mathbb{Z})}(G/U)$. Then Proposition 9 tells us that N/S is finitely generated and abelian-by-finite, that S/C is isomorphic to an arithmetic group, and that S is itself an arithmetic group (the latter statement will only be needed later on).

Since A acts faithfully by conjugation on G and in this way gives every automorphism of G, it follows that

$$N = C]A,$$

and this implies that $S = C](A \cap S)$. So putting

$$K = A \cap S,$$

we have $K \cong S/C$ and $A/K \cong AS/S \le N/S$. The result follows.

Exercise 10   Let G be a polycyclic group such that Fitt(G) is torsion-free and $G/$Fitt(G) is free abelian. Show that Aut G is isomorphic to an arithmetic group.
(*Hint*: Put $F = $ Fitt(G). Lemma 10 of Chapter 5 shows that $C_{\text{Aut}(G)}(G/F)$ has finite index in Aut G. Now show that in the proof of Theorem 2, above, we can arrange it so that $K = C_{\text{Aut}(G)}(G/F)$.)

Let us derive some purely group-theoretic consequences from Theorem 2. These will follow from the theorem of Borel and Harish-Chandra (see section E):

Theorem 3   If H is an arithmetic group, then

(i) H is finitely presented;
(ii) the finite subgroups of H lie in finitely many conjugacy classes in H.

Lemma 10   Let G be a group and N a normal subgroup. If N and G/N are finitely presented, then G is finitely presented.

Proof   Let $^-:G \to G/N$ be the natural map. N is generated by a finite set $\{y_1,\ldots,y_r\}$, say, subject to defining relations $u_i(y_1,\ldots,y_r)=1$, $i=1,\ldots,s$. G/N is generated by a finite set $\{\bar{x}_1,\ldots,\bar{x}_t\}$, say, subject to the defining relations $v_i(\bar{x}_1,\ldots,\bar{x}_t)=1$, $i=1,\ldots,m$. Then certainly G is generated by $\{y_1,\ldots,y_r,x_1,\ldots,x_t\}$, and these generators satisfy relations of the form

$$u_i(\mathbf{y})=1, \quad i=1,\ldots s$$
$$v_i(\mathbf{x})=\lambda_i(\mathbf{y}), \quad i=1,\ldots,m$$
$$\left.\begin{array}{l} y_i^{x_j}=\mu_{ij}(\mathbf{y}), \\ y_i^{x_j^{-1}}=v_{ij}(\mathbf{y}), \end{array}\right\} i=1,\ldots,s; \quad j=1,\ldots,m \tag{11}$$

where $\mathbf{x}=(x_1,\ldots,x_t)$ and $\mathbf{y}=(y_1,\ldots,y_r)$. I claim that (11) is a set of defining relations for G on the set $\{x_1,\ldots,x_t,y_1,\ldots,y_r\}$. To establish this, let $\{\xi_1,\ldots,\xi_t,\eta_1,\ldots,\eta_r\}$ be a set of symbols in 1:1 correspondence with $\{x_1,\ldots,y_r\}$ and consider the group Γ generated by $\{\xi_1,\ldots,\eta_r\}$ subject to the relations obtained from (11) on substituting ξ_i for x_i and η_j for y_j, for each i,j. We have a group epimorphism $\pi:\Gamma \to G$ sending ξ_i to x_i and η_j to y_j for each i and j, and we have to show that $\ker \pi = K$, say, is trivial. Put $\Delta = \langle \eta_1,\ldots,\eta_r \rangle$. The third and fourth sets of relations coming from (11) show that $\Delta \lhd \Gamma$, and the original presentation for G/N shows that π induces an isomorphism of Γ/Δ onto G/N. Therefore $K \leq \Delta$. So if $a \in K$, we can express a in the form $a=w(\eta_1,\ldots,\eta_r)$. Then

$$1 = a\pi = w(y_1,\ldots,y_r).$$

Since the relations $u_i(\mathbf{y}), i=1,\ldots,s$, are a set of defining relations for N on $\{y_i,\ldots,y_r\}$, the relation $w(y_1,\ldots,y_r)=1$ must be a consequence of the relations $u_i(\mathbf{y})=1, i=1,\ldots,s$. As $u_i(\eta_1,\ldots,\eta_r)=1$ for $i=1,\ldots,s$, it follows that $w(\eta_1,\ldots,\eta_r)=1$. Thus $a=1$ and $K=1$ as required.

This is all we need to deduce

Theorem 4   If G is a polycyclic-by-finite group, then G, Aut G, and Hol G are all finitely presented.

Proof   Cyclic groups, finite groups, and arithmetic groups are all finitely presented. So the result follows from Theorem 2 (and Theorem 3) by repeated applications of Lemma 10.

The other property of interest was to do with conjugacy of finite subgroups. For the remainder of this section, let us say that *a group G has the property* $\mathscr{P}$ *if the finite subgroups of G lie in finitely many conjugacy classes*. Now it is *not* true that this property is preserved by the formation of group extensions (see Exercise 11, below); however we can make do with

Lemma 11 Let B a group and A a normal subgroup. Suppose that B/A has $\mathscr{P}$ and that AF has $\mathscr{P}$ for every finite subgroup F of B. Then B has $\mathscr{P}$.

The proof is more or less obvious. A more noteworthy result is

Lemma 12 Let A be a free abelian group of finite rank and suppose $A \lhd_f B$. Then B has $\mathscr{P}$.

Proof Suppose F is a finite subgroup of B. Since A is torsion-free, $AF = A]F$, and since $|B:A|$ is finite there are only finitely many possibilities for the group AF. So we may assume that $B = A]H$ for some H, and it will suffice to show that the complements for A and B in lie in finitely many conjugacy classes. As we saw in section A of Chapter 3, this is equivalent to showing that

$$\text{Der}(H, A)/\text{Ider}(H, A) \text{ is finite.} \tag{12}$$

Since H is finite and A is finitely generated abelian, $\text{Der}\,(H, A)$ is also finitely generated abelian. So (12) will follow if we can show that

$$|H| \cdot \text{Der}\,(H, A) \subseteq \text{Ider}\,(H, A). \tag{13}$$

Take $\delta \in \text{Der}(H, A)$ and put $m = |H|$. Let $h \in H$. Writing A additively, we have

$$\sum_{x \in H} (xh)\delta = \left(\sum_{x \in H} x\delta \right)h + m \cdot h\delta.$$

As x runs over H, so does xh; so putting

$$a = \sum_{x \in H} x\delta$$

we see that

$$a = ah + m \cdot h\delta.$$

Thus $m\delta$ is the inner derivation $h \mapsto -a(h-1)$, and (13) follows.

We also need a sort of converse to Lemma 11:

Lemma 13 If a group N with property $\mathscr{P}$ is a semi-direct product $N = C]A$, then A also has $\mathscr{P}$.

Proof Since N has $\mathscr{P}$, it will suffice to show that two subgroups of A which are conjugate in N are already conjugate in A. So suppose $F \leq A$

and $F^x \leq A$ with $x \in N$. Let $\pi: N \to A$ denote the projection homomorphism $(ca \mapsto a)$. Then

$$F^x = (F^x)\pi = (F\pi)^{x\pi} = F^{x\pi}$$

which is what we want since $x\pi \in A$.

Putting the pieces together we now obtain

Theorem 5   If G is a polycyclic-by-finite group, then G, Aut G, and Hol G all have the property $\mathscr{P}$.

Proof   An obvious argument by induction on the Hirsch number, using Lemmas 12 and 11, shows that G has $\mathscr{P}$. Similarly, every finite extension of G has $\mathscr{P}$, so once we have shown that Aut G has $\mathscr{P}$, Lemma 11 will ensure that Hol $G = G]$ Aut G also has $\mathscr{P}$.

To see why Aut G has the property $\mathscr{P}$, recall the situation established in the proof of Theorem 2. We had $A = $ Aut G related to certain other groups as follows:

$$N = C]A \leq GL_n(\mathbb{Z})$$
$$C \leq S \lhd N$$

where N/S is finitely generated and abelian-by-finite, and S is arithmetic. By Lemma 12, N/S has $\mathscr{P}$. By Lemma 7 in section A, every subgroup of N which contains S as a subgroup of finite index is itself an arithmetic group, and therefore has $\mathscr{P}$ by Theorem 3. Thus Lemma 11 shows that N has $\mathscr{P}$. That A has $\mathscr{P}$ now follows by Lemma 13.

Exercise 11   Show that an extension of a group with $\mathscr{P}$ by another group with $\mathscr{P}$ need not itself have the property $\mathscr{P}$.
(*Hint*: Let N be free abelian of infinite rank and let X be cyclic of order 2, acting on N by inversion. Put $G = N]X$. Then compute $\mathrm{Der}(X, N)/\mathrm{Ider}(X, N)$. Try some different examples!)

C. Finite extensions

Let G be a polycyclic-by-finite group and suppose H is a finite extension of G; that is, $G \lhd H$ and H/G is finite. An embedding $\theta: G \to GL_n(\mathbb{Z})$ gives rise to an induced embedding $\bar{\theta}: H \to GL_{nd}(\mathbb{Z})$, where $d = |H:G|$, as described in Lemma 8 of Chapter 5. Let us compute the restriction of $\bar{\theta}$ to G. Recall that $\bar{\theta}$ is constructed using a transversal $\{x_1, \ldots, x_d\}$ to the cosets of G in H; if $\tau_i \in$ Aut G is the map $g \mapsto x_i g x_i^{-1}$, we find that

$$g\bar{\theta} = \mathrm{diag}(g^{\tau_1}\theta, \ldots, g^{\tau_d}\theta) \quad \text{for } g \in G.$$

Now suppose θ extends to an embedding of $G]$ Aut G into $GL_n(\mathbb{Z})$ – such an embedding exists, by Theorem 7 of Chapter 5. Then $g^{\tau_i}\theta = (g\theta)^{\tau_i\theta}$ for each $g \in G$ and each i, so we now have

$$g\bar{\theta} = (\text{diag}\,(g\theta, \ldots, g\theta))^{\text{diag}(\tau_1\theta, \ldots, \tau_d\theta)}$$

for each $g \in G$. Thus if we define

$$\theta^* : G \to GL_{nd}(\mathbb{Z}); \quad g\theta^* = \text{diag}\,(\underset{\longleftarrow \quad d \quad \longrightarrow}{g\theta, \ldots, g\theta})$$

and put

$$T = \text{diag}\,(\tau_1\theta, \ldots, \tau_d\theta) \in GL_{nd}(\mathbb{Z}),$$

we have

$$(G\theta^*)^T = G\bar{\theta} \lhd_f H\bar{\theta}.$$

Thus, finally,

$$H \cong H\bar{\theta} \cong (H\bar{\theta})^{T^{-1}},$$
$$G\theta^* \lhd_f (H\bar{\theta})^{T^{-1}} \leq GL_{nd}(\mathbb{Z}).$$

Now the map θ^* depends only on d, G and θ, not on the group H. So what we have established is

Lemma 14   Let G be a polycyclic-by-finite group and d a positive integer. Then there exists an embedding $\theta^* : G \to GL_m(\mathbb{Z})$, for some m, such that every group H containing G as a normal subgroup of index d is isomorphic to some subgroup K of $GL_m(\mathbb{Z})$ with $G\theta^* \lhd K$ and $|K : G\theta^*| = d$.

This shows that to obtain an over-view of the finite extensions of G, it suffices to consider the case where $G \leq GL_m(\mathbb{Z})$ and $G \lhd_f H \leq GL_m(\mathbb{Z})$. So what we are looking at is the collection of all finite subgroups of $N_{GL_n(\mathbb{Z})}(G)/G$, which is very similar to the topic discussed in the previous section.

Proposition 10   Let Γ be an arithmetic group and G a soluble-by-finite subgroup of Γ. Then the subgroups K of Γ such that $G \lhd_f K$ lie in finitely many conjugacy classes in Γ.

Before proving this, let us deduce the main result of this section.

Theorem 6   If G is a polycyclic-by-finite group and d is a positive integer, then the groups H which have a subgroup of index at most d isomorphic to G lie in finitely many isomorphism classes.

Proof   Of course, it is enough to consider groups H which actually contain G as a subgroup, with $|G : H| \leq d$. Then

$$G^0 = \bigcap_{h \in H} G^h$$

is normal in H, and

$$|G:G^0| \leq |H:G^0| \leq d!.$$

Now G has only finitely many subgroups of index at most $d!$, so there are only finitely many possibilities for G^0 and for the index $|H:G^0|$. Thus changing the notation, it will suffice to consider those groups H which contain G as a *normal* subgroup of index *exactly d*.

By Lemma 14, there is an embedding $\theta^*:G \to GL_m(\mathbb{Z})$, for some m, such that every such H is isomorphic to a group K with $G\theta^* \lhd_f K \leq GL_m(\mathbb{Z})$. Taking $\Gamma = GL_m(\mathbb{Z})$ and $G\theta^*$ in place of G in Proposition 10, we see that there are only finitely many possibilities for K up to conjugacy in $GL_m(\mathbb{Z})$. Hence there are only finitely many possibilities for H up to isomorphism, and the theorem follows.

Proof of Proposition 10   we have an arithmetic group Γ and a soluble-by-finite subgroup G of Γ. Let M be a triangularizable characteristic subgroup of finite index in G. If $G \lhd_f K \leq \Gamma$ then $M \lhd_f K$; so replacing G by M, we may as well assume that G is itself triangularizable. Then $N_\Gamma(G)$ is an arithmetic group, by Corollary 4 and by Lemma 7 in section A; since $G \lhd K$ implies that $K \leq N_\Gamma(G)$, we may also replace Γ by $N_\Gamma(G)$ and assume that $G \lhd \Gamma$.

Say $\Gamma \leq GL_n(\mathbb{Z})$ (this is a fair assumption, by Exercise 4 in section A). Let $\mathscr{G}$ be the closure of G and $\mathscr{H}$ the closure of Γ in $GL_n(\mathbb{C})$. Then $\mathscr{G}$ and $\mathscr{H}$ are $\mathbb{Q}$-groups by Proposition 7, and $\mathscr{G} \lhd \mathscr{H}$ by Proposition 3. Thus Corollary 3 in section A gives a $\mathbb{Q}$-rational homomorphism $\pi:\mathscr{H} \to \mathscr{K}$, for some $\mathbb{Q}$-group $\mathscr{K}$, with $\ker \pi = \mathscr{G}$, and by Proposition 5 we may take $\mathscr{K} = \mathscr{H}\pi$. Lemma 7 shows that Γ is an arithmetic subgroup in $\mathscr{H}$, so $\Gamma\pi$ is arithmetic in $\mathscr{K}$ by Theorem 1. Thus we conclude that $\Gamma/(\mathscr{G} \cap \Gamma)$ is isomorphic to an arithmetic group.

Now if $G \lhd_f K \leq \Gamma$ then $(\mathscr{G} \cap \Gamma)K/(\mathscr{G} \cap \Gamma)$ is a finite subgroup of $\Gamma/(\mathscr{G} \cap \Gamma)$; hence by Theorem 3 there are only finitely many possibilities up to conjugacy in Γ for the group $(\mathscr{G} \cap \Gamma)K$. It will suffice to show, therefore, that the groups K with $G \lhd_f K$ and $(\mathscr{G} \cap \Gamma)K = P$, for some fixed subgroup P of Γ, lie in finitely many conjugacy classes in P.

Of course P is a finite extension of $\mathscr{G} \cap \Gamma$. Also $\mathscr{G} \cap \Gamma$ is a polycyclic group, by Proposition 2 and the results of Chapter 2. Therefore P/G is polycyclic-by-finite. The required result now follows by Theorem 5, in the previous section.

D. An algorithm

In this section I will sketch, in outline only, an algorithm for deciding whether two soluble-by-finite subgroups of $GL_n(\mathbb{Z})$ are conjugate. This was a task left open in section D of Chapter 7. The interest of this lies not only in the metamathematical result – a decidability theorem – but also in the insight we thereby gain into the various factors which determine the conjugacy class of a soluble subgroup in $GL_n(\mathbb{Z})$. It is on such factors as these that any classification of polycyclic groups must ultimately depend; this point will be further illustrated in the following chapters.

For reasons which will become apparent, it is convenient to consider conjugacy in an arbitrary arithmetic group, rather than in $GL_n(\mathbb{Z})$; so our concern henceforth is to describe

Algorithm 1 Let Γ be an arithmetic group, P and Q two soluble-by-finite subgroups of Γ. The algorithm decides whether P and Q are conjugate in Γ.

As this is an outline, I am going to suppress a number of problems of a technical nature: about the precise way in which the various groups that arise are supposed to be explicitly given, and about the effective computability of various things which have to be computed in the course of the algorithm. The reader may take it as a challenge to (spot and) fill in the gaps.

Everything ultimately depends on two algorithms concerning arithmetic groups, which have already been mentioned in section D of Chapter 6. For references, see section E below.

Algorithm A This finds a finite set of generators for an explicitly given arithmetic group.

Algorithm B Let $\mathscr{G}$ be a $\mathbb{Q}$-group and $\rho:\mathscr{G} \to GL_m(\mathbb{C})$ a $\mathbb{Q}$-rational representation. Let Γ be an arithmetic subgroup of G. Given a and $b \in \mathbb{Q}^m$, the algorithm decides whether there exists $\gamma \in \Gamma$ with $a\rho(\gamma) = b$; if so, it also finds such a γ.

We deal first with some special cases of Algorithm 1. In what follows, Γ is supposed to be an arithmetic group contained in $GL_n(\mathbb{Z})$, P and Q are subsets of $GL_n(\mathbb{Q})$, and the task is to determine whether there exists $\gamma \in \Gamma$ such that $P^\gamma = Q$. I shall write

$$P \sim_\Gamma Q$$

to mean that such a γ exists, i.e. that P and Q are conjugate under Γ.

Case 1   Where P and Q are finite subsets of $GL_n(\mathbb{Q})$.

Discussion   Suppose $|P| = |Q| = d$; if $|P| \neq |Q|$ there can be no question of conjugacy. Define a representation

$$\rho:GL_n(\mathbb{C}) \to GL_{dn^2}(\mathbb{C})$$

by letting $\rho(g)$, for $g \in GL_n(\mathbb{C})$, represent the action of g by conjugation on the direct sum of d copies of $M_n(\mathbb{C})$. Say $P = \{p_1, \ldots, p_d\}$ and $Q = \{q_1, \ldots, q_d\}$. Then $P^\gamma = Q$, for $\gamma \in GL_n(\mathbb{Z})$, if and only if

$$p_i^\gamma = q_{i\sigma}, \ i = 1, \ldots, d$$

for some permutation σ of $\{1, \ldots, n\}$. Thus $P \sim_\Gamma Q$ if and only if there exists $\gamma \in \Gamma$ such that

$$(p_1, \ldots, p_d)\rho(\gamma) = (q_{1\sigma}, \ldots, q_{d\sigma})$$

for some $\sigma \in S_d$. This can be decided by $d!$ applications of Algorithm B.

Case 2   Where P and Q are unipotent subgroups of $GL_n(\mathbb{Z})$.

Discussion   Here we must refer back to sections A and B of Chapter 6. Put

$$L_1 = (n-1)!\,\mathbb{Z}\log P, L_2 = (n-1)!\,\mathbb{Z}\log Q;$$

then L_1 and L_2 are $\mathbb{Z}$-submodules of $M_n(\mathbb{Z})$. Certainly, if $P^\gamma = Q$ then $L_1^\gamma = L_2$; the first thing to decide is whether $\mathbb{Q}L_1^\gamma = \mathbb{Q}L_2$ for some $\gamma \in \Gamma$. Without loss of generality, $L_1 \cong L_2$ as $\mathbb{Z}$-modules. Say

$$L_1 = \overset{d}{\underset{i=1}{\bigoplus}} \, \alpha_i \mathbb{Z}, L_2 = \overset{d}{\underset{i=1}{\bigoplus}} \, \beta_i \mathbb{Z}.$$

Now we want to proceed as in Case 1 above: the difficulty here is that there will in general be infinitely many possible choices of $\mathbb{Z}$-basis for L_2, so even if $L_1^\gamma = L_2$ we cannot say, for example, that $\alpha_i^\gamma = \beta_i$ for each i. To get round the difficulty we take an *exterior power*; see Lang's *Algebra*, Chapter XVI section 6. We have

$$\wedge^d L_1 = (\alpha_1 \wedge \ldots \wedge \alpha_d)\mathbb{Z} \subseteq \wedge^d M_n(\mathbb{Z})$$
$$\wedge^d L_2 = (\beta_1 \wedge \ldots \wedge \beta_d)\mathbb{Z} \subseteq \wedge^d M_n(\mathbb{Z}),$$

and if

$$L_1^\gamma = L_2 \text{ with } \gamma \in GL_n(\mathbb{Z}), \text{ then}$$
$$\alpha_1^\gamma \wedge \ldots \wedge \alpha_d^\gamma = \pm \beta_1 \wedge \ldots \wedge \beta_d. \tag{14}$$

Defining $\rho: GL_n(\mathbb{C}) \to GL_{(n^2)}(\mathbb{C}) = \text{Aut}(\wedge^d M_n(\mathbb{C}))$ by putting

$$(x_1 \wedge \ldots \wedge x_d)\rho(g) = x_1^g \wedge \ldots \wedge x_d^g$$

for $x_1, \ldots, x_d \in M_n(\mathbb{C})$ and $g \in GL_n(\mathbb{C})$, we decide whether (14) holds for some $\gamma \in \Gamma$ by two applications of Algorithm B.

If (14) holds for no $\gamma \in \Gamma$, we conclude that $P \nsim_\Gamma Q$. If (14) holds for some $\gamma = \gamma_0 \in \Gamma$, we replace P by P^{γ_0}, and thereby reduce to the case where

$$\wedge^d L_1 = \wedge^d L_2. \tag{15}$$

This does not yet imply that $L_1 = L_2$: but it does imply that $\mathbb{Q}L_1 = \mathbb{Q}L_2$ ((15) implies that for each $i, \beta_1 \wedge \ldots \wedge \beta_d \wedge \alpha_i = 0$, whence each α_i is $\mathbb{Q}$-linearly dependent on L_2; similarly, each β_i is $\mathbb{Q}$-linearly dependent on L_1). Hence we have

$$\mathbb{Q}L_1 \cap M_n(\mathbb{Z}) = \mathbb{Q}L_2 \cap M_n(\mathbb{Z}) = M, \text{ say.}$$

Moreover, if $\gamma \in \Gamma$ satisfies $P^\gamma = Q$, then $M^\gamma = M$, so γ belongs to the arithmetic group $N_\Gamma(M)$ (*Exercise*: why is $N_\Gamma(M)$ arithmetic?). So replacing Γ by $N_\Gamma(M)$, we may suppose that Γ normalizes the ℤ-submodule M of $M_n(\mathbb{Z})$.

Now there exists a positive integer s such that

$$sM \subseteq L_1 \cap L_2;$$

Lemma 1 of Chapter 6, section A, shows that there is a positive integer m such that

$$m \cdot \mathbb{Z} \log P \subseteq \log P, m \cdot \mathbb{Z} \log Q \subseteq \log Q;$$

and Lemma 3 of Chapter 6, section B, gives us a positive integer t such that

$$g, h \in G \Rightarrow \log g + t \log h \in \log G$$

where G is either P or Q (these lemmas dealt with unitriangular groups, but, as remarked in Chapter 6, this implies no real loss of generality). Now put $r = tsm$ and let $\pi : GL_n(\mathbb{Z}) \to GL_n(\mathbb{Z}/r\mathbb{Z})$ be the natural map. I claim that

$$P \sim_\Gamma Q \Leftrightarrow P\pi \sim_{\Gamma\pi} Q\pi. \qquad (16)$$

Granting this for the moment, we can now decide whether $P \sim_\Gamma Q$ by calculating, in the finite group $\Gamma\pi$, whether the subgroups $P\pi$ and $Q\pi$ are conjugate. To do this, we have to know what the group $\Gamma\pi$ is, and this is where Algorithm A comes in. We find a finite generating set X for Γ, and observe that $\Gamma\pi$ is the subgroup of the finite group $GL_n(\mathbb{Z}/r\mathbb{Z})$ generated by the set $X\pi$.

Let us prove (16). Only the implication $\Leftarrow$ needs comment, so suppose $P\pi^{\gamma\pi} = Q\pi$ for some $\gamma \in \Gamma$. Let $g \in P$. Then $g^\gamma\pi = h\pi$ for some $h \in Q$; now

$$g^\gamma \equiv h \mod rM_n(\mathbb{Z})$$

$$\Rightarrow (n-1)! \log g^\gamma \equiv (n-1)! \log h \mod rM_n(\mathbb{Z}),$$

and since $M^\gamma = M$ this gives

$$(n-1)!(\log g^\gamma - \log h) \in M \cap rM_n(\mathbb{Z}) = rM = tsmM$$

$$\subseteq tmL_2 = (n-1)! tm\mathbb{Z} \log Q$$

$$\subseteq (n-1)! t \log Q.$$

So

$$\log g^\gamma = \log h + t \log k \qquad \text{with } k \in Q$$

$$\in \log Q$$

by the choice of t. Therefore $g^\gamma \in Q$. Thus $P^\gamma \subseteq Q$ and since, similarly, $Q^{\gamma^{-1}} \subseteq P$ we conclude that $P^\gamma = Q$.

Before going further with Algorithm 1, it will be as well to discuss a point which is easily overlooked. Supposing P and Q to be subgroups of $GL_n(\mathbb{Z})$, specified by finite generating sets, can we tell (even) whether or not $P = Q$? The answer is 'yes' if P and Q are soluble-by-finite (though not in general):

Algorithm 2   Given soluble-by-finite subgroups P and Q of $GL_n(\mathbb{Z})$, by finite generating sets, the algorithm decides whether or not $P = Q$.

Discussion   Since $P = Q$ if and only if each generator of P belongs to Q and *vice versa*, it will be enough if we can decide whether a given matrix x belongs to a given soluble-by-finite group $Y = \langle y_1, \ldots, y_d \rangle \le GL_n(\mathbb{Z})$. The idea is to carry out two procedures simultaneously. One of them (the 'yes procedure') will definitely confirm that $x \in Y$, after finitely many steps, provided that x does indeed lie in Y; the other (the 'no procedure') will definitely confirm that $x \notin Y$, after finitely many steps, if indeed $x \notin Y$. In either case, then, the question will be settled after finitely many steps.

The 'yes procedure' The mth step consists in evaluating all the group words of length $m - 1$ on $y_1, \ldots, y_d$, and seeing whether any of the (finitely many) resulting matrices is equal to x.

The 'no procedure' The mth step consists in listing the elements of the finite group $Y\pi_m \le GL_n(\mathbb{Z}/m\mathbb{Z})$, and seeing whether any of these elements is equal to $x\pi_m$.

It is clear that if $x \in Y$, then the 'yes procedure' will confirm this. If $x \notin Y$, then I claim that for some positive integer m we will find $x\pi_m \notin Y\pi_m$, so that the 'no procedure' in this case terminates with a definite answer. To see this, recall Theorem 5 of Chapter 4: this says that Y *is closed in the congruence topology on* $GL_n(\mathbb{Z})$; in other words,

$$Y = \bigcap_{m \in \mathbb{N}} Y \cdot \ker \pi_m,$$

and this makes it clear that $x \notin Y \Leftrightarrow x\pi_m \notin Y\pi_m$ for some m.

We return now to Algorithm 1. A subgroup G of $GL_n(\mathbb{Z})$ is called *unipotent-free* if G contains no unipotent elements $\ne 1$, i.e. if $G \cap G_u = 1$. If G is also abelian, then G_s is a subgroup of $GL_n(\mathbb{Q})$ and the map $g \mapsto g_s$ is an isomorphism of G onto G_s: we saw in section A of Chapter 7 that the map is a homomorphism, and its kernel is now $G \cap G_u = 1$. When G is abelian and unipotent-free, I shall denote the inverse isomorphism by

$$\lambda_G : G_s \to G;$$

thus $g_s \lambda_G = g$ for each $g \in G$.

Case 3 Where P and Q are abelian unipotent-free subgroups of $GL_n(\mathbb{Z})$.

Discussion   As before, we want to apply Algorithm B, and just as in Case 2 we have to get round the problem that a given generating set for P could, *a priori*, be conjugate to any one of infinitely many generating sets for Q. The answer in the present case depends on the following observation:

> given an n-tuple $(\xi_1, \dots, \xi_n)$ of algebraic numbers, there are at most $n!$ elements of Q having $\xi_1, \dots, \xi_n$ as eigenvalues.

To see this, note that Q_s^α is diagonal, for some matrix α over a suitable extension field of $\mathbb{Q}$ (because Q is abelian). If $q \in Q$ has $\xi_1, \dots, \xi_n$ as eigenvalues, then $q_s^\alpha = \operatorname{diag}(\xi_{1\sigma}, \dots, \xi_{n\sigma})$ for some $\sigma \in S_n$; thus

$$q = (\operatorname{diag}(\xi_{1\sigma}, \dots, \xi_{n\sigma}))^{\alpha^{-1}} \lambda_Q$$

for some $\sigma \in S_n$, and as $|S_n| = n!$ the claim follows.

Now suppose $P = \langle p_1, \dots, p_d \rangle$. Let $\xi_{i,1}, \dots, \xi_{i,n}$ be the eigenvalues of p_i, for $i = 1, \dots, d$. If $P^\gamma = Q$ with $\gamma \in \Gamma$, then the matrices $q_i = p_i^\gamma$, $i = 1, \dots, d$, generate Q. From the preceding discussion it follows that there exist $\sigma_1, \dots, \sigma_d \in S_n$ such that, for each i,

$$q_i = (\operatorname{diag}(\xi_{i,1\sigma_i}, \dots, \xi_{i,n\sigma_i}))^{\alpha^{-1}} \lambda_Q. \tag{17}$$

Having computed the $\xi_{i,j}$ and a suitable α, we proceed as follows.
Step 1   For each d-tuple $(\sigma_1, \dots, \sigma_d) \in S_n^d$, decide whether the group $\langle q_1, \dots, q_d \rangle$, where $q_1, \dots, q_d$ are defined by (17), is equal to Q.

This can be done by Algorithm 2 (since $Q \le GL_n(\mathbb{Z})$, we can immediately discount any q_i which is not in $GL_n(\mathbb{Z})$). If the answer is 'no' for every such d-tuple, conclude that $P \not\sim_\Gamma Q$. Otherwise, let $\mathcal{X}$ be the set of d-tuples $(\sigma_1, \dots, \sigma_d)$ for which the answer is 'yes', and go on to
Step 2   For each d-tuple $(\sigma_1, \dots, \sigma_d) \in \mathcal{X}$ decide whether there exists $\gamma \in \Gamma$ such that $p_i^\gamma = q_i$ for $i = 1, \dots, d$, where the q_i are given by (17).

For this, we use Algorithm B just as in Case 1 (only here we are dealing with *ordered* finite sets, so the procedure is even simpler). If the answer is 'no' for each d-tuple in $\mathcal{X}$, conclude that $P \not\sim_\Gamma Q$. Otherwise, Algorithm B finds $\gamma_0 \in \Gamma$ such that $p_i^{\gamma_0} = q_i$, $i = 1, \dots, d$, and then $P^{\gamma_0} = Q$.

For the final stage, we shall need

Lemma 15   Let $H \le K \triangleleft \Gamma \le GL_n(\mathbb{Z})$. Suppose that $H \triangleleft \Gamma$ and that K is soluble-by-finite. Let P/H and Q/H be finite subgroups of K/H. If $P\pi_m \sim_{\Gamma\pi_m} Q\pi_m$ for every positive integer m, then $P \sim_\Gamma Q$. (Here π_m denotes the natural map $GL_n(\mathbb{Z}) \to GL_n(\mathbb{Z}/m\mathbb{Z})$.)

Proof By Theorem 6 (section C, above), K/H has only finitely many conjugacy classes of finite subgroups. Let $F_1/H, \ldots, F_r/H$ be representatives for these classes. For each $m \in \mathbb{N}$, we are given $\gamma_m \in \Gamma$ such that

$$P^{\gamma_m} \pi_{m!} = Q \pi_{m!}. \tag{18}$$

Since H and K are normal in Γ, there exists for each m some index $i_m \in \{1, \ldots, r\}$ such that P^{γ_m} is conjugate to F_{i_m} in K. At least one value of i_m, say j, occurs for infinitely many values of m; then changing some of the γ_ms, without violating (18), we may suppose that $P^{\gamma_m} \sim_K F_j$ for every $m \in \mathbb{N}$. Let $y_m \in K$ satisfy $P^{\gamma_m} = F_j^{y_m}$. Then (18) gives

$$F_j^{y_m} \pi_{m!} = Q \pi_{m!} \quad \forall m \in \mathbb{N}.$$

By Theorem 5 of Chapter 4, every normal subgroup of finite index in K contains $K \cap \ker \pi_{m!}$ for some m. So F_j is conjugate to Q in every finite quotient of K; hence, by Theorem 7 of Chapter 4, F_j is conjugate to Q in K. Since P is conjugate to F_j in Γ, the result follows.

Algorithm 1 P and Q are soluble-by-finite subgroups of the arithmetic group $\Gamma \leq GL_n(\mathbb{Z})$. We are to decide whether $P \sim_\Gamma Q$.

Step 1   Find a positive integer e such that P^e and Q^e are triangularizable. Find $u(P^e)$ and $u(Q^e)$.

Step 2   Decide whether $u(P^e) \sim_\Gamma u(Q^e)$, by Case 2, above. If not, conclude that $P \nsim_\Gamma Q$. Otherwise, replace P by a suitable conjugate in Γ to reduce to the case where

$$u(P^e) = u(Q^e) = U, \quad \text{say}.$$

Now if $P^\gamma = Q$ then $U^\gamma = U$, so we may replace Γ by $N_\Gamma(U)$ and assume henceforth that $U \triangleleft \Gamma$; note that $N_\Gamma(U)$ is again arithmetic, by Lemmas 7 and 8.

Step 3   Let $\mathcal{U}$ be the closure of U in $GL_n(\mathbb{C})$ and put $\mathcal{N} = N_{GL_n(\mathbb{C})}(\mathcal{U})$. Then $\mathcal{U}$ and $\mathcal{N}$ are $\mathbb{Q}$-groups and $\mathcal{U} \triangleleft \mathcal{N}$. Let $\theta : \mathcal{N} \to GL_m(\mathbb{C})$ be a $\mathbb{Q}$-rational homomorphism with $\ker \theta = \mathcal{U}$ and $\mathcal{N}(\mathbb{Z})\theta \leq GL_m(\mathbb{Z})$. Then $\Gamma\theta$ is an arithmetic group, by Theorem 1; also $P^e\theta$ and $Q^e\theta$ are abelian, since $U \leq \ker \theta$ and P^e and Q^e are triangularizable; and $P^e\theta$ and $Q^e\theta$ are unipotent-free, by Proposition 4 and Exercise 9. So by Case 3, we may determine whether or not $P^e\theta \sim_{\Gamma\theta} Q^e\theta$.

If the answer is 'no', conclude that $P \nsim_\Gamma Q$. Otherwise, replacing P by a suitable conjugate in Γ we reduce to the case where $P^e\theta = Q^e\theta$. Then $P^e\mathcal{U} = Q^e\mathcal{U}$, so

$$P^e(\mathcal{U} \cap \Gamma) = Q^e(\mathcal{U} \cap \Gamma) = E, \quad \text{say}.$$

Step 4   Note that $\mathcal{U} \cap \Gamma \leq \mathcal{U}(\mathbb{Z})$ and $|\mathcal{U}(\mathbb{Z}):U|$ is finite, by Proposition 8. Since $U \leq P^e \cap Q^e$ it follows that $E^f \leq P^e \cap Q^e$ for some positive integer f. Also P and Q both normalize E (they normalize U, hence also $\mathcal{U}$, by Proposition 3); so $E^f \vartriangleleft P$ and $E^f \vartriangleleft Q$. Moreover, P/E^f and Q/E^f are both finite, since $(P^e)^f \leq E^f$ and $(Q^e)^f \leq E^f$.

If $\gamma \in \Gamma$ satisfies $P^\gamma = Q$, then $E^\gamma = E$ and so $\gamma \in N_\Gamma(E^f)$. As $N_\Gamma(E^f)$ is arithmetic (by Corollary 4) we may replace Γ by $N_\Gamma(E^f)$. Thus putting $H = E^f$, we have reduced to the following situation:

$$H \vartriangleleft_f P, \quad H \vartriangleleft_f Q, \quad H \vartriangleleft \Gamma.$$

Let $\mathcal{H}$ be the closure of H in $GL_n(\mathbb{C})$ and $\mathcal{M} = N_{GL_n(\mathbb{C})}(\mathcal{H})$. Then $\Gamma \leq \mathcal{H}(\mathbb{Z})$ and $P/(P \cap \mathcal{H})$, $Q/(Q \cap \mathcal{H})$ are finite. Just as in Step 3, using now Case 1 in place of Case 3, we determine whether or not there exists $\gamma \in \Gamma$ with

$$P^\gamma(\mathcal{H} \cap \Gamma) = Q(\mathcal{H} \cap \Gamma).$$

If the answer is 'no', conclude that $P \nsim_\Gamma Q$. Otherwise, replace P by a suitable conjugate in Γ to reduce to the case where

$$P(\mathcal{H} \cap \Gamma) = Q(\mathcal{H} \cap \Gamma) = K, \quad \text{say.}$$

As usual, we may now replace Γ by $N_\Gamma(K)$ and assume that $K \vartriangleleft \Gamma$.

Step 5   Lemma 15 now shows that $P \sim_\Gamma Q$ if and only if $P\pi_m \sim_{\Gamma\pi_m} Q\pi_m$ for every $m \in \mathbb{N}$. So to decide whether $P \sim_\Gamma Q$ we proceed as in Algorithm 2:

The 'yes procedure' Enumerate the elements of Γ, and for each $\gamma \in \Gamma$ decide (by Algorithm 2) whether $P^\gamma = Q$.

The 'no procedure' For $m = 1, 2, 3, \ldots$, in turn, decide whether $P\pi_m \sim_{\Gamma\pi_m} Q\pi_m$. For this, we use Algorithm A, just as in Case 2 above.

After finitely many steps, one of these two procedures will terminate with a definite answer. This concludes Algorithm 1.

Exercise 12   Describe algorithms which determine whether two subgroups of a polycyclic-by-finite group are (*a*) equal, (*b*) conjugate.

E. Notes

Zariski topology   Section A only gave the bare bones of what was needed for the rest of the chapter. For more information, in a similar spirit, see Wehrfritz (1973*b*), Chapters 5, 6 and especially 14. To get a better feeling for what lies behind these methods, one should also have a look at the introductory chapters of books on algebraic geometry. Ones which I have found helpful include Northcott (1980), Fogarty (1969), Shafarevich (1974), and Borel (1969*b*).

Arithmetic groups   Theorem 1 (that the image of an arithmetic group is again arithmetic) is from Borel (1966); the main part of the proof, dealing with the case of an 'isogeny', can be found in the book Borel (1969*a*): see Theoreme 8.9 on page 59. Theorem 3 is from Borel (1962); most of the proof appears in Borel & Harish-Chandra (1962), and the fact that arithmetic groups are finitely generated is Theorem 6.2 of the latter paper. An account of this work is also given in Chapters 7 and 9 of Borel's book on arithmetic groups (Borel (1969*a*)).

Soluble Z-linear groups   The ideas of section B were largely inspired by the paper Wehrfritz (1974); Theorem 2, above, is implicit in that paper. The fact that Aut G is finitely presented was proved by Auslander (1969). Our Theorem 5, on the other hand, seems previously to have escaped notice. The result on the 'isolator property', Exercise 8 in section B, is due to B. Wehrfritz. Results related to Exercise 7, but with different proofs, will be found in Wehrfritz (1971*b*). Wehrfritz (1980), Prop. 1.3, shows that if G is a polycyclic group and H is any subgroup, then there is an embedding $\rho : G \to GL_n(\mathbb{Z})$, for some n, such that $H\rho$ is Zariski-closed in $G\rho$. This is a striking result which may well have useful applications, though I know of none at present.

Proposition 10 and Theorem 2, about finite extensions, are from Segal (1978). These results will be useful in Chapters 9 and 10.

The algorithm   of section D is new (as far as I know); I worked it out with Fritz Grunewald when we were proving the main theorem of Chapter 9.

The effective undecidability of the question whether two finitely generated subgroups of $GL_n(\mathbb{Z})$ are equal, in general, can be established as follows. Let F_1 and F_2 be free groups of rank 2; it is known that there can be no uniform algorithm for deciding whether given finitely generated subgroups of $F_1 \times F_2$ are equal (see Lyndon & Schupp (1977), Theorem 4.3 in Chapter IV). Our claim then follows from the fact that

$$F_1 \times F_2 \cong \left\langle \begin{bmatrix} 1 & 0 & & \\ 2 & 1 & & \\ & & 1 & \\ & & & 1 \end{bmatrix}, \begin{bmatrix} 1 & 2 & & \\ 0 & 1 & & \\ & & 1 & \\ & & & 1 \end{bmatrix}, \begin{bmatrix} 1 & & & \\ & 1 & & \\ & & 1 & 0 \\ & & 2 & 1 \end{bmatrix}, \begin{bmatrix} 1 & & & \\ & 1 & & \\ & & 1 & 2 \\ & & 0 & 1 \end{bmatrix} \right\rangle$$

$$\leqslant GL_4(\mathbb{Z}).$$

9
A finiteness theorem

If Γ is a polycyclic group, then subgroups of Γ which have conjugate images in every finite quotient of Γ must be conjugate in Γ: this was Theorem 7 of Chapter 4. We have now seen several ways in which arithmetic groups resemble polycyclic groups, and the present chapter is devoted to an analogue of that theorem, where Γ is taken to be an arbitrary arithmetic group. As stated, the result is false for such a Γ; so we must be content with a *finiteness* theorem. The precise statement will have to wait until section B, below; a special case which exhibits all the main features is as follows: *If $\mathscr{C}$ is a family of soluble-by-finite subgroups of $GL_n(\mathbb{Z})$ and if the images of the groups in $\mathscr{C}$ are all conjugate in $GL_n(\mathbb{Z}/m\mathbb{Z})$ for every $m \neq 0$, then $\mathscr{C}$ is contained in the union of finitely many conjugacy classes of subgroups in $GL_n(\mathbb{Z})$.* This result is the key to the more interesting finiteness theorem which forms the subject of Chapter 10.

The theorem is the hardest one in this book, and I shall give nothing like the full proof of it. The purpose of this chapter is merely to introduce the most important concepts and methods which lie behind the theorem. Some of these, particularly those belonging to the 'arithmetical' theory of algebraic groups, have interest and applications which go far beyond group theory; a simple account of some major results in this area is given in section A. Sections B and C deal with special cases of the theorem, and section D explains in broad outline how the general case is tackled.

A. Local and global orbits of algebraic groups

The situation we are going to discuss has already been encountered in connection with Algorithm B in section D of Chapter 8. Consider a $\mathbb{Q}$-group $\mathscr{G}$, of degree n, and a $\mathbb{Q}$-rational representation

$$\rho : \mathscr{G} \to GL_m(\mathbb{C}).$$

The group $\mathscr{G}$ acts via ρ on the vector space $\mathbb{C}^m$, and we are going to study the orbits of this action; in particular, the orbits of $\mathscr{G}(\mathbb{Z})$ on $\mathbb{Q}^m$.

The first question is: to what extent is an orbit of $\mathscr{G}(\mathbb{Z})$ determined by congruences? Before formulating this question precisely let us consider a slightly special case. Assume that the polynomial equations defining $\mathscr{G}$ as a subset of $GL_n(\mathbb{C})$ all have integer coefficients; for a positive integer r, define $\mathscr{G}(\mathbb{Z}/r\mathbb{Z})$ be the set of all matrices in $GL_n(\mathbb{Z}/r\mathbb{Z})$ whose entries satisfy these equations over $\mathbb{Z}/r\mathbb{Z}$. Supposing also that the rational expressions which define the map ρ are in fact polynomials with integer coefficients, we may formally define a map

$$\bar{\rho} : \mathscr{G}(\mathbb{Z}/r\mathbb{Z}) \to M_m(\mathbb{Z}/r\mathbb{Z})$$

by means of those same polynomials, reduced mod r. Under these circumstances, we now say of two vectors a and b in $\mathbb{Z}^m$ that they are $\mathscr{G}$-*equivalent mod* r if there exists a matrix $g \in \mathscr{G}(\mathbb{Z}/r\mathbb{Z})$ such that

$$\bar{a}\bar{\rho}(g) = \bar{b}, \tag{1}$$

where $\bar{a} = a \bmod r$ etc. If this holds for every positive integer r, we say that a and b belong to the same *local orbit* of $\mathscr{G}$.

Evidently a local orbit of G is the union of a number of orbits of $\mathscr{G}(\mathbb{Z})$: the question is, how many? Having reformulated our definitions so that they make sense in general, we shall be in a position to discuss the first main result,

Theorem 1   Every local orbit of $\mathscr{G}$ in $\mathbb{Q}^m$ is the union of finitely many orbits of $\mathscr{G}(\mathbb{Z})$.

This result, due essentially to Borel and Serre, is extremely wide-ranging in its implications. A straightforward example is given in

Exercise 1   Show that a set of matrices in $SL_n(\mathbb{Z})$ whose images in $SL_n(\mathbb{Z}/r\mathbb{Z})$ are conjugate for every $r \in \mathbb{N}$ contains only finitely many non-conjugate matrices in $SL_n(\mathbb{Z})$.

To define 'local orbits' in general, we need the language of *p-adic numbers*. I shall assume familiarity with the basic properties of the p-adic number field $\mathbb{Q}_p$ and the ring of p-adic integers $\mathbb{Z}_p$; there are several good introductions to this material and some suitable references are given in section E.

The statement that (1) holds for some $g \in \mathscr{G}(\mathbb{Z}/r\mathbb{Z})$ is equivalent to the statement that a certain set of polynomial congruences, in the entries of an $n \times n$ matrix over $\mathbb{Z}$, have a simultaneous solution modulo r. If this holds with r running over all powers of a fixed prime p, the corresponding set of simultaneous polynomial equations has a simultaneous solution over the p-adic integers. Thus there exists $g \in \mathscr{G}(\mathbb{Z}_p)$ such that $a\rho(g) = b$ (we really do get $g \in \mathscr{G}(\mathbb{Z}_p)$, because a matrix $g \in M_n(\mathbb{Z}_p)$ satisfying $g\pi_{p^i} \in GL_n(\mathbb{Z}/p^i\mathbb{Z})$ for $i > 0$ has det g coprime to p, whence det g is a unit in $\mathbb{Z}_p$ and g is invertible).

Hence if a and b are in the same local $\mathscr{G}$-orbit, then a and b lie in the same orbit of $\mathscr{G}(\mathbb{Z}_p)$ for every prime p.

Exercise 2 Suppose $\mathscr{G}$ and ρ are defined by polynomials over $\mathbb{Z}$, as before. Show that if a and b in $\mathbb{Z}^m$ lie in the same orbit of $\mathscr{G}(\mathbb{Z}_p)$, for every prime p, then they lie in the same local orbit of $\mathscr{G}$ as defined above. (*Hint*: Use the Chinese Remainder Theorem.)

The discussion so far motivates our general

Definition Let $a, b \in \mathbb{Q}^m$. Then a and b lie in the same *local orbit* of $\mathscr{G}$, or are *locally $\mathscr{G}$-equivalent*, if for every prime p there exists $g_p \in \mathscr{G}(\mathbb{Z}_p)$ such that $a\rho(g_p) = b$.

For a field k, we denote an algebraic closure of k by $\bar{k}$. There are three stages in the proof of Theorem 1, and the first consists in relating local orbits to orbits of $\mathscr{G}(\bar{\mathbb{Q}})$:

Proposition 1 If a and b in $\mathbb{Q}^m$ lie in the same orbit of $\mathscr{G}(\mathbb{Q}_p)$, for some prime p, then they lie in the same orbit of $\mathscr{G}(\bar{\mathbb{Q}})$.

Proof This is an application of *Hilbert's Nullstellensatz*: the version that we need says that *a finitely generated $\mathbb{Q}$-algebra which is a field is a finite extension of $\mathbb{Q}$*. For this, see e.g. Atiyah & Macdonald: *Introduction to Commutative Algebra*, Corollary 5.24 on page 67. Now let k be any extension field of $\mathbb{Q}$. The statement

$$a\rho(g) = b \text{ for some } g \in \mathscr{G}(k)$$

is equivalent to the solubility, in k^{n^2+1}, of the following system of equations over $\mathbb{Q}$, where $\mathbf{X} = (X_{st})$ is an $n \times n$ matrix with indeterminate entries and Y is the (n^2+1)th indeterminate (and $a = (a_1, \ldots, a_m)$, $b = (b_1, \ldots, b_m)$):

$$(\det \mathbf{X})Y - 1 = 0 \tag{2}$$

$$\sum_i a_i P_{ij}(\mathbf{X}) - b_j Q(\mathbf{X}) = 0, \quad j = 1, \ldots, m \tag{3}$$

(where $\rho(\mathbf{X})_{ij} = P_{ij}(\mathbf{X})/Q(\mathbf{X})$ for $1 \le i, j \le m$)

$$\varphi_\lambda(\mathbf{X}) = 0, \quad \lambda \in \Lambda \tag{4}$$

(where the φ_λ, $\lambda \in \Lambda$, are the polynomials defining the $\mathbb{Q}$-group $\mathscr{G}$). Let R be the polynomial ring $\mathbb{Q}[X_{11}, \ldots, X_{nn}, Y]$, and let I be the ideal of R generated by the left-hand sides of (2), (3) and (4). Now suppose $a\rho(g) = b$ with $g \in \mathbb{Q}_p$. The map sending X_{st} to g_{st} for each s, t, and sending Y to $(\det g)^{-1}$, extends to a $\mathbb{Q}$-algebra homomorphism $\theta : R \to \mathbb{Q}_p$; and since

$(g,(\det g)^{-1})$ furnishes a solution to the equations (2)–(4), I is contained in ker θ. Let M be a maximal ideal of R containing ker θ (note that $1\notin\ker\theta$). By the Nullstellensatz, R/M is a finite field extension of $\mathbb{Q}$; call it k and let $\pi:R\to k$ be the natural map. Put $y=Y\pi$ and for each s,t, put $h_{st}=X_{st}\pi$. Then $h=(h_{st})\in M_n(k)$ satisfies

$$h\in\mathscr{G}(k),\quad a\rho(h)=b;$$

for the left-hand sides of equations (2)–(4) lie inside $I\subseteq\ker\pi$, so the image under π of Y and of the X_{st} do indeed satisfy these equations. Since $k\subseteq\bar{\mathbb{Q}}$, the proposition follows.

The next stage belongs to the theory of *Galois Cohomology*. Whenever a group Γ acts on a group G, one can form the so-called *first cohomology set* $H^1(\Gamma,G)$. To define this, we need *cocycles* (strictly, '1-cocycles'); these are just derivations in a non-commutative setting. Thus a cocycle of Γ in G is a map $\sigma:\Gamma\to G$ such that

$$(\alpha\beta)\sigma=(\alpha\sigma)^\beta\cdot\beta\sigma\quad\text{for all}\quad\alpha,\beta\in\Gamma;$$

for technical reasons which I don't wish to go into, we also insist that σ factor through some finite quotient of Γ. The set of all cocycles of Γ in G is denoted $Z^1(\Gamma,G)$. Two cocycles σ and τ are *equivalent*, $\sigma\sim\tau$, if there exists $g\in G$ such that

$$\alpha\tau=g^\alpha\cdot\alpha\sigma\cdot g^{-1}\quad\text{for all}\quad\alpha\in\Gamma.$$

Exercise 3 Check that $\sim$ is an equivalence relation on $Z^1(\Gamma,G)$. The set of equivalence classes $Z^1(\Gamma,G)/\sim$ is the cohomology set $H^1(\Gamma,G)$.

Now let $\mathscr{H}$ be a $\mathbb{Q}$-group, and for a field k write

$$\Gamma_k=\mathrm{Gal}(\bar{k}/k).$$

If $k\supseteq\mathbb{Q}$, then Γ_k operates on the group $\mathscr{H}(\bar{k})$, by acting on the entries of the matrices: since $\mathscr{H}$ is defined by equations over $\mathbb{Q}$, we have $g^\gamma\in\mathscr{H}(\bar{k})$ whenever $g\in\mathscr{H}(\bar{k})$ and $\gamma\in\Gamma_k$. Thus we have cohomology sets $H^1(\Gamma_k,\mathscr{H}(\bar{k}))$. For these we use the abbreviated notation

$$H^1(\Gamma_k,\mathscr{H}(\bar{k}))=H^1(k,\mathscr{H}).$$

Let us see how $H^1(\mathbb{Q},\mathscr{H})$ is related to $H^1(\mathbb{Q}_p,\mathscr{H})$. We suppose that for each prime p, the algebraic closure $\bar{\mathbb{Q}}_p$ of $\mathbb{Q}_p$ is chosen to contain $\bar{\mathbb{Q}}$. Then $\Gamma_{\mathbb{Q}_p}$ acts by restriction on $\bar{\mathbb{Q}}$, and we have the restriction map

$$r_p:\Gamma_{\mathbb{Q}_p}\to\Gamma_{\mathbb{Q}}.$$

Also $\mathscr{H}(\bar{\mathbb{Q}})\le\mathscr{H}(\bar{\mathbb{Q}}_p)$. So to each cocycle $\sigma\in Z^1(\Gamma_{\mathbb{Q}},\mathscr{H}(\bar{\mathbb{Q}}))$ we may associate a cocycle

$$\sigma_p=r_p\circ\sigma\in Z^1(\Gamma_{\mathbb{Q}_p},\mathscr{H}(\bar{\mathbb{Q}}_p)).$$

Moreover, it is a trivial matter to verify that $\sigma \sim \tau \Rightarrow \sigma_p \sim \tau_p$, for σ and τ in $Z^1(\Gamma_{\mathbb{Q}}, \mathscr{H}(\overline{\mathbb{Q}}))$. Hence there is a canonical map

$$\omega_p : H^1(\mathbb{Q}, \mathscr{H}) \to H^1(\mathbb{Q}_p, \mathscr{H})$$

$$[\sigma]\omega_p = [\sigma_p], \quad \sigma \in Z^1(\Gamma_{\mathbb{Q}}, \mathscr{H}(\overline{\mathbb{Q}}))$$

where [ ] denotes equivalence class in the respective H^1. Putting these maps together, for all primes p, gives a map

$$\omega : H^1(\mathbb{Q}, \mathscr{H}) \to \prod_p H^1(\mathbb{Q}_p, \mathscr{H}),$$

the object on the right being the Cartesian product taken over all primes p.

We now quote a beautiful theorem of Borel and Serre (see section E):

Theorem 2 The map ω has finite fibres. ('Fibre' = inverse image of a point.)

How does this help with Theorem 1? For $c \in \mathbb{Q}^m$ let $\mathscr{H}_c$ be the stabilizer of c in $\mathscr{G}$, i.e.

$$\mathscr{H}_c = \{g \in \mathscr{G} \,|\, c\rho(g) = c\}.$$

Then $\mathscr{H}_c$ is again a $\mathbb{Q}$-group. Let $\mathscr{C}$ be a local orbit of $\mathscr{G}$ in $\mathbb{Q}^m$. By Proposition 1, $\mathscr{C}$ is contained in a single orbit of $\mathscr{G}(\overline{\mathbb{Q}})$; so fixing an element $c \in \mathscr{C}$, we have for each $a \in \mathscr{C}$ an element $g_a \in \mathscr{G}(\overline{\mathbb{Q}})$ with $a\rho(g_a) = c$. If $\gamma \in \Gamma_{\mathbb{Q}}$, then also $a\rho(g_a^\gamma) = c$, because the components of a and c, and the coefficients of the rational expressions defining ρ, all lie in $\mathbb{Q}$. Consequently

$$c\rho(g_a^{-\gamma} \cdot g_a) = c\rho(g_a^\gamma)^{-1}\rho(g_a) = a\rho(g_a) = c,$$

and so $g_a^{-\gamma} g_a \in \mathscr{H}_c$. Thus we may define a map

$$\Delta_a : \Gamma_{\mathbb{Q}} \to \mathscr{H}_c(\overline{\mathbb{Q}})$$

by

$$\gamma\Delta_a = g_a^{-\gamma} g_a, \quad \gamma \in \Gamma_{\mathbb{Q}},$$

for each $a \in \mathscr{C}$. Write $\mathscr{H} = \mathscr{H}_c$ henceforth.

Claim 1 $\Delta_a \in Z^1(\Gamma_{\mathbb{Q}}, \mathscr{H}(\overline{\mathbb{Q}}))$.

Proof Routine (note that Δ_a factors through the finite group $\mathrm{Gal}(k/\mathbb{Q})$, where k is a finite normal extension of $\mathbb{Q}$ such that $g_a \in \mathscr{G}(k)$).

Claim 2 The class of Δ_a in $H^1(\mathbb{Q}, \mathscr{H})$ is independent of the choice of $g_a \in \mathscr{G}(\overline{\mathbb{Q}})$.

Proof Suppose $g'_a \in \mathcal{G}(\bar{\mathbb{Q}})$ also satisfies $a\rho(g'_a) = c$. Put $h = g'^{-1}_a g_a$. Then for $\gamma \in \Gamma_{\mathbb{Q}}$,

$$h^\gamma \cdot \gamma \Delta_a \cdot h^{-1} = h^\gamma g_a^{-\gamma} g_a h^{-1}$$
$$= g'^{-\gamma}_a g'_a$$
$$= \gamma \Delta'_a,$$

where Δ'_a is defined like Δ_a, using g'_a in place of g_a. Thus $\Delta'_a \sim \Delta_a$.

The two claims show that we have a well defined map

$$\theta : \mathcal{C} \to H^1(\mathbb{Q}, \mathcal{H})$$
$$a \mapsto [\Delta_a].$$

Let us compute the fibres of θ.

Claim 3 The fibres of θ are just the orbits of $\mathcal{G}(\mathbb{Q})$ in $\mathcal{C}$ (strictly speaking, I mean the intersections of $\mathcal{C}$ with the orbits of $\mathcal{G}(\mathbb{Q})$ in $\mathbb{Q}^m$).

Proof Suppose $b = a\rho(x)$ with $x \in \mathcal{G}(\mathbb{Q})$. To define Δ_b, we may, in view of Claim 2, choose g_b equal to xg_a. Then for $\gamma \in \Gamma_{\mathbb{Q}}$,

$$\gamma \Delta_b = g_b^{-\gamma} g_b = g_a^{-\gamma} x^{-\gamma} x g_a = g_a^{-\gamma} g_a = \gamma \Delta_a,$$

since $x \in \mathcal{G}(\mathbb{Q})$ is invariant under $\Gamma_{\mathbb{Q}}$. Therefore $a\theta = b\theta$ and a, b belong to the same fibre of θ.

Suppose conversely that $a\theta = b\theta$ for some $a, b \in \mathcal{C}$. Then $\Delta_b \sim \Delta_a$, so there exists $h \in \mathcal{H}(\bar{\mathbb{Q}})$ such that

$$g_b^{-\gamma} g_b = h^\gamma \cdot g_a^{-\gamma} g_a \cdot h^{-1} \text{ for all } \gamma \in \Gamma_{\mathbb{Q}}.$$

Then

$$g_a h^{-1} g_b^{-1} = g_a^\gamma h^{-\gamma} g_b^{-\gamma} = (g_a h^{-1} g_b^{-1})^\gamma \text{ for all } \gamma \in \Gamma_{\mathbb{Q}}.$$

Since an element of $\bar{\mathbb{Q}}$ fixed by the whole of $\Gamma_{\mathbb{Q}}$ must lie in $\mathbb{Q}$, it follows that

$$y = g_a h^{-1} g_b^{-1} \in \mathcal{G}(\mathbb{Q}).$$

But

$$a\rho(y) = c\rho(h)^{-1} \rho(g_b)^{-1} = c\rho(g_b)^{-1} = b,$$

so a and b lie in the same orbit of $\mathcal{G}(\mathbb{Q})$.

We are now in a position to deduce

Proposition 2 Every local $\mathcal{G}$-orbit in $\mathbb{Q}^m$ is contained in the union of finitely many orbits of $\mathcal{G}(\mathbb{Q})$.

Proof What we shall show is that the set

$$\mathscr{C}\theta\omega \subseteq \prod_p H^1(\mathbb{Q}_p, \mathscr{H})$$

has just one element. Granting this, it will follow that $\mathscr{C}\theta$ is contained in a single fibre of ω, hence by Theorem 2 that $\mathscr{C}\theta$ is finite. The proposition then follows by Claim 3.

Let $a, b \in \mathscr{C}$. We must show that $a\theta\omega = b\theta\omega$, i.e. that for each prime p,

$$a\theta\omega_p = b\theta\omega_p.$$

Fix a prime p. By hypothesis, there exists $g \in \mathscr{G}(\mathbb{Z}_p) \leq \mathscr{G}(\mathbb{Q}_p)$ such that $a\rho(g) = b$. Let $g_a, g_b \in \mathscr{G}(\bar{\mathbb{Q}})$ satisfy $a\rho(g_a) = c = b\rho(g_b)$, and put

$$h = g_a^{-1}gg_b.$$

Since $\bar{\mathbb{Q}} \subseteq \bar{\mathbb{Q}}_p$, and by the choice of g_a and g_b, we have $h \in \mathscr{H}(\bar{\mathbb{Q}}_p)$. From the definitions, we have, for $\beta \in \Gamma_{\mathbb{Q}_p}$,

$$\beta(\Delta_a)_p = g_a^{-\beta}g_a, \quad \beta(\Delta_b)_p = g_b^{-\beta}g_b.$$

So

$$\begin{aligned} h^\beta \cdot \beta(\Delta_b)_p \cdot h^{-1} &= g_a^{-\beta}g^\beta g_b^\beta \cdot g_b^{-\beta}g_b \cdot g_b^{-1}g^{-1}g_a \\ &= g_a^{-\beta}g^\beta g^{-1}g_a \\ &= g_a^{-\beta}g_a = \beta(\Delta_a)_p, \end{aligned}$$

using here the fact that $g \in \mathscr{G}(\mathbb{Q}_p) \Rightarrow g^\beta = g$. Thus

$$(\Delta_b)_p \sim (\Delta_a)_p$$

and so

$$a\theta\omega_p = [(\Delta_a)_p] = [(\Delta_b)_p] = b\theta\omega_p$$

as required. This completes the proof.

The reader will note that we in fact established a bit more than was stated in Proposition 2: *a set of vectors in $\mathbb{Q}^m$ which is contained in a single orbit of $\mathscr{G}(\mathbb{Q}_p)$, for every prime p, is contained in the union of finitely many orbits of $\mathscr{G}(\mathbb{Q})$.* A typical application is given in the following exercise:

Exercise 4 Let $\mathscr{X}$ be a family of finite-dimensional $\mathbb{Q}$-algebras (not necessarily associative). Show that if the algebras in $\mathscr{X}$ all become isomorphic when tensored with $\mathbb{Q}_p$, for every prime p, then $\mathscr{X}$ contains only finitely many isomorphism types of algebras.
(*Hint*: The algebras have a common dimension, say n. Consider the action of $GL_n(\mathbb{C})$ on the n^3-dimensional space of structure constants (compare Chapter 6, Exercise 18 in section D, and the discussion preceding it).)

Let us return to Theorem 1. In view of Proposition 2, what remains to be shown is this: if $\mathscr{C}$ is a subset of $\mathbb{Q}^m$ and if for every b and c in $\mathscr{C}$ there exist $g \in \mathscr{G}(\mathbb{Q})$ and $g_p \in \mathscr{G}(\mathbb{Z}_p)$, for all primes p, such that

$$b\rho(g) = c, \quad b\rho(g_p) = c \quad \forall p, \tag{5}$$

then $\mathscr{C}$ meets only finitely many orbits of $\mathscr{G}(\mathbb{Z})$. To express the hypotheses in a succinct form, one introduces the so-called (finite) *adele group* of $\mathscr{G}$. This is a subgroup $\mathscr{G}(\mathbb{A}_f)$ of the Cartesian product, over all primes p, of the groups $\mathscr{G}(\mathbb{Q}_p)$, and is defined thus:

$$\mathscr{G}(\mathbb{A}_f) = \left\{ \hat{g} = (g_p) \in \prod_p \mathscr{G}(\mathbb{Q}_p) \,\middle|\, g_p \in \mathscr{G}(\mathbb{Z}_p) \text{ for almost all primes } p \right\}.$$

('Almost all' means 'all but finitely many'.) It is easy to see that $\mathscr{G}(\mathbb{A}_f)$ is indeed a subgroup, and that it contains the smaller subgroup

$$\mathscr{G}^\infty = \prod_{\text{all } p} \mathscr{G}(\mathbb{Z}_p).$$

The group $\mathscr{G}(\mathbb{A}_f)$ also contains a copy of $\mathscr{G}(\mathbb{Q})$: if $g \in \mathscr{G}(\mathbb{Q})$, then $g \in \mathscr{G}(\mathbb{Z}_p)$ for every prime p except for those primes which divide the denominator of some entry of g, or divide either the numerator or denominator of $\det g$. As this set of exceptional primes is finite, the element

$$\hat{g} = (g_p) \text{ with } g_p = g \text{ for every } p$$

in $\prod \mathscr{G}(\mathbb{Q}_p)$ actually belongs to $\mathscr{G}(\mathbb{A}_f)$. Thus we may, and henceforth shall, identify each $g \in \mathscr{G}(\mathbb{Q})$ with the corresponding $\hat{g} \in \mathscr{G}(\mathbb{A}_f)$; in this way $\mathscr{G}(\mathbb{Q})$ is identified with a subgroup of $\mathscr{G}(\mathbb{A}_f)$.

What is the intersection, in $\mathscr{G}(\mathbb{A}_f)$, of $\mathscr{G}^\infty$ with $\mathscr{G}(\mathbb{Q})$? Certainly $\mathscr{G}(\mathbb{Z}) \leq \mathscr{G}^\infty \cap \mathscr{G}(\mathbb{Q})$. Conversely, suppose

$$\hat{g} = (g_p) \in \mathscr{G}^\infty \cap \mathscr{G}(\mathbb{Q});$$

then there exists $g \in \mathscr{G}(\mathbb{Q})$ with $g_p = g$ for all p, and since $g_p \in \mathscr{G}(\mathbb{Z}_p)$ for every p, we have $g \in \mathscr{G}(\mathbb{Z}_p)$ for every p. Now a rational number which lies in $\mathbb{Z}_p$ for every p belongs to $\mathbb{Z}$, so we conclude that $g \in M_n(\mathbb{Z})$ and that $\det g \in \mathbb{Z}$, $(\det g)^{-1} \in \mathbb{Z}$. Hence $g \in GL_n(\mathbb{Z}) \cap \mathscr{G}(\mathbb{Q}) = \mathscr{G}(\mathbb{Z})$. This shows that

$$\mathscr{G}^\infty \cap \mathscr{G}(\mathbb{Q}) = \mathscr{G}(\mathbb{Z}).$$

Now we can extend the map ρ in the obvious way (i.e. componentwise) to a map of $\mathscr{G}(\mathbb{A}_f)$ into $\prod GL_m(\mathbb{Q}_p)$. Then the hypothesis at (5) above can be formulated in the following way:

for each b and $c \in \mathscr{C}$, there exist $g \in \mathscr{G}(\mathbb{Q})$ and $\hat{g} \in \mathscr{G}^\infty$ such that $b\rho(g) = c = b\rho(\hat{g})$;

here we tacitly identify each vector $v \in \mathbb{Q}^m$ with the 'vector'

$$\hat{v} = (v_p), \quad v_p = v \text{ for all } p,$$

in $\prod_p \mathbb{Q}_p^m$, the natural space on which $\prod_p GL_m(\mathbb{Q}_p)$ is acting.

As before, we fix $c \in \mathscr{C}$, for $b \in \mathscr{C}$ let $g_b \in \mathscr{G}(\mathbb{Q})$ and $\hat{g}_b \in \mathscr{G}^\infty$ play the roles of g, $\hat{g}$ respectively in (6), and put

$$\hat{h}_b = g_b^{-1} \hat{g}_b \in \mathscr{G}(\mathbb{A}_f).$$

Defining $\mathscr{H}$ to be the stabilizer of c in $\mathscr{G}$ as before, it is clear that then

$$\hat{h}_b \in \mathscr{H}(\mathbb{A}_f)$$

for each $b \in \mathscr{C}$. Suppose we make a different choice of g_b and of $\hat{g}_b$, say $g_b' \in \mathscr{G}(\mathbb{Q})$ and $\hat{g}_b' \in \mathscr{G}^\infty$. Then $g_b' = g_b y$ and $\hat{g}_b' = \hat{g}_b \hat{y}$ for some $y \in \mathscr{H}(\mathbb{Q})$ and $\hat{y} \in \mathscr{H}^\infty$, so we have

$$\hat{h}_b' = g_b'^{-1} \hat{g}_b' = y^{-1} \hat{h}_b \hat{y}.$$

Thus $\hat{h}_b'$ and $\hat{h}_b$ belong to the same *double coset*

$$\mathscr{H}(\mathbb{Q}) \cdot \hat{h}_b \cdot \mathscr{H}^\infty$$

of $\mathscr{H}(\mathbb{A}_f)$ w.r.t. the subgroups $\mathscr{H}(\mathbb{Q})$ and $\mathscr{H}^\infty$. Denoting the set of all such double cosets by

$$\mathscr{D} = \mathscr{H}(\mathbb{Q}) \backslash \mathscr{H}(\mathbb{A}_f) / \mathscr{H}^\infty,$$

we thus have a well defined map

$$\psi : \mathscr{C} \to \mathscr{D}$$
$$b \to \mathscr{H}(\mathbb{Q}) \cdot \hat{h}_b \cdot \mathscr{H}^\infty.$$

Claim 4 The fibres of ψ are the orbits of $\mathscr{G}(\mathbb{Z})$ in $\mathscr{C}$.

Proof Suppose $a, b \in \mathscr{C}$ and $a\rho(x) = b$ with $x \in \mathscr{G}(\mathbb{Z})$. In view of the preceding discussion, to compute $a\psi$ and $b\psi$ we may choose $g_a = x g_b$, $\hat{g}_a = x \hat{g}_b$. Then

$$\hat{h}_a = g_a^{-1} \hat{g}_a = g_b^{-1} \hat{g}_b = \hat{h}_b,$$

so certainly $a\psi = b\psi$.

Suppose conversely that $a, b \in \mathscr{C}$ with $a\psi = b\psi$. Then there exist $y \in \mathscr{H}(\mathbb{Q})$ and $\hat{y} \in \mathscr{H}^\infty$ such that

$$\hat{h}_a = y \hat{h}_b \hat{y}.$$

Thus

$$g_a^{-1} \hat{g}_a = y g_b^{-1} \hat{g}_b \hat{y}$$

and so

$$g_b y^{-1} g_a^{-1} = \hat{g}_b \hat{y} \hat{g}_a^{-1} \in \mathscr{G}(\mathbb{Q}) \cap \mathscr{G}^\infty = \mathscr{G}(\mathbb{Z}).$$

Since
$$b\rho(g_b y^{-1} g_a^{-1}) = c\rho(y)^{-1}\rho(g_a)^{-1} = a,$$
it follows that a and b lie in the same orbit of $\mathscr{G}(\mathbb{Z})$.

It is clear now that Theorem 1 will follow from

Theorem 3 For every $\mathbb{Q}$-group $\mathscr{H}$, the set
$$\mathscr{H}(\mathbb{Q})\backslash\mathscr{H}(\mathbb{A}_f)/\mathscr{H}^\infty$$
is finite.

This equally remarkable companion result to Theorem 2 is due to Borel; see section E for references.

This completes the discussion of Theorem 1, and we turn next to a different question: how are the orbits of $\mathscr{G}(\mathfrak{o})$ related to those of $\mathscr{G}(\mathbb{Z})$, when $\mathfrak{o}$ is the ring of integers in some algebraic number field k? The answer again depends on Galois cohomology, this time that of arithmetic groups. The result we shall obtain is

Theorem 4 Every orbit of $\mathscr{G}(\mathfrak{o})$ meets only finitely many orbits of $\mathscr{G}(\mathbb{Z})$ in $\mathbb{Z}^m$.

If we enlarge the ring $\mathfrak{o}$, the orbits of $\mathscr{G}(\mathfrak{o})$ only get bigger; so we may assume for the purposes of the proof that k is normal over $\mathbb{Q}$. Let Γ be the Galois group of k over $\mathbb{Q}$. Then Γ operates on $\mathscr{G}(\mathfrak{o})$ in the usual way, and since $\mathfrak{o} \cap \mathbb{Q} = \mathbb{Z}$ the elements of $\mathscr{G}(\mathfrak{o})$ fixed by Γ comprise exactly $\mathscr{G}(\mathbb{Z})$. Let
$$\mathscr{C} = c\rho(\mathscr{G}(\mathfrak{o})) \cap \mathbb{Z}^m,$$
with $c \in \mathbb{Z}^m$, be the intersection of an orbit of $\mathscr{G}(\mathfrak{o})$ with $\mathbb{Z}^m$. We want to show that $\mathscr{C}$ consists of finitely many orbits of $\mathscr{G}(\mathbb{Z})$. Let $\mathscr{H}$ be the stabilizer of c in $\mathscr{G}$. Just as above, in Claims 1 and 2, we can define a map
$$\theta : \mathscr{C} \to H^1(\Gamma, \mathscr{H}(\mathfrak{o}))$$
$$a \mapsto [\Delta_a];$$
if $a \in \mathscr{C}$ and $a\rho(g) = c$ with $g \in \mathscr{G}(\mathfrak{o})$, then $\Delta_a : \Gamma \to \mathscr{H}(\mathfrak{o})$ sends $\gamma \in \Gamma$ to $g^{-\gamma}g$. Just as in Claim 3, the fibres of θ will be exactly the orbits of $\mathscr{G}(\mathbb{Z})$ in $\mathscr{C}$. So theorem 4 will be a consequence of

Theorem 5 Let $\mathscr{H}$ be a $\mathbb{Q}$-group, k a finite normal extension field of $\mathbb{Q}$ with ring of integers $\mathfrak{o}$, and $\Gamma = \mathrm{Gal}(k/\mathbb{Q})$. Then $H^1(\Gamma, \mathscr{H}(\mathfrak{o}))$ is finite.

This result is also due to Borel and Serre. It will be instructive to see how it is proved.

Lemma 1 Let H be a group and Γ a finite group acting on H. Then there is a $1:1$ correspondence between the set of conjugacy classes of complements to H in $H]\Gamma$ and the set $H^1(\Gamma, H)$.

If H is abelian, this is just a restatement of Propositions 1 and 2 of Chapter 3, section A. The proof in the general case is identical, and is left to the reader: only take care that whenever products in H are considered, they are written in the correct order!

Now we know that in an arithmetic group, the finite subgroups lie in finitely many conjugacy classes: this theorem of Borel was stated as Theorem 3 in Chapter 8. So Theorem 5 will follow from Lemma 1, once we have established that the semi-direct product $\mathscr{H}(\mathfrak{o})]\Gamma$ is isomorphic to an arithmetic group. This we shall do in Proposition 3 below, after some preliminary discussion.

Let $(u_1, \ldots, u_d)$ be a $\mathbb{Z}$-basis for $\mathfrak{o}$. Then $k = \oplus u_i \mathbb{Q}$, and we may represent an element α of k by a matrix $\alpha^\dagger = (\alpha_{ij}) \in M_d(\mathbb{Q})$, where

$$u_i \alpha = \sum_j \alpha_{ij} u_j, \quad i = 1, \ldots, d.$$

The map † thus embeds k as a subring in $M_d(\mathbb{Q})$, and $\mathfrak{o}^\dagger \subseteq M_d(\mathbb{Z})$. This embedding then gives an embedding of $M_n(k)$ into $M_{nd}(\mathbb{Q})$:

$$g = (g_{ij}) \mapsto g^\dagger = \begin{bmatrix} g_{11}^\dagger & \cdots & g_{1n}^\dagger \\ \vdots & & \vdots \\ g_{n1}^\dagger & \cdots & g_{nn}^\dagger \end{bmatrix}.$$

It is easy to see that $^\dagger : M_n(k) \to M_{nd}(\mathbb{Q})$ is an injective ring homomorphism; its restriction to $GL_n(k)$ is therefore an injective group homomorphism into $GL_{nd}(\mathbb{Q})$, and maps $GL_n(\mathfrak{o})$ into $GL_{nd}(\mathbb{Z})$.

Lemma 2 (i) $GL_n(\mathfrak{o})^\dagger = GL_n(k)^\dagger \cap GL_{nd}(\mathbb{Z})$.

(ii) Suppose $k = \mathbb{Q}(\lambda)$. Then $GL_n(k)^\dagger$ is the centralizer in $GL_{nd}(\mathbb{Q})$ of the matrix $(\lambda 1_n)^\dagger$.

Proof (i) An element $\alpha \in k$ satisfies $\alpha^\dagger \in M_d(\mathbb{Z})$ if and only if $u_i \alpha \in \mathfrak{o}$ for $i = 1, \ldots, d$, which holds if and only if $\alpha \in \mathfrak{o}$. It follows that

$$M_n(k)^\dagger \cap M_{nd}(\mathbb{Z}) = M_n(\mathfrak{o})^\dagger.$$

A moment's thought about inverses shows that this implies (i).

(ii) The scalar matrix $\lambda 1_n$ commutes with every matrix in $GL_n(k)$; since † is a homomorphism it follows that $GL_n(k)^\dagger$ is contained in the centralizer, C say, of $(\lambda 1_n)^\dagger$ in $GL_{nd}(\mathbb{Q})$. Suppose conversely that $x \in C$. Partition the matrix

x into $d \times d$ blocks thus:

$$x = \begin{bmatrix} \xi_{11}(x) & \cdots & \xi_{1n}(x) \\ \vdots & & \vdots \\ \xi_{n1}(x) & \cdots & \xi_{nn}(x) \end{bmatrix} \tag{7}$$

with each $\xi_{ij}(x) \in M_d(\mathbb{Q})$. Evidently $x \in M_n(k)^\dagger$ if and only if each $\xi_{ij}(x) \in k^\dagger$.

By hypothesis, x commutes with $(\lambda 1_n)^\dagger$. Therefore $\xi_{ij}(x)$ commutes with $\lambda^\dagger$ in $M_d(\mathbb{Q})$, for each i and j. If $\xi \in M_d(\mathbb{Q})$ commutes with $\lambda^\dagger$, then ξ commutes with every element of $k^\dagger$, by the choice of λ; if $\alpha \in k$ is the image of $1 \in k$ under the endomorphism represented by ξ, one verifies immediately that then $\alpha^\dagger = \xi$. Thus each $\xi_{ij}(x)$ does indeed belong to $k^\dagger$ and so $x \in M_n(k)^\dagger$. Consideration of x^{-1} again shows now that $x \in GL_n(k)^\dagger$ as claimed. This proves the Lemma.

The next step is to represent $\Gamma = \text{Gal}(k/\mathbb{Q})$ in $M_{nd}(\mathbb{Q})$. If $\gamma \in \Gamma$ we have $u_i^\gamma = \sum_j \gamma_{ij} u_j$ where $\bar{\gamma} = (\gamma_{ij}) \in GL_d(\mathbb{Z})$. Moreover, if $\alpha \in k$ and $\gamma \in \Gamma$

$$(\alpha^\gamma)^\dagger = (\alpha^\dagger)^{\bar{\gamma}}, \tag{8}$$

(an equation in $GL_d(\mathbb{Q})$). To see this, let $\beta \in k$ and observe that

$$\beta \cdot \alpha^\gamma = (\beta^{\gamma^{-1}} \cdot \alpha)^\gamma;$$

thus multiplication by α^γ has the same effect on k as

$$\gamma^{-1} \circ (\text{multiply by } \alpha) \circ \gamma,$$

and (8) is just the matrix expression of this identity, w.r.t. the basis $(u_1, \ldots, u_d)$ of k. Now for $\gamma \in \Gamma$ define

$$\gamma^* = \text{diag}(\underbrace{\bar{\gamma}, \ldots, \bar{\gamma}}_{n}) \in GL_{nd}(\mathbb{Z}).$$

It follows from (8) that the map

$$GL_n(k)] \Gamma \to GL_{nd}(\mathbb{Q})$$
$$g \cdot \gamma \mapsto g^\dagger \cdot \gamma^* \tag{9}$$

is a group homomorphism. In fact it gives an isomorphism of $GL_n(k)]\Gamma$ onto $GL_n(k)^\dagger]\Gamma^*$: this will follow if we check that $GL_n(k)^\dagger \cap \Gamma^* = 1$, and this in turn holds because the only element of Γ which fixes λ is the identity. The following is now clear:

Lemma 3 Let H be a Γ-invariant subgroup of $GL_n(k)$. Then the map (9) restricts to an isomorphism of $H]\Gamma$ onto $H^\dagger]\Gamma^*$; if also $H \le GL_n(\mathfrak{o})$, then $H^\dagger]\Gamma^* \le GL_{nd}(\mathbb{Z})$.

We can now prove

Proposition 3　　Let $\mathscr{H}$ be a $\mathbb{Q}$-group, of degree n, and let k, $\mathfrak{o}$ and Γ be as in Theorem 5. Then the semi-direct product $\mathscr{H}(\mathfrak{o})]\Gamma$ is isomorphic to an arithmetic group.

Proof　　In view of Lemma 3, it will suffice to prove that the group $\mathscr{H}(\mathfrak{o})^{\dagger}]\Gamma^{*} \leq GL_{nd}(\mathbb{Z})$ is an arithmetic group. As Γ^{*} is finite, this will follow, by Lemma 7 in section A of Chapter 8, if we show that $\mathscr{H}(\mathfrak{o})^{\dagger}$ is a closed subgroup of $GL_{nd}(\mathbb{Z})$.

Now there is a set S of polynomials, in n^2 variables, over $\mathbb{Q}$ such that for $y \in GL_n(k)$,

$$y \in \mathscr{H}(k) \Leftrightarrow P(y) = 0 \quad \forall P \in S$$

(as usual, $P(y)$ denotes the polynomial P evaluated at the entries of the matrix y). It follows that for $x \in GL_n(k)^{\dagger} \leq GL_{nd}(\mathbb{Q})$,

$$x \in \mathscr{H}(k)^{\dagger} \Leftrightarrow P(\xi_{11}(x), \ \xi_{12}(x), \ldots, \xi_{nn}(x)) = 0 \quad \forall P \in S,$$

where the $d \times d$ matrices $\xi_{ij}(x)$ are given in (7), above. Hence by Lemma 2(ii), for $x \in GL_{nd}(\mathbb{Q})$ we have

$$x \in \mathscr{H}(k)^{\dagger} \Leftrightarrow x \in GL_n(k)^{\dagger} \ \text{and} \ P(\xi_{ij}(x)) = 0 \quad \forall P \in S$$
$$\Leftrightarrow x \cdot (\lambda 1_n)^{\dagger} - (\lambda 1_n)^{\dagger} \cdot x = 0 \ \text{and}$$
$$P(\xi_{ij}(x)) = 0 \quad \forall P \in S.$$

Thus $\mathscr{H}(k)^{\dagger}$ is a closed subset of $GL_{nd}(\mathbb{Q})$, and Lemma 2 (i) then shows that

$$\mathscr{H}(\mathfrak{o})^{\dagger} = \mathscr{H}(k)^{\dagger} \cap GL_n(\mathfrak{o})^{\dagger} = \mathscr{H}(k)^{\dagger} \cap GL_{nd}(\mathbb{Z})$$

is closed in $GL_{nd}(\mathbb{Z})$. This completes the proof of Proposition 3, and therewith that of Theorem 5.

It is worth remarking that the technique just described provides a supply of new arithmetic groups, and of new $\mathbb{Q}$-groups. Whenever $\mathscr{H}$ is a $\mathbb{Q}$-group of degree n and k is a finite normal extension field of $\mathbb{Q}$, of degree d, the group $\mathscr{H}(k)^{\dagger}$ is a closed subgroup of $GL_{nd}(\mathbb{Q})$, and its closure in $GL_{nd}(\mathbb{C})$ is then a $\mathbb{Q}$-group $\mathscr{H}_1$, say, such that

$$\mathscr{H}_1(\mathbb{Q}) \cong \mathscr{H}(k), \quad \mathscr{H}_1(\mathbb{Z}) \cong \mathscr{H}(\mathfrak{o}).$$

Thus for example $\mathfrak{o}^{*} = GL_1(\mathfrak{o})$ is in a natural way an arithmetic group, sitting inside $GL_d(\mathbb{Z})$. This shows how the Dirichlet Units Theorem (at least the qualitative statement) is generalized by the theorem of Borel and Harish-Chandra that every arithmetic group is finitely generated. (The quantitative statement of the Units Theorem also generalizes to arithmetic groups, but in a rather sophisticated way; see section E for references.)

B. The main theorem: the unipotent case

For each prime p let α_p be a p-adic integer, and let m be a positive integer. Let $p_1, \ldots, p_s$ be the primes dividing m, so that $m = p_1^{e_1}, \ldots, p_s^{e_s}$ say. By the Chinese Remainder Theorem, there is an integer a, unique modulo m, such that

$$\alpha_{p_i} \equiv a \bmod p_i^{e_i} \mathbb{Z}_{p_i}, \quad i = 1, \ldots, s.$$

We then have $\alpha_p \equiv a \bmod m\mathbb{Z}_p$ for every prime p. Putting

$$(\alpha_p)\pi_m = a + m\mathbb{Z},$$

we thus get a surjective ring homomorphism

$$\pi_m : \prod_p \mathbb{Z}_p \to \mathbb{Z}/m\mathbb{Z}.$$

This extends in the obvious way to a group homomorphism

$$\pi_m : GL_n^\infty \to GL_n(\mathbb{Z}/m\mathbb{Z}).$$

Now let $\mathscr{G}$ be a $\mathbb{Q}$-group, of degree n. We define an equivalence relation on the collection of all subsets of $GL_n(\mathbb{Z})$ as follows: for $P, Q \subseteq GL_n(\mathbb{Z})$, $P \approx _{\mathscr{G}} Q$ *if and only if there exists* $g \in \mathscr{G}^\infty$ *such that* $P^g\pi_m = Q\pi_m$ *for every* $m \in \mathbb{N}$. (Remember that $GL_n(\mathbb{Z})$ is considered as a subset of GL_n^∞, via the 'diagonal' embedding.) If we had chosen to define the sets $\mathscr{G}(\mathbb{Z}/m\mathbb{Z}) \subseteq GL_n(\mathbb{Z}/m\mathbb{Z})$, as we did at the beginning of section A, we would equally well say that $P \approx _{\mathscr{G}} Q$ if P and Q have conjugate images in $\mathscr{G}(\mathbb{Z}/m\mathbb{Z})$, for every positive integer m; the present definition has the merit of bypassing uncomfortable questions about whether $\mathscr{G}(\mathbb{Z}/m\mathbb{Z})$ is even a group.

The main result of the chapter can now be stated:

Theorem 6   Let $\mathscr{G}$ be a $\mathbb{Q}$-group. Then every $\approx _{\mathscr{G}}$-class of soluble-by-finite subgroups of $\mathscr{G}(\mathbb{Z})$ is the union of finitely many conjugacy classes of subgroups in $\mathscr{G}(\mathbb{Z})$.

The relationship between Theorem 6 and Theorem 1 is just like that between Algorithm 1 and Algorithm B in section D of Chapter 8. The argument which reduces the one to the other, however, is a lot more complicated. In this section we deal with the special case of Theorem 6 where the subgroups in question are all unipotent.

Lemma 4   Let P and Q be soluble-by-finite subgroups of $\mathscr{G}(\mathbb{Z})$.

(i) If $P \approx _{\mathscr{G}} Q$ then $P^e \approx _{\mathscr{G}} Q^e$ for every positive integer e.

(ii) Given P, and given that Q^e is conjugate to P^e in $\mathscr{G}(\mathbb{Z})$, there are only finitely many possibilities for Q up to conjugacy in $\mathscr{G}(\mathbb{Z})$.

Proof (i) is clear from the definition. For (ii), we may suppose that $Q^e = P^e$, and the result then follows from Proposition 10 in section C of Chapter 8.

We shall repeatedly use this lemma as a reduction step for Theorem 6, and do so for the first time right now. Thus suppose $\mathscr{C}$ is a family of unipotent subgroups of $\mathscr{G}(\mathbb{Z})$, contained in a single $\approx_{\mathscr{G}}$-class. We want to show that $\mathscr{C}$ contains only finitely many non-conjugate subgroups in $\mathscr{G}(\mathbb{Z})$. Theorem 5 of Chapter 6 shows that for a suitable positive integer e, the groups P^e, $P\in\mathscr{C}$ are all lattice groups; Lemma 1 now shows that we may safely replace each $P\in\mathscr{C}$ by its subgroup P^e. Thus we may assume that $\mathscr{C}$ consists of *lattice groups*.

We interrupt the argument at this point, for another definition. Let $\rho:\mathscr{G}\to GL_r(\mathbb{C})$ be a $\mathbb{Q}$-rational representation, with the property that $\rho(\mathscr{G}^\infty)\le GL_r^\infty$. For $\mathbb{Z}$-submodules A and B of $\mathbb{Z}^r$, write

$$A \sim_\rho B$$

if there exists $g\in\mathscr{G}^\infty$ such that

$$A\rho(g)\pi_m = B\pi_m \quad \forall m\in\mathbb{N},$$

where π_m is now the natural map from $\prod_p \mathbb{Z}_p^r$ onto $(\mathbb{Z}/m\mathbb{Z})^r$.

Returning to Theorem 6, we consider a family $\mathscr{C}$ of unipotent lattice subgroups of $\mathscr{G}(\mathbb{Z})$, contained in a single $\approx_{\mathscr{G}}$-class. Let $\rho:\mathscr{G}\to GL_{n^2}(\mathbb{C})$ be the representation of $\mathscr{G}$ acting by conjugation on $M_n(\mathbb{C}) = \mathbb{C}^{n^2}$; it is easy to see that then $\rho(\mathscr{G}^\infty)\le GL_{n^2}^\infty$. For $P\in\mathscr{C}$ put $L_P = (n-1)!\log P$. Then each L_P is a $\mathbb{Z}$-submodule of $M_n(\mathbb{Z})$, and it is clear from the definitions that the modules L_P, $P\in\mathscr{C}$, lie in a single $\sim_\rho$-class in $M_n(\mathbb{Z}) = \mathbb{Z}^{n^2}$. Thus the 'unipotent case' of Theorem 6 will follow once we have proved

Proposition 4   Let $\mathscr{G}$ be a $\mathbb{Q}$-group and $\rho:\mathscr{G}\to GL_r(\mathbb{C})$ a $\mathbb{Q}$-rational representation, with $\rho(G^\infty)\le GL_r^\infty$. Then each $\sim_\rho$-class of $\mathbb{Z}$-submodules of $\mathbb{Z}^n$ is the union of finitely many orbits of $\mathscr{G}(\mathbb{Z})$, acting via ρ on the set of all $\mathbb{Z}$-submodules of $\mathbb{Z}^r$.

(Note that $\rho(\mathscr{G}(\mathbb{Z}))\le GL_r^\infty\cap GL_r(\mathbb{Q}) = GL_r(\mathbb{Z})$, so $\mathscr{G}(\mathbb{Z})$ really does act on $\mathbb{Z}^r$.)

Clearly, the thing to do is to reduce Proposition 4 to Theorem 1, somehow. The reduction comes in two stages; it is very similar to the proof of Proposition 2 in section D of Chapter 4.

First stage   Supposing Proposition 4 is known for each $\sim_\rho$-class of *cyclic* $\mathbb{Z}$-modules (and every ρ and r), we deduce the general case. Let $\mathscr{L}$ be a $\sim_\rho$-class of $\mathbb{Z}$-submodules in $\mathbb{Z}^r$. It is easy to see that the modules in $\mathscr{L}$ all have the same rank, say d. Define

$$\psi : \mathscr{G} \to GL_{\binom{r}{d}}(\mathbb{C}) = \operatorname{Aut} \wedge^d(\mathbb{C}^r)$$

by $\psi(g) = \wedge^d \rho(g)$ for $g \in \mathscr{G}$, where $\wedge^d \rho(g)$ maps the vector $v_1 \wedge \ldots \wedge v_d$ to $v_1 \rho(g) \wedge \ldots \wedge v_d \rho(g)$, for $v_1, \ldots, v_d \in \mathbb{C}^r$. Then ψ is a $\mathbb{Q}$-rational representation and $\psi(\mathscr{G}^\infty) \leq GL_{\binom{r}{d}}^\infty$ (provided we take the usual basis for $\wedge^d(\mathbb{C}^r)$, namely the vectors $e_{i_1} \wedge \ldots \wedge e_{i_d}$ with $i_1 < i_2 < \ldots < i_d$, where $(e_1, \ldots, e_r)$ is the standard basis of $\mathbb{C}^r$). It is quite straightforward to see that if $A, B \leq \mathbb{Z}^r$ and if $A \sim_\rho B$, then $\wedge^d A \sim_\psi \wedge^d B$ in $\wedge^d(\mathbb{Z}^r) = \mathbb{Z}^{\binom{r}{d}}$. Moreover, if $A \cong B \cong \mathbb{Z}^d$, then $\wedge^d A$ and $\wedge^d B$ are cyclic $\mathbb{Z}$-modules. By hypothesis, then, the $\mathbb{Z}$-modules $\wedge^d A$ for $A \in \mathscr{L}$ lie in finitely many orbits of $\mathscr{G}(\mathbb{Z})$, acting via ψ on $\wedge^d(\mathbb{Z}^r)$. Just as in section D of Chapter 8, we may infer from this that the subspaces $\mathbb{Q}A \leq \mathbb{Q}^r$, for $A \in \mathscr{L}$, lie in finitely many orbits of $\mathscr{G}(\mathbb{Z})$, acting via ρ.

At this point we need the following simple observation; recall that for $A \leq \mathbb{Z}^r$, the *isolator* $i(A)$ of A is defined by $i(A)/A = \tau(\mathbb{Z}^r/A)$.

Lemma 5 Let $A, B \leq \mathbb{Z}^r$ and suppose that $i(A) = i(B) = I$, say. If $A\pi_m$ and $B\pi_m$ lie in the same orbit of $GL_r(\mathbb{Z}/m\mathbb{Z})$, for all positive integers m, then $|I:A| = |I:B|$.

Proof Choose a positive integer m so that $mI \leq A \cap B$. Since $m\mathbb{Z}^r \cap I = mI \subseteq B$, we have

$$(I + m\mathbb{Z}^r)/(B + m\mathbb{Z}^r) \cong I/(I \cap (B + m\mathbb{Z}^r)) = I/(B + mI) = I/B.$$

Therefore

$$|I:B| = |I\pi_m/B\pi_m| = |I\pi_m| \cdot |B\pi_m|^{-1}.$$

Similarly, $|I:A| = |I\pi_m| \cdot |A\pi_m|^{-1}$. The hypotheses imply that $|A\pi_m| = |B\pi_m|$, hence $|I:B| = |I:A|$ as claimed.

We may now complete the first stage of the proof. By the result of the previous paragraph, it will suffice now to show that given $A \in \mathscr{L}$, there are only finitely many $B \in \mathscr{L}$ such that $\mathbb{Q}B = \mathbb{Q}A$. But if $\mathbb{Q}B = \mathbb{Q}A$ then

$$i(B) = \mathbb{Q}B \cap \mathbb{Z}^r = \mathbb{Q}A \cap \mathbb{Z}^r = i(A) = I, \text{ say.}$$

Put $m = |I:A|$. Then by Lemma 5, we have

$$mI \leq B \leq I.$$

Since there are only finitely many groups lying between mI and I, there are only finitely many possibilities for B, and the result follows.

Second stage   We must now prove Proposition 4 for a $\sim_p$-class of cyclic $\mathbb{Z}$-modules. To do this we shall need

> **Lemma 6**   Let $\mathscr{H} \leq GL_r(\mathbb{C})$ be a $\mathbb{Q}$-group. Then
> $$\mathscr{H}_1 = \mathbb{C}^* \mathscr{H}$$
> is again a $\mathbb{Q}$-group. Moreover, $|\mathscr{H}_1(\mathbb{Z}) : \mathscr{H}(\mathbb{Z})|$ is finite, and $\mathscr{H}_1(\mathbb{Z}_p) \geq \mathbb{Z}_p^* \cdot 1_r$ for each prime p.

> *Proof*   Identify $GL_1(\mathbb{C}) \times \mathscr{H}$ with the set of all matrices
> $$\begin{bmatrix} x & 0 \dots 0 \\ 0 & \\ \vdots & h \\ 0 & \end{bmatrix}, x \in \mathbb{C}^*, h \in \mathscr{H}$$
> in $GL_{r+1}(\mathbb{C})$. The map τ which sends such a matrix to $xh \in \mathscr{H}_1$ is obviously a surjective $\mathbb{Q}$-rational homomorphism of $GL_1(\mathbb{C}) \times \mathscr{H}$ onto $\mathscr{H}_1$. Proposition 5 of chapter 8 shows that $\mathscr{H}_1$ is then a closed subgroup of $GL_r(\mathbb{C})$ and Proposition 7 of that chapter ensures that in fact $\mathscr{H}_1$ is a $\mathbb{Q}$-group. By Theorem 1 of Chapter 8, $(GL_1(\mathbb{Z}) \times \mathscr{H}(\mathbb{Z}))\tau$ is arithmetic in $\mathscr{H}_1$; but $GL_1(\mathbb{Z}) = \{\pm 1\}$, so $(GL_1(\mathbb{Z}) \times \mathscr{H}(\mathbb{Z}))\tau = \pm \mathscr{H}(\mathbb{Z})$, and consequently $\mathscr{H}(\mathbb{Z})$ is commensurable with $\mathscr{H}_1(\mathbb{Z})$. Since $\mathscr{H}(\mathbb{Z}) \leq \mathscr{H}_1(\mathbb{Z})$ we have $|\mathscr{H}_1(\mathbb{Z}):\mathscr{H}(\mathbb{Z})| < \infty$ as claimed. For the final claim, embed $\mathbb{Z}_p$ into $\mathbb{C}$.

Now in Proposition 4, we have a representation $\rho : \mathscr{G} \to GL_r(\mathbb{C})$. Take $\mathscr{H} = \rho(\mathscr{G}) \leq GL_r(\mathbb{C})$. By hypothesis, $\rho(\mathscr{G}^\infty) \leq GL_r^\infty$, so we have
$$\rho(\mathscr{G}^\infty) \leq \mathscr{H}^\infty. \tag{10}$$
Suppose $a, b \in \mathbb{Z}^r \setminus \{0\}$ satisfy $a\mathbb{Z} \sim_p b\mathbb{Z}$. Then from (10) we know that there exists $\hat{h} \in \mathscr{H}^\infty$ such that
$$(a\hat{h}\mathbb{Z})\pi_m = (b\mathbb{Z})\pi_m \quad \forall m \in \mathbb{N}.$$
In particular, if we fix a prime p and denote the p-component of $\hat{h}$ by $h_p \in \mathscr{H}(\mathbb{Z}_p)$, we see that
$$(ah_p)\pi_{p^j} = (\lambda_j b)\pi_{p^j} \quad \forall j \in \mathbb{N} \tag{11}$$
where each λ_j is a suitable integer coprime to p. It follows from (11) that the sequence (λ_j) converges to some p-adic unit λ, say, such that
$$ah_p = \lambda b.$$
Now let $\mathscr{H}_1$ be the $\mathbb{Q}$-group constructed in Lemma 6. Then $\lambda^{-1} h_p \in \mathscr{H}_1(\mathbb{Z}_p)$, and so a and b lie in the same orbit of $\mathscr{H}_1(\mathbb{Z}_p)$ acting in the

natural way on $\mathbb{Z}_p^r$. Since p was arbitrary, we have established that a and b are locally $\mathscr{H}_1$-equivalent.

Proposition 4 now follows easily. Let $\mathscr{L}$ be a $\sim_p$-class of cyclic $\mathbb{Z}$-submodules of $\mathbb{Z}^r$; say

$$\mathscr{L} = \{a_\alpha \mathbb{Z} \,|\, \alpha \in A\}.$$

We have just seen that the generators $a_\alpha, \alpha \in A$, all belong to a single local orbit of the group $\mathscr{H}_1$. By Theorem 1, they lie in finitely many orbits of $\mathscr{H}_1(\mathbb{Z})$, and hence, in view of Lemma 6, in finitely many orbits of $\mathscr{H}(\mathbb{Z})$. Now Theorem 1 of Chapter 8 shows that $\rho(\mathscr{G}(\mathbb{Z}))$ has finite index in $\mathscr{H}(\mathbb{Z})$: so the $a_\alpha, \alpha \in A$, lie in finitely many orbits of $\mathscr{G}(\mathbb{Z})$, acting via ρ; and the result follows.

C. The diagonalizable case

In this section we consider the special case of Theorem 6 where the $\approx_{\mathscr{G}}$-class consists of diagonalizable abelian subgroups of $\mathscr{G}(\mathbb{Z})$. The main ideas in this case are quite different and belong to algebraic number theory. I shall concentrate on the special case $\mathscr{G} = GL_n(\mathbb{C})$, which already exhibits the more interesting features of the argument.

Suppose, then, that we have a $\approx$-class $\mathscr{C}$ of diagonalizable abelian subgroups of $GL_n(\mathbb{Z})$ (where '$\approx$' stands for '$\approx_{GL_n}$'). We are to show that $\mathscr{C}$ breaks up into finitely many conjugacy conjugacy classes in $GL_n(\mathbb{Z})$. The first step is to show that all the groups in $\mathscr{C}$ can be diagonalized over the *same* finite extension field of $\mathbb{Q}$. For a matrix group A, denote the group generated by the eigenvalues of all matrices in A by X_A. We can then state

Proposition 5 Let A and B be diagonalizable abelian subgroups of $GL_n(\mathbb{Z})$. If A and B have conjugate images in $GL_n(\mathbb{Z}/p\mathbb{Z})$ for every prime p, then the fields $\mathbb{Q}(X_A)$ and $\mathbb{Q}(X_B)$ are equal.

Proof Say $A = \langle a_1, \ldots, a_d \rangle$. Let $\lambda_{i1}, \ldots, \lambda_{in}$ be the eigenvalues of a_i; then

$$X_A = \langle \lambda_{ij} \,|\, 1 \le i \le d, 1 \le j \le n \rangle,$$

and $\mathbb{Q}(X_A)$ is a finite normal extension of $\mathbb{Q}$. Let $\mathfrak{o}$ be its ring of algebraic integers, and note that $X_A \le \mathfrak{o}^*$.

Now take $b \in B$ and let p be a prime number. Then $b\pi_p$ is conjugate to $a\pi_p$ in $GL_n(\mathbb{Z}/p\mathbb{Z})$ for some $a \in A$. So if χ_b, χ_a denote the characteristic polynomials of b and of a, respectively, and $\bar{\chi}_b, \bar{\chi}_a$ their reductions modulo p, then $\bar{\chi}_b = \bar{\chi}_a$. Since the eigenvalues of a lie in $\mathfrak{o}$, $\bar{\chi}_a$ splits into linear factors over the ring $\mathfrak{o}/\mathfrak{p}\mathfrak{o}$. Consequently $\bar{\chi}_b$ splits likewise over $\mathfrak{o}/\mathfrak{p}\mathfrak{o}$. We now appeal to a number-theoretic result:

Theorem 7 Let k be an algebraic number field, normal over $\mathbb{Q}$, with ring of integers $\mathfrak{o}$. Let $f(X)\in\mathfrak{o}[X]$ be a monic irreducible polynomial. If, for all but finitely many primes $\mathfrak{p}$ of $\mathfrak{o}$, the image of f splits into linear factors over $\mathfrak{o}/\mathfrak{p}$, then f splits into linear factors over $\mathfrak{o}$.

The case $k = \mathbb{Q}$ of this theorem played a key role in section A of Chapter 4 (see also section E of Chapter 4); for the general case, see section E below.

Returning to Proposition 5, we apply Theorem 7 to each irreducible factor of χ_b over $\mathfrak{o}$, noting that every prime $\mathfrak{p}$ of $\mathfrak{o}$ contains $p\mathfrak{o}$ for the corresponding (rational) prime number p. The conclusion is that χ_b splits into linear factors over $\mathfrak{o}$. Thus all the eigenvalues of b belong to $\mathfrak{o}$, and as $b\in B$ was arbitrary we see that $X_B \subseteq \mathfrak{o} \subseteq \mathbb{Q}(X_A)$. The proposition follows by symmetry.

If A is a diagonalizable abelian subgroup of $GL_n(\mathbb{Z})$, then A is diagonalizable over the field $\mathbb{Q}(X_A)$: this is a simple piece of linear algebra. So applying Proposition 5 to the family of groups $\mathscr{C}$, we see that *there exists an algebraic number field k (finite and normal over $\mathbb{Q}$) such that every group in $\mathscr{C}$ is diagonalizable over k*. We keep this field k fixed, and denote its ring of integers by $\mathfrak{o}$.

The next step hinges on a simple but crucial fact:

Lemma 7 Let F be a field and X, Y subgroups of $D_n(F)$. If $Y = X^h$ for some $h\in GL_n(F)$, then there is a permutation matrix σ such that $h\sigma^{-1}$ centralizes X.

Proof The eigenspaces of X and of Y are spanned by subsets of the standard basis of F^n, since X and Y are diagonal. Since $Y = X^h$, the matrix h maps the eigenspaces of X to those of Y; because of the particular shape of these eigenspaces, we can find a permutation matrix σ such that $h\sigma^{-1}$ fixes each eigenspace of X. But X acts as scalars on each of its eigenspaces, so $h\sigma^{-1}$ centralizes X.

Returning to our problem, for each $A\in\mathscr{C}$ let $\alpha_A\in GL_n(k)$ be a matrix such that A^{α_A} is diagonal. Take A and B in $\mathscr{C}$, and let $\mathfrak{p}$ be a prime ideal of $\mathfrak{o}$ such that α_A and α_B are $\mathfrak{p}$-integral (i.e. such that α_A and α_B lie in $GL_n(\mathfrak{o}_{(\mathfrak{p})})$, where $\mathfrak{o}_{(\mathfrak{p})}$ is the localisation of $\mathfrak{o}$ w.r.t. $\mathfrak{p}$); only finitely many primes of $\mathfrak{o}$ are excluded by this condition. Let p be the rational prime number in $\mathfrak{p}$. Let $\hat{g}\in GL_n^{\infty}$ be such that $A^{\hat{g}}\pi_m = B\pi_m$ for all m; thus in particular

$$A\pi_p^{\hat{g}\pi_p} = B\pi_p. \tag{12}$$

Extending the notation in the obvious way, put

$$h = (\alpha_A \pi_{\mathfrak{p}})^{-1} \cdot \hat{g} \pi_{\mathfrak{p}} \cdot (\alpha_B \pi_{\mathfrak{p}}) \in GL_n(\mathfrak{o}/\mathfrak{p}),$$

$$X = A^{\alpha_A} \pi_{\mathfrak{p}}, \quad Y = B^{\alpha_B} \pi_{\mathfrak{p}} \le D_n(\mathfrak{o}/\mathfrak{p}),$$

and note that $X^h = Y$. Lemma 7 now shows that there is a permutation matrix $\sigma = \sigma(\mathfrak{p})$ such that $h\sigma^{-1}$ centralizes X; in other words, then,

$$\alpha_A^{-1} \hat{g} \alpha_B \sigma^{-1} \text{ centralizes } A^{\alpha_A} \text{ modulo } \mathfrak{p}. \tag{13}$$

With (12), this implies that

$$A^{\alpha_A \sigma} \pi_{\mathfrak{p}} = B^{\alpha_B} \pi_{\mathfrak{p}}. \tag{14}$$

Now suppose we knew that $\sigma(\mathfrak{p})$ were independent of $\mathfrak{p}$. Then (14) would almost almost tell us that the congruence closure of B^{α_B} in $D_n(\mathfrak{o})$ is equal to the congruence closure of $A^{\alpha_A \sigma}$; since $D_n(\mathfrak{o}) = \mathfrak{o}^* \times \ldots \times \mathfrak{o}^*$ (n factors), Theorem 4 of Chapter 4 shows that the congruence topology on $D_n(\mathfrak{o})$ is the same as the profinite topology, and as $D_n(\mathfrak{o})$ is a finitely generated abelian group, all its subgroups are closed w.r.t. the latter topology. Thus we could (almost) infer that $B^{\alpha_B} = A^{\alpha_A \sigma}$, hence that B is conjugate to A in $GL_n(k)$.

Unfortunately, this won't quite do; there is no reason to suppose that the same permutation matrix σ will work for every prime $\mathfrak{p}$ of $\mathfrak{o}$. However, the principle is sound, provided we can improve Theorem 4 of Chapter 4 sufficiently for the purpose. To describe the requisite improvement, we make a

Definition Let $\mathfrak{o}$ be a ring of algebraic integers and $\mathscr{Q}$ a set of primes of $\mathfrak{o}$. The set $\mathscr{Q}$ is called *ample* if for every subgroup H of finite index in $\mathfrak{o}^*$, there exist finitely many primes $\mathfrak{p}_1, \ldots, \mathfrak{p}_s$ in $\mathscr{Q}$ such that

$$\mathfrak{o}^* \cap \left(1 + \bigcap_1^s \mathfrak{p}_i \right) \le H.$$

Though this was not stated explicitly, we in fact proved, in section E of Chapter 4, that the set of *all* primes of $\mathfrak{o}$ is ample. The improved version is as follows:

Theorem 8 Let $\mathfrak{o}$ be a ring of algebraic integers and $\mathscr{P}$ the set of all primes of $\mathfrak{o}$. Suppose

$$\mathscr{P} = \mathscr{Q}_1 \cup \ldots \cup \mathscr{Q}_r \cup \mathscr{F}$$

with r finite and $\mathscr{F}$ a finite subset of $\mathscr{P}$. Then at least one of the sets $\mathscr{Q}_1, \ldots, \mathscr{Q}_r$ is ample.

The proof is similar to that given in section E of Chapter 4, though technically more complicated. See section E, below for the reference.

Thus equipped, let us return to the argument where we left it, at (13) above. With A and B as before, let $\mathscr{F}$ be the finite set of primes $\mathfrak{p}$ of $\mathfrak{o}$ for which α_A or α_B is not $\mathfrak{p}$-integral, and for each $\mathfrak{p} \notin \mathscr{F}$, let $\sigma(\mathfrak{p})$ be a permutation matrix satisfying (13). Denote the set of all primes of $\mathfrak{o}$ by $\mathscr{P}$: thus we have a map

$$\mathscr{P} \backslash \mathscr{F} \to S_n; \quad \mathfrak{p} \mapsto \sigma(\mathfrak{p}).$$

The fibres of this map partition $\mathscr{P} \backslash \mathscr{F}$ into at most $n!$ subsets, and by Theorem 7 at least one of these subsets must be ample. Thus there exists $\tau \in S_n$ such that the set

$$\mathscr{Q} = \{\mathfrak{p} \in \mathscr{P} \backslash \mathscr{F} \,|\, \sigma(\mathfrak{p}) = \tau\}$$

is ample. I claim now that

$$A^{\alpha_A \tau} = B^{\alpha_B}. \tag{15}$$

For if $\mathfrak{p}_1, \ldots, \mathfrak{p}_s \in \mathscr{Q}$, then (13) shows that $\alpha_A^{-1} \hat{g} \alpha_B \tau^{-1}$ centralizes A^{α_A} modulo $\mathfrak{p}_1 \cap \ldots \cap \mathfrak{p}_s = \mathfrak{a}$, say, and so

$$A^{\alpha_A \tau} \pi_{\mathfrak{a}} = B^{\alpha_B} \pi_{\mathfrak{a}}. \tag{16}$$

As $\mathscr{Q}$ is ample, the equation (16) for all such ideals $\mathfrak{a}$ shows that $A^{\alpha_A \tau}$ and B^{α_B} have the same profinite closure in $D_n(\mathfrak{o})$; and this gives (15) since $D_n(\mathfrak{o})$ is a finitely generated abelian group.

What we have proved so far is

Proposition 6   Let $\mathscr{C}$ be a $\approx$-class of abelian subgroups of $GL_n(\mathbb{Z})$, all diagonalizable over an algebraic number field k. Then the groups in $\mathscr{C}$ are all conjugate in $GL_n(k)$.

This completes the subtlest part of our argument. To deduce Theorem 6 for diagonalizable abelian subgroups in $GL_n(\mathbb{Z})$, we need two further results, both of which are relatively straightforward:

Proposition 7   Let k be an algebraic number field. Then k has a finite extension k_1 with the following property: if $\mathscr{C}$ is a family of abelian subgroups of $GL_n(\mathbb{Z})$, all conjugate in $GL_n(k)$ and all diagonalizable over k, than $\mathscr{C}$ breaks up into finitely many classes under conjugation in $GL_n(\mathfrak{o}_1)$, where $\mathfrak{o}_1$ is the ring of integers of k_1.

Proposition 8   Let k_1 be an algebraic number field, with ring of integers $\mathfrak{o}_1$. Let $\mathscr{C}$ be a family of finitely generated subgroups of $GL_n(\mathbb{Z})$ which are all conjugate in $GL_n(\mathfrak{o}_1)$. Then $\mathscr{C}$ breaks up into finitely many classes under conjugation in $GL_n(\mathbb{Z})$.

The special case of Theorem 6 under consideration clearly follows from Propositions 5, 6, 7 and 8. Let us prove Proposition 8 first:

Proof of Proposition 8   This is a direct application of Theorem 4. Choose $A = \langle a_1, \ldots, a_d \rangle \in \mathscr{C}$. For each $B \in \mathscr{C}$ we are given $g_B \in GL_n(\mathfrak{o}_1)$ with $A^{g_B} = B$; put

$$v_B = (a_1^{g_B}, \ldots, a_d^{g_B}) \in \underbrace{M_n(\mathbb{Z}) \oplus \ldots \oplus M_n(\mathbb{Z})}_{d}.$$

Define

$$\rho : GL_n(\mathbb{C}) \to GL_{dn^2}(\mathbb{C}) = \mathrm{Aut}\,(\underbrace{M_n(\mathbb{C}) \oplus \ldots \oplus M_n(\mathbb{C})}_{d})$$

to be the representation of $GL_n(\mathbb{C})$ acting by conjugation on each summand $M_n(\mathbb{C})$. The vectors v_B, $B \in \mathscr{C}$, lie in a single orbit of $GL_n(\mathfrak{o}_1)$, acting via ρ, and each $v_B \in \mathbb{Z}^{dn^2}$. By Theorem 4, the v_B lie in finitely many orbits of $GL_n(\mathbb{Z})$. Since each group B is generated by the d component matrices of v_B, it follows that $\mathscr{C}$ breaks up into finitely many classes under conjugation in $GL_n(\mathbb{Z})$.

Proposition 7 depends on three lemmas.

Lemma 8   Let k be an algebraic number field with ring of integers $\mathfrak{o}$. Then there is a finite extension field k_1 of k, with ring of integers $\mathfrak{o}_1$, such that for every finitely generated $\mathfrak{o}$-submodule E of k^n, the $\mathfrak{o}_1$-module $E\mathfrak{o}_1$ $\subseteq k_1^n$ is free as an $\mathfrak{o}_1$-module.

Proof   The group of ideal classes of $\mathfrak{o}$ is finite (see, e.g., Stewart & Tall (1979), Chapter 9). Let $\mathfrak{a}_1, \ldots, \mathfrak{a}_h$ be representatives for these classes. Then for each i, $\mathfrak{a}_i^h$ is a principal ideal $c_i\mathfrak{o}$, say. We take

$$k_1 = k(c_1^{1/h}, \ldots, c_h^{1/h}).$$

Then every ideal of $\mathfrak{o}$ will generate a principal ideal of $\mathfrak{o}_1$, and the lemma follows from the structure theory of torsion-free modules over Dedekind domains: see P.M. Cohn's *Algebra*, vol. 2, Chapter 11 section 6.

Lemma 9   Let $k \subseteq k_1$ be as in Lemma 8. Let $A = \langle a_1, \ldots, a_d \rangle$ be a subgroup of $GL_n(\mathfrak{o})$ which is diagonalizable over k. Then there exists a positive integer m, depending only on the eigenvalues of $a_1, \ldots, a_d$, such that $A^\alpha \le D_n(\mathfrak{o})$ for some $\alpha \in GL_n(k_1)$ with $\alpha \in M_n(\mathfrak{o}_1)$ and $m\alpha^{-1} \in M_n(\mathfrak{o}_1)$.

Proof   Suppose the distinct eigenvalues of a_i are $\lambda_{i1}, \ldots, \lambda_{ir_i}$. Put

$$\Delta_i = \prod_{1 \le j \le l \le r_i} (\lambda_{ij} - \lambda_{il})$$

($\Delta_i = 1$ if $r_i = 1$); define a positive integer m_i by

$$m_i \mathbb{Z} = \mathbb{Z} \cap \Delta_i \mathfrak{o},$$

and put

$$m = \prod_{i=1}^{d} m_i.$$

Now let $U_1, \ldots, U_r$ be the eigenspaces of A in k^n. A familiar argument of linear algebra (left to the reader!) shows that

$$m\mathfrak{o}^n \subseteq \bigoplus_{i=1}^{r} (\mathfrak{o}^n \cap U_i).$$

Embed $\mathfrak{o}^n$ in $\mathfrak{o}_1^n$: then

$$m\mathfrak{o}_1^n \subseteq \bigoplus_{i=1}^{r} (\mathfrak{o}^n \cap U_i)\mathfrak{o}_1 \subseteq \mathfrak{o}_1^n. \tag{17}$$

By the choice of $\mathfrak{o}_1$, each of the modules $(\mathfrak{o}^n \cap U_i)\,\mathfrak{o}_1$ is free as an $\mathfrak{o}_1$-module; let $(u_{i1}, \ldots, u_{in_i})$ be a basis. Then

$$(u_{ij} | 1 \leq i \leq r, \quad 1 \leq j \leq n_i)$$

is a basis for k_1^n w.r.t. which the action of A is represented by diagonal matrices; and (17) shows that the corresponding base-change matrix α satisfies $\alpha \in M_n(\mathfrak{o}_1)$, $m\alpha^{-1} \in M_n(\mathfrak{o}_1)$.

Lemma 10   Let k_1 be an algebraic number field with ring of integers $\mathfrak{o}_1$, and let m be a positive integer. Then the matrices $\alpha \in GL_n(k_1)$ with $\alpha \in M_n(\mathfrak{o}_1)$ and $m\alpha^{-1} \in M_n(\mathfrak{o}_1)$ lie in finitely many right cosets of $GL_n(\mathfrak{o}_1)$.

Proof   If α is such a matrix, we have

$$m\mathfrak{o}_1^n \subseteq \mathfrak{o}_1^n \alpha \subseteq \mathfrak{o}_1^n.$$

This gives only finitely many possibilities for the module $\mathfrak{o}_1^n \alpha$. If α, $\beta \in GL_n(k_1)$ satisfy $\mathfrak{o}_1^n \alpha = \mathfrak{o}_1^n \beta$, then $\alpha\beta^{-1} \in GL_n(\mathfrak{o}_1)$, so α and β belong to the same right coset of $GL_n(\mathfrak{o}_1)$. This establishes the lemma.

Proof of Proposition 7   Given k, take k_1 as in Lemma 8. Now $\mathscr{C}$ is supposed to be a family of abelian subgroups of $GL_n(\mathbb{Z})$, all conjugate in $GL_n(k)$ and diagonalizable over k: we are to show that $\mathscr{C}$ breaks up into finitely many classes under conjugation in $GL_n(\mathfrak{o}_1)$. Let $A = \langle a_1, \ldots, a_d \rangle \in \mathscr{C}$, and take m as in Lemma 9. Since conjugation does not alter the eigenvalues of $a_1, \ldots, a_d$, Lemma 9 shows that for each $B \in \mathscr{C}$ there exists $\alpha_B \in M_n(\mathfrak{o}_1)$, with $m\alpha_B^{-1} \in M_n(\mathfrak{o}_1)$, such that B^{α_B} is diagonal. Then the groups $B^{\alpha_B} \leq D_n(k_1)$ are all conjugate in $GL_n(k_1)$, so by Lemma 7 there are at most $n!$ distinct groups among the B^{α_B}, $B \in \mathscr{C}$. Hence it will suffice to show that for a given $D \leq D_n(k_1)$, the groups $B \in \mathscr{C}$ such that $B^{\alpha_B} = D$ lie in finitely many conjugacy classes in $GL_n(\mathfrak{o}_1)$. This now follows without more ado from Lemma 10.

This finishes the Proof of Theorem 6 for diagonalizable subgroups in $\mathscr{G} = GL_n$. The transition to the case where $\mathscr{G}$ is an arbitrary $\mathbb{Q}$-group depends on one further theorem:

Theorem 9 If $\mathscr{G}$ is a $\mathbb{Q}$-group and k is an algebraic number field, then the maximal k-split tori of $\mathscr{G}$ are conjugate under the action of $\mathscr{G}(k)$.

A 'k-split torus' is an abelian subgroup of $\mathscr{G}$ which is diagonalizable over k, and both k-closed and connected in the Zariski topology. To prove Theorem 6, we consider a family $\mathscr{C}$ of diagonalizable abelian subgroups of $\mathscr{G}(\mathbb{Z})$. By what we have already done, we may suppose that the groups in $\mathscr{C}$ are all conjugate in $GL_n(\mathbb{Z})$, and diagonalizable over some algebraic number field k. Replacing each group in $\mathscr{C}$ by a subgroup of finite index, we can then assume that each $B \in \mathscr{C}$ is contained in some maximal k-split torus T_B in G. If A and B are in $\mathscr{C}$, Theorem 9 then tells us that $T_A = T_B^x$ for some $x \in \mathscr{G}(k)$, so A and B^x are both contained in T_A. Keeping A fixed, we deduce from Lemma 7 that there are at most $n!$ possibilities for B^x as B runs through $\mathscr{C}$. Thus we may as well assume that the groups in $\mathscr{C}$ are all conjugate in $\mathscr{G}(k)$ as well as in $GL_n(\mathbb{Z})$. Together with the original hypothesis that $\mathscr{C}$ is contained in a single $\approx_{\mathscr{G}}$-class, this provides enough of a handle with which to apply Borel's Theorem (Theorem 3 in section A); only one has to use a more general version of that result, concerning the 'adele group over k' of the $\mathbb{Q}$-group $\mathscr{G}$.

For the details of this argument, the reader is referred to the original paper (see section E).

D. The general case

Before discussing, in outline, how this is done, I want to mention some limits to further generalisations of Theorem 6. First of all, one cannot deduce in Theorem 6 that the given $\approx_{\mathscr{G}}$-class of subgroups consists of a *single* conjugacy class in $\mathscr{G}(\mathbb{Z})$. Examples demonstrating this will be described in Chapter 11. Secondly, the restriction to soluble-by-finite subgroups is not superfluous, as the following exercise shows:

Exercise 5 There exists an infinite family $\mathscr{C}$ of pairwise non-isomorphic finitely generated free subgroups of $GL_2(\mathbb{Z})$, whose images in $GL_2(\mathbb{Z}/m\mathbb{Z})$ are all equal for every positive integer m.

Corollary 1   For every $n \geq 2$, there exists a $\approx_{GL_n}$-class of finitely generated subgroups of $GL_n(\mathbb{Z})$ which contains infinitely many pairwise non-conjugate subgroups.

(*Hint:* For the Corollary, embed $GL_2(\mathbb{Z})$ in $GL_n(\mathbb{Z})$. For the Exercise, let F be a free subgroup of rank 2 in $GL_2(\mathbb{Z})$, e.g.

$$F = \left\langle \begin{pmatrix} 1 & 2 \\ 0 & 1 \end{pmatrix}, \begin{pmatrix} 1 & 0 \\ 2 & 1 \end{pmatrix} \right\rangle.$$

For each r, let θ_r be an epimorphism of F onto the alternating group A_r, and put $F_r = \ker \theta_r$. By Schreier's Theorem (see Magnus, Karrass & Solitar: *Combinatorial Group Theory*, section (2.4)), F_r is free of rank $1 + \frac{1}{2}r!$ By considering the ranks of elementary abelian subgroups, show that A_r is not isomorphic to a section of $GL_2(\mathbb{Z}/p\mathbb{Z})$, for any prime p, provided r is large enough. Deduce that $F_r \pi_m = F \pi_m$ for all $m \in \mathbb{N}$, provided r is large (note that the kernel of the natural map $GL_2(\mathbb{Z}/m\mathbb{Z}) \to \prod_{p|m} GL_2(\mathbb{Z}/p\mathbb{Z})$ is nilpotent, while A_r is simple for $r \geq 5$).)

We turn now to Theorem 6. The first step is the

Reduction to the abelian case. Let $\mathscr{G}$ be a $\mathbb{Q}$-group and $\mathscr{C}$ a $\approx_{\mathscr{G}}$-class of soluble-by-finite subgroups of $\mathscr{G}(\mathbb{Z})$. Assuming Theorem 6 proved for the case where $\mathscr{C}$ consists of abelian groups, we deduce that $\mathscr{C}$ consists of finitely many conjugacy classes in $\mathscr{G}(\mathbb{Z})$.

Step 1   By Lemma 4, we may assume that $\mathscr{C}$ consists of triangularizable groups.

Step 2   We may assume that the derived groups A' for $A \in \mathscr{C}$ are all equal. For after Step 1, we have each A' unipotent, and clearly whenever $A \approx_{\mathscr{G}} B$ then $A' \approx_{\mathscr{G}} B'$; so by the result of section B, the groups A', $A \in \mathscr{C}$, lie in finitely many conjugacy classes in $\mathscr{G}(\mathbb{Z})$. Dividing $\mathscr{C}$ into finitely many subsets and replacing each group in $\mathscr{C}$ by a suitable conjugate in $\mathscr{G}(\mathbb{Z})$, we achieve the required reduction.

Step 3   We now have a unipotent subgroup U of $\mathscr{G}(\mathbb{Z})$ with $A' = U$ for every $A \in \mathscr{C}$. Let $\bar{U}$ be the closure of U in $\mathscr{G}$ and put $\mathscr{H} = N_{\mathscr{G}}(\bar{U})$. By Corollary 3 in section A of Chapter 8, there is a $\mathbb{Q}$-group $\mathscr{K}$ and a surjective $\mathbb{Q}$-rational homomorphism $\varphi : \mathscr{H} \to \mathscr{K}$ with $\ker \varphi = \bar{U}$ and $\mathscr{H}(\mathbb{Z})\varphi \leq \mathscr{K}(\mathbb{Z})$. It has to be checked that the groups $A\varphi$, $A \in \mathscr{C}$, lie in a single $\approx_{\mathscr{K}}$-class in $\mathscr{K}(\mathbb{Z})$; then as $A\varphi \cong A/(A \cap \bar{U})$ is abelian for each $A \in \mathscr{C}$, we deduce by the 'abelian case' of Theorem 6 that the groups $A\varphi$ lie in finitely many conjugacy classes in $\mathscr{K}(\mathbb{Z})$. By Theorem 1 of Chapter 8, they therefore lie in finitely many conjugacy classes in $\mathscr{H}(\mathbb{Z})\varphi$. Hence the groups $A \cdot \bar{U}(\mathbb{Z})$ lie in finitely many conjugacy classes in $\mathscr{H}(\mathbb{Z})$, and our conclusion now follows easily from the fact that $|\bar{U}(\mathbb{Z}) : U|$ is finite. This completes the reduction to the abelian case.

Henceforth, we assume that $\mathscr{C}$ consists of abelian subgroups of $\mathscr{G}(\mathbb{Z})$. For $A \in \mathscr{C}$, put

$$A_u = \{a_u \mid a \in A\}$$
$$A_s = \{a_s \mid a \in A\}$$
$$A_N = A \cap A_u$$

(see section A of Chapter 7). Thus for $A \in \mathscr{C}$ we have

$$A_N \leq A \leq A_u \times A_s \leq \mathscr{G}(\mathbb{Q}).$$

Replacing each $A \in \mathscr{C}$ by a suitable subgroup of finite index, one reduces to the case where

$$A_u \times A_s \leq \mathscr{G}(\mathbb{Z})$$

for each $A \in \mathscr{C}$. The next major step is to show that as A runs through $\mathscr{C}$, the groups A_u lie in a single $\approx_{\mathscr{g}}$-class, as do the groups A_s and the groups A_N. In fact we need something a little stronger, namely

Lemma 11 Let A, B be abelian subgroups of $GL_n(\mathbb{Z})$ with A_u, A_s, B_u and B_s all contained in $GL_n(\mathbb{Z})$. Suppose $\hat{g} \in GL_n^\infty$ satisfies

$$A^{\hat{g}} \pi_m = B \pi_m \quad \forall m \in \mathbb{N}.$$

Then

$$A_u^{\hat{g}} \pi_m = B_u \pi_m, \ A_s^{\hat{g}} \pi_m = B_s \pi_m \text{ and } A_N^{\hat{g}} \pi_m = B_N \pi_m$$

for every $m \in \mathbb{N}$.

I shall prove this lemma in section C of the next chapter, where a general framework for the relevant kind of argument will be ready to hand.

Using the third claim of Lemma 11, we argue just as in the 'reduction to the abelian case' to reduce to the situation where $A_N = 1$ for each $A \in \mathscr{C}$ (also needed is Proposition 4 of Chapter 8). So we may assume henceforth that $\mathscr{C}$ consists of *unipotent-free* groups.

Lemma 11 shows that the groups A_s, $A \in \mathscr{C}$, lie in a single $\approx_{\mathscr{g}}$-class. By the 'diagonalizable case' of Theorem 6, we may therefore suppose that there exists a diagonalizable subgroup D of $\mathscr{G}(\mathbb{Z})$ such that

$$A_s = D \quad \forall A \in \mathscr{C}. \tag{18}$$

Now put

$$\Xi = C_{GL_n(\mathbb{C})}(D), \ \mathscr{G}_1 = N_{\mathscr{g}}(\Xi).$$

One can prove, using Lemma 11, that $\mathscr{C}$ is contained in a single $\approx_{\mathscr{g}_1}$-class of $\mathscr{G}_1(\mathbb{Z})$. Having done this, we deduce from Lemma 11 and the 'unipotent case' of Theorem 6 that the groups A_u, $A \in \mathscr{C}$, lie in finitely many conjugacy classes in $\mathscr{G}_1(\mathbb{Z})$. Putting

$$\mathscr{G}_2 = \mathscr{G} \cap \Xi$$

we have

$$|\mathcal{G}_1(\mathbb{Z}):\mathcal{G}_2(\mathbb{Z})| < \infty,$$

essentially by Lemma 7, in section C above. Therefore the groups A_u for $A \in \mathcal{C}$ lie in finitely many conjugacy classes in $\mathcal{G}_2(\mathbb{Z})$. Dividing $\mathcal{C}$ into finitely many subsets and conjugating each group in $\mathcal{C}$ by a suitable element of $\mathcal{G}_2(\mathbb{Z})$, we may then assume that there exists a unipotent subgroup U of $\mathcal{G}(\mathbb{Z})$ such that

$$A_u = U \quad \forall A \in \mathcal{C};$$

the point of introducing $\mathcal{G}_2$ into the picture is that we may effect this reduction without violating (18).

Thus far, we have reduced the proof of Theorem 6 to the following very special situation: where $\mathcal{C}$ is a family of subgroups of $U \times D \le \mathcal{G}(\mathbb{Z})$, with

$$A_u = U, A_s = D, A_N = 1 \quad \forall A \in \mathcal{C}.$$

In particular, the projection of each A onto D is an isomorphism. It would seem that not much room for manoeuvre remains; after all, $U \times D$ is a finitely generated abelian group, and the subgroups A of $U \times D$ satisfying the condition described correspond in a 1:1 manner with homomorphisms of D onto U. However, a rather subtle argument is still required to complete the final stage of the proof, and I shall not give it here; the reader might like to think about it before looking up the original paper.

E. Notes

Suitable references for p-adic numbers are Serre (1973), Chapter 1; Borevich & Shafarevich (1966), Chapter 1; Cassells (1978), Chapter 3. The general reference for Galois cohomology of algebraic groups is Serre (1964). Our Theorem 2 is Théorème 7.1 of Borel & Serre (1964), and our Proposition 2 is essentially Corollaire 7.12 of that paper.

The whole discussion in section A really belongs to the 'theory of descent'; an interesting introducing to this theory, in a very general setting, will be found in Part V of Waterhouse (1979). See also Serre (1979) Chapter 10.

The theorem on conjugacy classes of finite subgroups in an arithmetic group is from Borel (1962).

For a general discussion of 'adele groups', with detailed examples, see Chapter V of Humphreys (1980). Chapter III of these notes also explains how the Dirichlet Units Theorem and the finiteness of the class number of an algebraic number field can be viewed as properties of the algebraic group GL_1.

Theorem 5 in section A is a special case of Borel & Serre (1964), Proposition 3.8.

The 'double cosets theorem', Theorem 3 in section A, is from Borel (1963).

The main result of the chapter, Theorem 6, is from Grunewald & Segal (1982). A special case appeared earlier, in Grunewald & Segal (1978*b*); this paper is perhaps worth looking at as an introduction to some of the main ideas in a rather more explicit form.

Proposition 4 appeared in Grunewald & Segal (1978*c*), and Theorem 8 in Grunewald & Segal (1979*a*).

Theorem 7 is not hard to deduce from Čebotarev's Theorem; or see Janusz (1973), Chapter IV Exercise 5 (stated there for $k = \mathbb{Q}$, but equally valid in general).

Theorem 9, on the conjugacy of maximal k-split tori, is Theorem 15.9 in Chapter V of Borel (1969*b*).

10
Polycyclic groups with isomorphic finite quotients

Let G be a polycyclic-by-finite group. We saw in Chapter 4 (and in Chapter 1) that certain properties of G are determined by the finite quotient groups of G. Perhaps the most obvious question in the same circle of ideas is 'do the finite quotients of G determine G up to isomorphism?' If the answer were 'yes', one could rather easily (in principle) solve the isomorphism problem for polycyclic groups, by the usual 'yes procedure' and 'no procedure' method.

However, the answer is 'no'; counterexamples are not hard to come by, if you know where to look, and several different kinds are described in the next chapter. Instead, we shall devote this chapter to proving a *finiteness theorem*. Let $\mathcal{F}(G)$ denote the set of isomorphism types of finite quotients of G; then *the polycyclic-by-finite groups H such that $\mathcal{F}(H) = \mathcal{F}(G)$ lie in finitely many isomorphism classes.*

When G is a $\mathfrak{T}$-group, the result is now an easy consequence of work done in earlier chapters; this case is proved in section A. To go further, we need the formalism of *profinite completions*: the profinite completion $\hat{G}$ of G is a group which is built out of the finite quotients of G, just as the p-adic integers are constructed from the family of finite rings $\mathbb{Z}/p^n\mathbb{Z}$. A sketchy and utilitarian account of profinite completions is given in section B, where we show, in particular, that for polycyclic-by-finite groups G and H, $\mathcal{F}(G) = \mathcal{F}(H)$ if and only if $\hat{G} \cong \hat{H}$.

As important step in the proof of the main theorem is established in section C; we show there that

$$\mathrm{Fitt}\,(\hat{G}) = (\mathrm{Fitt}\,(G))\hat{}\,.$$

Thus the finite quotients of G determine the finite quotients of Fitt (G) up to isomorphism, a striking result which already justifies the method of profinite completions (I know of no other way to prove it).

The proof is completed in sections D and E. The main result on which the theorem depends is the finiteness theorem of Chapter 9; but results from all the previous chapters are needed in the course of the argument, and in particular the theory of semi-simple splitting of Chapter 7 plays a major role.

A. Pickel's first theorem

This is

Theorem 1 Let G be a $\mathfrak{X}$-group. Then the $\mathfrak{X}$-groups H such that $\mathscr{F}(H) = \mathscr{F}(G)$ lie in finitely many isomorphism classes.

To prove it, recall from Chapter 5 that we have a canonical embedding

$$\beta_G : G \to \mathrm{Tr}_1(n, \mathbb{Z})$$

for a certain $n = n(G)$. If H is another $\mathfrak{X}$-group, then $H \cong G$ if and only if $n(H) = n(G)$ and $H\beta_H$ is conjugate to $G\beta_G$ in $GL_n(\mathbb{Z})$. Now we saw in Chapter 9, section B, that every $\approx$ -class of unipotent subgroups of $GL_n(\mathbb{Z})$ breaks up into finitely many conjugacy classes (where $\approx$ stands for $\approx_{GL_n}$). Hence to establish Theorem 1 it will suffice to prove

Lemma 1 Let G and H be $\mathfrak{X}$-groups with $\mathscr{F}(H) = \mathscr{F}(G)$. Then $n(H) = n(G) = n$, say, and $H\beta_H \approx G\beta_G$ in $GL_n(\mathbb{Z})$.

Proof Let c be the nilpotency class of G, $\mathfrak{g}$ the augmentation ideal of $\mathbb{Z}G$, and

$$I_G/\mathfrak{g}^{c+1} = \tau(\mathbb{Z}G/\mathfrak{g}^{c+1}).$$

The representation β_G gives the action of G, by right multiplication, on the factor ring $\mathbb{Z}G/I_G$ of $\mathbb{Z}G$. Write

$$V(G) = \mathbb{Z}G/I_G$$

and note that $V(G) \cong \mathbb{Z}^{n(G)}$. What we have to do is to compare the G-module $V(G)$ with H-module $V(H)$. Note first of all that H has the same nilpotency class as G, namely c; for since G is residually finite, c is just the maximum of the classes of the finite quotients of G.

Now let m be a positive integer. The next step is to describe $V(G)/mV(G)$ in terms of a suitable finite quotient group of G. Choose a positive integer q such that $qI_G \subseteq \mathfrak{g}^{c+1}$. Since $\mathbb{Z}G/\mathfrak{g}^{c+1}$ is finitely generated as a $\mathbb{Z}$-module, $\mathbb{Z}G/(qm\mathbb{Z}G + \mathfrak{g}^{c+1})$ is a finite ring, so the kernel of the natural homomorphism of G into the unit group of this ring has finite index in G. Thus G^k is contained in this kernel for some positive integer k, which means that

$$G^k - 1 \subseteq qm\mathbb{Z}G + \mathfrak{g}^{c+1}. \tag{1}$$

Write $\bar{G} = G/G^k$ and denote the obvious ring epimorphism by

$$^- : \mathbb{Z}G \to \mathbb{Z}\bar{G}.$$

The kernel of this epimorphism is $(G^k - 1)\mathbb{Z}G$, so because of (1), the map $-$ induces an isomorphism

$$\mathbb{Z}G/(qm\mathbb{Z}G + \mathfrak{g}^{c+1}) \xrightarrow{\sim} \mathbb{Z}\bar{G}/(qm\mathbb{Z}\bar{G} + \bar{\mathfrak{g}}^{c+1}),$$

under which the image of $q\mathbb{Z}G + \mathfrak{g}^{c+1}$ is mapped onto the image of $q\mathbb{Z}\bar{G} + \bar{\mathfrak{g}}^{c+1}$. The choice of q ensures that

$$V(G)/mV(G) \cong (q\mathbb{Z}G + \mathfrak{g}^{c+1})/(qm\mathbb{Z}G + \mathfrak{g}^{c+1})$$

(*Exercise*: verify this!). We may therefore conclude that

$$V(G)/mV(G) \cong (q\mathbb{Z}\bar{G} + \bar{\mathfrak{g}}^{c+1})/(qm\mathbb{Z}\bar{G} + \bar{\mathfrak{g}}^{c+1}). \tag{2}$$

Now choose q and k so that we also have $qI_H \subseteq \mathfrak{h}^{c+1}$ and $H^k - 1 \subseteq qm\mathbb{Z}H + \mathfrak{h}^{c+1}$. Since $\mathcal{F}(G) = \mathcal{F}(H)$, we have $H/H^k \cong G/G^k$, because G/G^k is the biggest finite quotient of G whose exponent divides k, and similarly for H/H^k. Let

$$\theta_m : \bar{G} = G/G^k \to H/H^k = \bar{H}$$

be an isomorphism. Then θ_m induces an isomorphism

$$(q\mathbb{Z}\bar{G} + \bar{\mathfrak{g}}^{c+1})/(qm\mathbb{Z}\bar{G} + \bar{\mathfrak{g}}^{c+1}) \stackrel{\sim}{\to} (q\mathbb{Z}\bar{H} + \mathfrak{h}^{c+1})/(qm\mathbb{Z}\bar{H} + \bar{\mathfrak{h}}^{c+1}),$$

which together with (2) yields an isomorphism

$$\varphi_m : V(G)/mV(G) \stackrel{\sim}{\to} V(H)/mV(H). \tag{3}$$

The first conclusion to be drawn from this is that $n(H) = n(G)$. For $V(G)/mV(G)$ has order $m^{n(G)}$ and $V(H)/mV(H)$ has order $m^{n(H)}$; provided $m > 1$ this forces $n(H) = n(G)$. So we may put $n = n(G) = n(H)$, and choose $\mathbb{Z}$-bases in $V(G)$ and $V(H)$, giving isomorphisms

$$V(G) \stackrel{\lambda}{\to} \mathbb{Z}^n \stackrel{\mu}{\leftarrow} V(H),$$

and defining the maps $\beta_G : G \to GL_n(\mathbb{Z})$ and $\beta_H : H \to GL_n(\mathbb{Z})$. We keep λ and μ fixed (independent of m) for the rest of the argument.

Now let m be an arbitrary positive integer. Then λ and μ induce isomorphisms λ_m and μ_m:

$$
\begin{array}{ccc}
V(G)/mV(G) & \stackrel{\phi_m}{\longrightarrow} & V(H)/mV(H) \\
\lambda_m \downarrow & & \downarrow \mu_m \\
\mathbb{Z}^n/m\mathbb{Z}^n & \stackrel{\gamma_m}{\dashrightarrow} & \mathbb{Z}^n/m\mathbb{Z}^n
\end{array}
$$

Define $\gamma_m \in GL_n(\mathbb{Z}/m\mathbb{Z})$ by

$$\gamma_m = \lambda_m^{-1} \varphi_m \mu_m,$$

where φ_m is the map given at (3) above. We complete the proof of Lemma 1 by showing that

$$((G\beta_G)\pi_m)^{\gamma_m} = (H\beta_H)\pi_m, \tag{4}$$

where $\pi_m : GL_n(\mathbb{Z}) \to GL_n(\mathbb{Z}/m\mathbb{Z})$ is the natural map.

To prove (4), let $g \in G$ and $h \in H$ satisfy $\bar{g}\theta_m = \bar{h}$, and write $\cdot g$ (respectively $\cdot h$) for the action of g (respectively h) on $V(G)/mV(G)$ (respectively $V(H)/mV(H)$).

Now consider the following diagram:

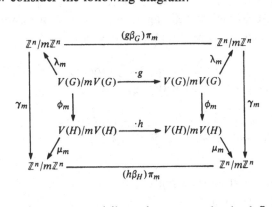

The top and bottom quadrilaterals commute by the definition of β_G and β_H. The two side quadrilaterals commute by definition of γ_m. The central square commutes because of the way φ_m was defined: the reader should check this for himself/herself. It follows that

$$(h\beta_H)\pi_m = \gamma_m^{-1} \cdot (g\beta_G)\pi_m \cdot \gamma_m.$$

This gives (4), since θ_m was an isomorphism of $\bar{G}$ onto $\bar{H}$.

Thus Lemma 1 is proved, and with it also Theorem 1.

Let us extend Theorem 1 a little. If G is a finite extension of a finitely generated nilpotent group, then G^k is a $\mathfrak{T}$-group for some positive integer k, and G/G^k is finite. Suppose H is another such group, with $\mathscr{F}(H) = \mathscr{F}(G)$. Then $H/H^k \cong G/G^k$; if we knew that $H^k \cong G^k$, we could deduce from Theorem 6 of Chapter 8 that there are only finitely many possibilities for H up to isomorphism.

Exercise 1 Let G and H be polycyclic-by-finite groups with $\mathscr{F}(G) = \mathscr{F}(H)$, and let k be a positive integer. Show that $\mathscr{F}(G^k) = \mathscr{F}(H^k)$. Show also that H is nilpotent if and only if G is nilpotent.

Exercise 2 Show that Theorem 1 holds if we replace '$\mathfrak{T}$-group' by 'torsion-free finitely generated nilpotent-by-finite group'.

It is not hard, but it is not trivial either, to get rid of the 'torsion-free' hypothesis here. The reader might like to try this before reading the next section, where we introduce the proper machinery.

B. Profinite completions

The language of profinite completions is a convenient formalism for discussing the finite quotients of a group; it generalises the method whereby families of congruences over $\mathbb{Z}$ can be replaced by equations over $\mathbb{Z}_p$ for various primes p. Before defining profinite completions, we must discuss a bit of general nonsense about families of sets.

Definition   A *directed set* is a partially ordered set Λ such that for every λ and $\mu \in \Lambda$ there exists $\nu \in \Lambda$ with $\nu \geq \lambda$ and $\nu \geq \mu$.

Definition   An *inverse system* is a family of sets $(A_\lambda)_{\lambda \in \Lambda}$, where Λ is a directed set, and a family of maps $(\varphi_{\lambda\mu} : A_\lambda \to A_\mu | \lambda, \mu \in \Lambda, \lambda \geq \mu)$, such that

$$\varphi_{\lambda\lambda} = \mathrm{Id}_{A_\lambda} \quad \forall \lambda$$

$$\varphi_{\lambda\mu}\varphi_{\mu\nu} = \varphi_{\lambda\nu} \quad \text{whenever } \lambda \geq \mu \geq \nu.$$

The whole system is denoted

$$(A, \varphi; \Lambda).$$

Definition   Let $(A, \varphi; \Lambda)$ be an inverse system. The *inverse limit* of $(A, \varphi; \Lambda)$ is the set

$$\varprojlim_{\lambda \in \Lambda} A_\lambda = \{(a_\lambda) \in \prod_{\lambda \in \Lambda} A_\lambda \, | \, a_\lambda \varphi_{\lambda\mu} = a_\mu \text{ whenever } \lambda \geq \mu\}.$$

Thus $\varprojlim A_\lambda$ is a subset of the Cartesian product of all the sets A_λ, consisting of all the 'vectors' whose components 'match up' properly according to the maps $\varphi_{\lambda\mu}$. For each $\mu \in \Lambda$ let $\varphi_\mu : \varprojlim A_\lambda \to A_\mu$ be the 'co-ordinate projection' sending (a_λ) to a_μ.

Exercise 3   Put $\hat{A} = \varprojlim A_\lambda$. Whenever $\nu \geq \mu$ in Λ, the diagram

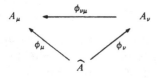

commutes, Suppose B is a set and $(\psi_\lambda)_{\lambda \in \Lambda}$ a family of maps, with $\psi_\lambda : B \to A_\lambda$ for each λ, such that the diagram

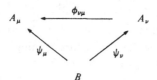

commutes whenever $v \geq \mu$. Show that there is a unique map $\beta : B \to \hat{A}$ such that $\psi_\lambda = \beta \varphi_\lambda$ for all $\lambda \in \Lambda$. (This *universal property* characterises $\varprojlim A_\lambda$ up to canonical bijection; the reader who is unfamiliar with such ideas should work out what I mean by this, and prove it.)

Let Λ be a directed set and Λ' a *cofinal* subset: i.e. for each $\lambda \in \Lambda$ there exists $\lambda' \in \Lambda'$ with $\lambda' \geq \lambda$. Then Λ' is also directed, and if $(A, \varphi; \Lambda)$ is an inverse system then there is an inverse system $(A, \varphi; \Lambda')$, obtained by leaving out the sets A_λ with $\lambda \in \Lambda \backslash \Lambda'$. It is easy to see that $\hat{A}' = \varprojlim_{\lambda \in \Lambda'} A_\lambda$ is just the image of $\hat{A} = \varprojlim_{\lambda \in \Lambda} A_\lambda$ under the natural projection map of $\prod_{\lambda \in \Lambda} A_\lambda$ onto $\prod_{\lambda \in \Lambda'} A_\lambda$ (delete all components A_λ for $\lambda \in \Lambda \backslash \Lambda'$). As Λ' is cofinal in Λ, the resulting map from $\hat{A}$ onto $\hat{A}'$ is bijective; indeed its inverse is given by the last exercise, for we may define $\psi_\lambda : \hat{A}' \to A_\lambda$ by

$$\psi_\lambda = \varphi'_v \varphi_{v\lambda}, \quad \text{some } v \geq \lambda \text{ with } v \in \Lambda',$$

where $\varphi'_v : \hat{A}' \to A_v$ is the usual co-ordinate projection. The upshot is that we may, whenever convenient, identify $\varprojlim_{\lambda \in \Lambda'} A_\lambda$ with $\varprojlim_{\lambda \in \Lambda} A_\lambda$, provided Λ' is a cofinal subset of Λ.

If the sets A_λ have some additional structure and the maps $\varphi_{\lambda\mu}$ preserve that structure, then $\varprojlim A_\lambda$ inherits the structure, and the maps φ_λ preserve it. Two cases interest us here.

(i) Suppose each A_λ is a group and each $\varphi_{\lambda\mu}$ is a homomorphism. Then $\prod A_\lambda$ is a group, and it is easy to see that $\varprojlim A_\lambda$ is a subgroup. The co-ordinate projections φ_λ are obviously homomorphisms in this case. Also the universal property, described in Exercise 3, holds within the category of groups: thus if B is a group and the maps $\psi_\lambda : B \to A_\lambda$ are homomorphisms, then the unique map $\beta : B \to \varprojlim A_\lambda$ given in the exercise is also a homomorphism. These statements are all easy to verify and this is left to the reader.

(ii) Suppose each A_λ is a Hausdorff topological space and each map $\varphi_{\lambda\mu}$ is continuous. Then $\prod A_\lambda$ is a Hausdorff space, in the product topology, and giving $\varprojlim A_\lambda$ the subspace topology we find that the maps φ_λ are all continuous.

Lemma 2 $\varprojlim A_\lambda = \hat{A}$, say, is closed in $\prod A_\lambda$.

Proof Suppose $\hat{a} = (a_\lambda) \in \prod A_\lambda \backslash \hat{A}$. Then for some $\lambda \geq \mu$ we have $a_\lambda \varphi_{\lambda\mu} \neq a_\mu$. Since A_μ is a Hausdorff space, A_μ has open subsets U and V with $a_\lambda \varphi_{\lambda\mu} \in U$, $a_\mu \in V$, and $U \cap V = \emptyset$. For $v \neq \lambda, \mu$, put $T_v = A_v$; put $T_\lambda =$

$U\varphi_{\lambda\mu}^{-1} \subseteq A_\lambda$ and $T_\mu = V$. The set $\prod T_\nu$ is then open in $\prod A_\nu$, it contains $\hat{a}$, and its intersection with $\hat{A}$ is empty. This shows that the complement of $\hat{A}$ in $\prod A_\lambda$ is open and so proves the lemma.

An important consequence of this lemma, or rather of its proof, is

Proposition 1   Suppose $(A, \varphi; \Lambda)$ is an inverse system of non-empty finite sets. Then $\varprojlim A_\lambda$ is non-empty.

Proof   Give each set A_λ the discrete topology. Then each A_λ is compact and Hausdorff. By Tychonoff's theorem (see J.L. Kelley's *Topology*, Chapter 5), the product space $\prod A_\lambda$ is compact. Now suppose $\varprojlim A_\lambda$ is empty. Then the open sets $\prod T_\nu$, of the form described in the above proof, cover $\prod A_\nu$; by compactness, a finite collection of such sets covers $\prod A_\nu$. Thus we have finitely many pairs (λ_i, μ_i), with $\lambda_i \ge \mu_i$, and subsets $U_i, V_i \subseteq A_{\mu_i}$, with $U_i \cap V_i = \varnothing$, such that

$$\prod A_\nu = \bigcup_{i=1}^{n} \prod_\nu T_\nu(i) \tag{5}$$

where $T_\nu(i) = A_\nu$ for $\nu \ne \lambda_i, \mu_i$; $T_{\lambda_i}(i) = U_i\varphi_{\lambda_i\mu_i}^{-1}$, and $T_{\mu_i}(i) = V_i$. Since Λ is directed, there is an element $\sigma \in \Lambda$ with $\sigma \ge \lambda_i$ for $i = 1,\dots,n$. Since A_σ is non-empty, we can choose an element $a_\sigma \in A_\sigma$. Now define

$$a_\nu = a_\sigma\varphi_{\sigma\nu} \quad \text{for} \quad \nu \le \sigma$$
$$a_\nu \in A_\nu \text{ arbitrary} \quad \text{for} \quad \nu \not\le \sigma.$$

It is then easy to see that the element $(a_\nu) \in \prod A_\nu$ belongs to none of the sets $\prod_\nu T_\nu(i)$, and we have a contradiction to (5).

Let us return to group theory. Let G be a group and $\mathscr{S}$ a family of normal subgroups of finite index in G, cofinal with the set of all normal subgroups of finite index (w.r.t. *reverse inclusion*: i.e. for each $N \lhd_f G$ there exists $K \in \mathscr{S}$ with $K \le N$). For $M, N \in \mathscr{S}$ with $M \le N$, let $\pi_{MN}: G/M \to G/N$ be the natural epimorphism. The *profinite completion* of G is the group

$$\hat{G} = \varprojlim_{N \in \mathscr{S}} G/N,$$

the inverse limit of the system $(G/N, \pi; \mathscr{S})$. In view of our previous discussion, it makes no material difference with family one takes for $\mathscr{S}$. In particular, assuming now that G is polycyclic-by-finite, the subgroups

$$G^{m!} = \langle x^{m!} | x \in G \rangle, \quad m \in \mathbb{N}$$

form a suitable cofinal family. This family has the useful feature of being *linearly ordered*, which gives another way of looking at $\hat{G}$.

Definition A sequence $(g_n)_{n \in \mathbb{N}}$ in G is a *Cauchy sequence* if

$$G^{n!} g_m = G^{n!} g_n \text{ whenever } m \geq n.$$

Two Cauchy sequences (g_n), (h_n) in G are *equivalent* if

$$G^{n!} g_n = G^{n!} h_n \quad \forall n \in \mathbb{N}.$$

A *null sequence* is a Cauchy sequence equivalent to the constant sequence (1).

It is easy to see that the Cauchy sequences in G form a group (with componentwise multiplication), and that the map

$$(g_n) \to (G^{n!} g_n)$$

is an epimorphism of this group onto $\hat{G} = \varprojlim G/G^{n!}$, whose kernel consists of the null sequences.

Let $\pi_m : G \to G/G^{m!}$ be the natural map, for each m. By Exercise 3, there is a unique homomorphism $\iota : G \to \hat{G}$ such that $\pi_m = \iota \pi'_m$ for each m, where $\pi'_m : \hat{G} \to G/G^{m!}$ is the co-ordinate projection. Evidently $g\iota$ is just the element of $\hat{G}$ represented by the constant Cauchy sequence (g). Now $g\iota = 1$ if and only if (g) is a null sequence, i.e. if and only if $g \in G^{n!}$ for every n. Assuming still that G is polycyclic-by-finite we know that G is residually finite, so our element $g \in \ker \iota$ must be the identity. Thus ι embeds G into $\hat{G}$; we shall usually identify G with its image in $\hat{G}$, and so think of G as a subgroup of $\hat{G}$.

We give each finite quotient group of G the discrete topology, and $\hat{G}$ the subspace topology induced by the product topology on $\prod G/G^{m!}$. The main properties of $\hat{G}$ can then be summarised as follows.

Proposition 2 Let G be a polycyclic-by-finite group, and consider G as a subgroup of $\hat{G}$.

(i) $\hat{G}$ is a compact Hausdorff topological group.

(ii) Let $\pi_m : \hat{G} \to G/G^{m!}$ denote the co-ordinate projection. Then the subgroups $\ker \pi_m$, $m \in \mathbb{N}$, form a base for the neighbourhoods of 1 in $\hat{G}$.

(iii) Let S be a subset of $\hat{G}$. Then the closure of S consists of all $x \in \hat{G}$ such that $x\pi_m \in S\pi_m$ for every $m \in \mathbb{N}$.

(iv) G is dense in $\hat{G}$.

Proof (i) By Lemma 2, $\hat{G}$ is a closed subset of $\prod G/G^{m!}$. This latter is a compact Hausdorff space by Tychonoff's theorem, and a topological group because each $G/G^{m!}$ is a (discrete) topological group. The claim follows because a closed subset of a compact Hausdorff space is compact.

(ii) follows from the definition of the product topology,

(iii) follows from (ii), and (iv) follows from (iii).

Proposition 3   Let $\theta:H \to G$ be a homomorphism of polycyclic-by-finite groups. Then there is a unique continuous homomorphism $\hat\theta:\hat H \to \hat G$ extending θ. Moreover.

(i) $\hat H\hat\theta$ is the closure of $H\theta$ in $\hat G$;

(ii) The assignment $\theta \mapsto \hat\theta$ is functorial, i.e. $\widehat{\mathrm{Id}}_H = \mathrm{Id}_{\hat H}$, and given homomorphisms $K \xrightarrow{\varphi} H \xrightarrow{\theta} G$, we get $(\varphi\theta)\hat{} = \hat\varphi\hat\theta:\hat K \to \hat G$;

(iii) If θ is injective (respectively, surjective), then so is $\hat\theta$.

Proof   For each n, let θ_n be the map of $H/H^{n!} \to G/G^{n!}$ induced by θ, and compose θ_n with the usual map $\hat H \to H/H^{n!}$ to give $\bar\theta_n:\hat H \to G/G^{n!}$. These maps evidently fit together so that we may apply Exercise 3, and obtain a map $\hat\theta:\hat H \to \hat G$ such that $\bar\theta_n = \hat\theta\pi_n$ for each n, where $\pi_n:\hat G \to G/G^{n!}$ is as usual. The reader may easily verify that $\hat\theta$ extends θ and is a continuous homomorphism; it is unique with these properties because H is dense in $\hat H$.

(i) Since $\hat H$ is compact and $\hat\theta$ is continuous, $\hat H\hat\theta$ is a compact subset of $\hat G$. Since $\hat G$ is a Hausdorff space, $\hat H\hat\theta$ is closed in $\hat G$. So it will suffice to show that $\hat H\hat\theta$ is contained in the closure of $H\theta$. This follows now from part (iii) of Proposition 2, since for each n we have

$$\hat H\hat\theta\pi_n = \hat H\bar\theta_n = (H/H^{n!})\theta_n = (H\theta)\pi_n.$$

(ii) $(\varphi\theta)\hat{}$ and $\hat\varphi\hat\theta$ both extend the map $\varphi\theta:K \to G$. By the uniqueness property, the two maps must be equal.

(iii) Suppose θ is injective. Let $x\in\hat H$ be represented by the Cauchy sequence (h_n) in H. For each n, there exists $m \geq n$ such that $G^{m!} \cap H\theta \leq (H\theta)^{n!} = H^{n!}\theta$ (Chapter 1, Exercise 11). If $x\hat\theta = 1$ then $h_m\theta\in G^{m!} \cap H\theta$ for each m, so given n we can find $m \geq n$ such that $h_m\theta\in H^{n!}\theta$; since θ is injective $h_m\in H^{n!}$, and since $h_n \equiv h_m$ modulo $H^{n!}$, it follows that (h_n) is a null sequence. Thus $x = 1$, and so $\hat\theta$ is injective.

Suppose θ is surjective. Let $y\in\hat G$. For each $n\in\mathbb N$ put

$$S_n = \{x\in H/H^{n!}|x\theta_n = y\pi_n\}.$$

Each S_n is a finite set, and is non-empty since θ_n is surjective. If $m \geq n$, the natural map $\pi_{mn}:H/H^{m!} \to H/H^{n!}$ sends S_m into S_n, and if $t \geq m \geq n$ then $\pi_{tn} = \pi_{tm}\pi_{mn}$. So we have an inverse system $(S, \pi; \mathbb N)$ of non-empty finite sets. By Proposition 1, the inverse limit is not empty. Let $\hat x = (x_n)_{n\in\mathbb N}$ belong to this inverse limit. Then $\hat x\in\hat H$, and

$$\hat x\hat\theta\pi_n = \hat x\bar\theta_n = x_n\theta_n = y\pi_n$$

for each n. Therefore $\hat x\hat\theta = y$, and we have shown that $\hat\theta$ is surjective.

Further basic properties of our construction are given in the following exercise.

Exercise 4 Let G be a polycyclic-by-finite group. For each n, put

$$\hat{G}(n) = \ker \pi_n$$

where $\pi_n : \hat{G} \to G/G^{n!}$ is as usual. (i) $\hat{G}(n) \cdot G = \hat{G}$ for every $n \in \mathbb{N}$. (ii) If G is finite then $\hat{G} = G$. (iii) If $H \leq G$, then $\hat{H}$ may be identified with the closure of H in $\hat{G}$. (iv) Let $N \lhd G$ and $\pi : G \to G/N$ the natural map. Then $\hat{\pi} : \hat{G} \to (G/N)\hat{\;}$ is surjective and $\ker \hat{\pi} = \hat{N}$. (v) If $N \lhd G$ and $C \leq G$ then $(NC)\hat{\;} = \hat{N}\hat{C}$ (considered as subgroups of $\hat{G}$). If also G/N is finite then $(NC)\hat{\;} = \hat{N}C$. (vi) If $H \leq_f G$ and $C \leq G$ then $(H \cap C)\hat{\;} = \hat{H} \cap \hat{C}$. (vii) Every automorphism of G extends uniquely to a continuous automorphism of $\hat{G}$.

(*Hint*: For (iii), apply Proposition 3 to the inclusion $H \to G$. For (iv), show that $\ker \hat{\pi}$ is equal to the closure of N in $\hat{G}$. For (v), assume w.l.o.g. that $G = NC$, and then use (iv). For (vi), let (c_n), (h_n) be Cauchy sequences in G with $c_n \in C$, $h_n \in H$ for all n, and suppose these sequences are equivalent. Show that $c_n \in C \cap H$ for all sufficiently large n.)

Corollary 1 The functor $\hat{\;}$ on the category of polycyclic-by-finite groups is exact.

Proof The claim is that if $K \xrightarrow{\varphi} H \xrightarrow{\theta} G$ are homomorphisms with $\ker \theta = K\varphi$, then $\ker \hat{\theta} = \hat{K}\hat{\varphi}$. Equivalently, by Proposition 3(i) and Exercise 4 (iii), we can state this as

$$\ker \hat{\theta} = (\ker \theta)\hat{\;}. \tag{6}$$

To prove (6), put $N = \ker \theta$ and factor θ as

$$H \xrightarrow{\pi} H/N \xrightarrow{\psi} G,$$

with ψ injective. Then $\hat{\psi}$ is injective (Proposition 3 (iii)) and $\hat{\theta} = \hat{\pi}\hat{\psi}$, so $\ker \hat{\theta} = \ker \hat{\pi}$ which is just $\hat{N}$ by (iv) of Exercise 4. Thus (6) holds.

The next thing we must do is to compare the natural topology on $\hat{G}$ with its own profinite topology. For arbitrary groups G the situation is unclear, but when the groups are polycyclic-by-finite the result is as simple as it could be. The main step is

Lemma 3 Let G be a polycyclic-by-finite group and k a positive integer. Then

$$\widehat{G^k} = \hat{G}^k.$$

Proof Let (g_n) be a Cauchy sequence in G representing an element $x \in \hat{G}$. Then (g_n^k) is a Cauchy sequence and it represents x^k, so $x^k \in \widehat{G^k}$ (which we think of as the closure of G^k in $\hat{G}$). Since the elements x^k generate $\hat{G}^k$, it follows that $\hat{G}^k \leq \widehat{G^k}$.

For the reverse inclusion, we argue by induction on the Hirsch number of G. If G is finite there is nothing to prove. Suppose G is free abelian. Then the map $g \mapsto g^k$ is an endomorphism of G; call it θ. Now $\hat{G}$ is abelian so the map $x \mapsto x^k$ is an endomorphism of $\hat{G}$, it is certainly continuous, and it extends θ. It must therefore be the map $\hat{\theta}$. It follows that

$$\hat{G}^k = \hat{G}\hat{\theta} = (G\theta)\hat{} = \widehat{G^k}.$$

Now consider the general case where G is infinite. Then G has a free abelian normal subgroup $K \neq 1$. There exists $l \in \mathbb{N}$ such that $G^l \cap K \leq K^k$, and we choose l to be a multiple of k. By inductive hypothesis we may suppose that

$$((G/K)\hat{})^l = ((G/K)^l)\hat{},$$

and this implies that

$$\hat{G}^l \hat{K} = (G^l K)\hat{} = \widehat{G^l}K$$

(by Exercise 4, (iv) and (v)). So (by Exercise 4(vi) and the first paragraph, above)

$$\widehat{G^l} = \widehat{G^l} \cap \hat{G}^l \hat{K} = \hat{G}^l(\widehat{G^l} \cap \hat{K}) = \hat{G}^l(G^l \cap K)\hat{} \leq \hat{G}^l \hat{K}^k.$$

Since K is free abelian, $\hat{K}^k = \hat{K}^k \leq \hat{G}^k$. Therefore

$$\widehat{G^l} \leq \hat{G}^l \hat{G}^k = \hat{G}^k.$$

But $\hat{G} = \widehat{G^l}G$, so $\hat{G} = \hat{G}^k G$; consequently

$$|\hat{G}:\hat{G}^k| = |G:G \cap \hat{G}^k| \leq |G:G^k| = |\hat{G}:\widehat{G^k}|,$$

since $\hat{G}/\widehat{G^k} \cong (G/G^k)\hat{}$. As $\hat{G}^k \leq \widehat{G^k}$, it follows that $\hat{G}^k = \widehat{G^k}$.

Corollary 2 The natural topology on $\hat{G}$ is the same as the profinite topology.

For the subgroups $\widehat{G^k}$ form a base for the neighbourhoods of 1 w.r.t. the natural topology; while the subgroups $\hat{G}^k$ form a base for the neighbourhoods of 1 w.r.t. the profinite topology, since they each have finite index in $\hat{G}$ by Lemma 3.

Corollary 3 If H and G are polycyclic-by-finite groups, then every homomorphism $\hat{H} \to \hat{G}$ is continuous, and has a closed image.

Proof Every homomorphism is continuous w.r.t. the profinite topologies, so the first claim follows from Corollary 2. The second claim then follows, because a continuous image of a compact space is compact, and a compact subset of a Hausdorff space is closed.

We can now, belatedly, justify all this machinery by proving

Proposition 4 For polycyclic-by-finite groups G and H, $\mathcal{F}(G) = \mathcal{F}(H)$ if and only if $\hat{G} \cong \hat{H}$.

Proof Suppose $\hat{G} \cong \hat{H}$. Then, by Lemma 3,
$$H/H^k \cong \hat{H}/\hat{H}^k \cong \hat{G}/\hat{G}^k \cong G/G^k$$
for each positive integer k. Therefore
$$\mathcal{F}(G) = \bigcup_{k \in \mathbb{N}} \mathcal{F}(G/G^k) = \bigcup_{k \in \mathbb{N}} \mathcal{F}(H/H^k) = \mathcal{F}(H).$$

Conversely, suppose $\mathcal{F}(G) = \mathcal{F}(H)$. For each n, let S_n be the set of all isomorphisms of $G/G^{n!}$ onto $H/H^{n!}$. Each S_n is finite and (by hypothesis) non-empty. There is an obvious map $\sigma_{mn} : S_m \to S_n$ whenever $m \geq n$; if $\alpha \in S_m$, $\alpha^{\sigma_{mn}} \in S_n$ makes the following square commute:

$$
\begin{array}{ccc}
G/G^{m!} & \xrightarrow{\ \ \alpha\ \ } & H/H^{m!} \\
\pi_{mn} \downarrow & & \downarrow \pi_{mn} \\
G/G^{n!} & \xrightarrow{\ \alpha^{\sigma_{mn}}\ } & H/H^{n!}
\end{array}
\qquad (7)
$$

Thus we have an inverse system $(S, \sigma ; \mathbb{N})$; its inverse limit is non-empty by Proposition 1. Let $(\alpha_n) \in \varprojlim S_n$. Then α_n is an isomorphism $G/G^{n!} \to H/H^{n!}$, for each n, and for $m \geq n$ we have $\alpha_m \pi_{mn} = \pi_{mn} \alpha_n$ (where the two maps π_{mn} are as in (7)). This shows that the isomorphism $(\alpha_n) : \prod_n G/G^{n!} \to \prod_n H/H^{n!}$ maps the subgroup $\hat{G}$ into the subgroup $\hat{H}$. Similarly, $(\alpha_n^{-1}) = (\alpha_n)^{-1}$ maps $\hat{H}$ into $\hat{G}$, whence we conclude that $\hat{H} \cong \hat{G}$.

We saw in Chapter 4 that in a soluble-by-finite subgroup of $GL_n(\mathbb{Z})$, the profinite topology is the same as the congruence topology. This provides yet another description for the profinite completion. Recall that

$$GL_n^\infty = \prod_{\text{primes } p} GL_n(\mathbb{Z}_p).$$

Define the *congruence topology* on GL_n^∞ to be that induced by the profinite topology on the additive group $\prod_p M_n(\mathbb{Z}_p)$ (which is the same as the product topology coming from the p-adic topology on each $M_n(\mathbb{Z}_p)$: for if $m \in \mathbb{N}$ then $mM_n(\mathbb{Z}_p) = M_n(\mathbb{Z}_p)$ for every $p \nmid m$, and $mM_n(\mathbb{Z}_p) = p^e M_n(\mathbb{Z}_p)$ if p^e exactly divides m). A base for the neighbourhoods of 1 w.r.t. the congruence topology in GL_n^∞ is given by the family of subgroups

$$K_m^\infty = GL_n^\infty \cap (1 + mM_n(\mathbb{Z}_p)), \ m \in \mathbb{N}.$$

Note that if $m = \prod p_i^{e_i}$ (with each $e_i \geq 1$) then

$$K_m^\infty = \prod_{p \nmid m} GL_n(\mathbb{Z}_p) \times \prod_i (1 + p_i^{e_i} M_n(\mathbb{Z}_{p_i})),$$

so K_n^∞ has finite index in GL_n^∞. If we consider $GL_n(\mathbb{Z})$ embedded in GL_n^∞ as usual (i.e. 'diagonally'), we also have

$$K_m^\infty \cap GL_n(\mathbb{Z}) = (1 + mM_n(\mathbb{Z})) \cap GL_n(\mathbb{Z}),$$

so the new congruence topology does extend the one originally defined in Chapter 4.

Proposition 5   Let $G \leq GL_n(\mathbb{Z})$ be polycyclic-by-finite. Then $\hat{G}$ may be identified with the closure of G, w.r.t. the congruence topology, in GL_n^∞.

Proof   By Theorem 5 of Chapter 4, the groups $K_{m!}^\infty \cap G$ form a cofinal family of normal subgroups of finite index in G. We may therefore identify $\hat{G}$ with the group of all Cauchy sequences in G, w.r.t. this family, modulo null sequences. Such a Cauchy sequence (q_m) converges p-adically for each prime p; so (q_m) converges in the congruence topology to some $\hat{g} \in GL_n^\infty$, and certainly $\hat{g} \in \bar{G}$, the congruence closure of G in GL_n^∞. Moreover, $\hat{g} = 1$ if and only if (q_m) is a null sequence; so $\hat{G}$ appears as a subgroup of $\bar{G}$.

Suppose $x \in \bar{G}$. Then for each $m \in \mathbb{N}$ there exists $g_m \in G$ such that $x \equiv q_m \bmod K_{m!}^\infty$. The sequence (g_m) is then a Cauchy sequence (w.r.t. to the family $(K_{m!}^\infty \cap G)$) in G, and its limit is x; so $x \in \hat{G}$. Thus $\hat{G}$ is the whole of $\bar{G}$ as claimed.

Note that

$$\hat{\mathbb{Z}} = \prod_p \mathbb{Z}_p;$$

for a Cauchy sequence in $\mathbb{Z}$ will converge p-adically for each prime p; while a family of p-adic integers (one for each prime p) can be simultaneously approximated by a Cauchy sequence in $\mathbb{Z}$, because of the Chinese Remainder Theorem. It follows that if $A = \mathbb{Z}^n$ then $\hat{A} = \prod_p \mathbb{Z}_p^n$.

Lemma 4   If $A = \mathbb{Z}^n$ then $\text{Aut } \hat{A} = GL_n^\infty$.

Proof   Certainly $GL_n^\infty \leq \text{Aut } \hat{A}$. Suppose $\alpha \in \text{Aut } \hat{A}$. Then α induces an automorphism $\alpha(p^m)$ on $\hat{A}/p^m\hat{A} \cong A/p^mA$, for each prime-power p^m, and for a fixed p, the $\alpha(p^m)$ together induce an automorphism α_p on $\varprojlim_m A/p^mA = \mathbb{Z}_p^n$. Thus $\alpha_p \in GL_n(\mathbb{Z}_p)$ for each p, and it is clear that the element

$$(\alpha_p) \in \prod_p GL_n(\mathbb{Z}_p) = GL_n^\infty$$

is equal to α. Thus GL_n^∞ is the whole of $\text{Aut } \hat{A}$.

From Proposition 5 we can deduce

Corollary 4 Let $A = \mathbb{Z}^n$ and let G be a polycyclic-by-finite group acting on A. Then $\hat{G}$ acts on $\hat{A}$ and

$$C_G(\hat{A}) = (C_G(A))\hat{}.$$

Proof Put $K = C_G(A)$. We have an injective homomorphism $\theta: G/K \to \text{Aut } A = GL_n(\mathbb{Z})$. Then $\hat{\theta}: (G/K)\hat{} \to ((G/K)\theta)\hat{}$ is injective, by Proposition 3 (iii). By Proposition 5 we may identify $((G/K)\theta)\hat{}$ with a subgroup of $GL_n^\infty = \text{Aut } \hat{A}$; this gives the action of $\hat{G}$ on $\hat{A}$, via

$$\hat{G} \to \hat{G}/\hat{K} \cong (G/K)\hat{} \xrightarrow{\hat{\theta}} \text{Aut } \hat{A},$$

and shows that the kernel of this action is exactly $\hat{K}$.

Remark The material of this section up to Proposition 4 has been quite formal; Proposition 5, on the other hand, invoked results from Chapter 4, and lies much deeper. In fact, the innocuous-looking corollary above is yet another equivalent formulation of Theorem 4 of Chapter 4:

Exercise 5 Assuming Corollary 4, deduce Theorem 4 of Chapter 4 (Chevalley's theorem about algebraic integer units).
(*Hint*: We saw in Chapter 4 that the said theorem follows from the fact that every abelian subgroup of $GL_n(\mathbb{Z})$ is closed in the congruence topology. Let G be such a subgroup, with congruence closure $\bar{G}$, and let $x \in \bar{G}$. Show that $H = \langle G, x \rangle$ is abelian, and that there exists $\hat{g} \in \hat{G}$ such that $x^{-1}\hat{g} \in C_H(\hat{A})$, where $A = \mathbb{Z}^n$. Deduce that $x \in H \cap \bar{G} = G$.)

Quite generally, if a polycyclic-by-finite group G acts on another such group A, we can extend the action to make $\hat{G}$ act on $\hat{A}$, in a natural way. The simplest way to describe this is to consider the semi-direct product $E = A]G$. Then $\hat{A}$ is a normal subgroup and $\hat{G}$ is a subgroup of the group $\hat{E}$ and the action of $\hat{G}$ on $\hat{A}$ is simply conjugation in $\hat{E}$.

Exercise 6 Let H be a polycyclic-by-finite group with three normal subgroups $Q < P$ and $K \leq C_H(P/Q)$. Put $A = P/Q$ and $G = H/K$, and let G act on A via conjugation in H. Verify that the action of $\hat{G}$ on $\hat{A}$ described above is the same as the action coming from conjugation within the group $\hat{H}$, when we identify $\hat{G}$ with $\hat{H}/\hat{K}$ and $\hat{A}$ with $\hat{P}/\hat{Q}$.
(*Hint*: though messy, this is essentially trivial,. For each $m \in \mathbb{N}$ take $n \in \mathbb{N}$ such that $G^{n!}$ centralizes $A/A^{m!}$, and consider the action of $G/G^{n!}$ on $A/A^{m!}$ (it will be the same whether we fit these two finite groups together into some

finite quotient of $A]G$, or into some finite quotient of H).)

Finally, let us extend Lemma 4 to $\mathfrak{X}$-groups. Recall from Chapter 5 that a $\mathfrak{X}$-group N has a canonical embedding β_N into some $\mathrm{Tr}_1(n, \mathbb{Z})$. Call N a *lattice $\mathfrak{X}$-group* if $N\beta_N$ is a lattice group (see section B of Chapter 6). We saw in Chapter 6, section C, that there is a $\mathbb{Q}$-group $\mathscr{G}$, which we shall denote $\mathscr{A}ut(N)$, such that

$$\mathscr{G}(\mathbb{Q}) = \mathrm{Aut}\,\mathscr{L}(N\beta_N)$$

where $\mathscr{L}(N\beta_N)$ is the Lie algebra $\mathbb{Q}\log N\beta_N$. When N is a lattice $\mathfrak{X}$-group, $\mathscr{G}(\mathbb{Z})$ may be identified with $\mathrm{Aut}\,N$.

Proposition 6   Let N be a lattice $\mathfrak{X}$-group and put $\mathscr{G} = \mathscr{A}ut\,N$. The identification of $\mathscr{G}(\mathbb{Z})$ with $\mathrm{Aut}\,N$ can be extended to an identification of $\mathscr{G}^\infty = \prod_p \mathscr{G}(\mathbb{Z}_p)$ with $\mathrm{Aut}\,\hat{N}$.

Proof   As we did for $\mathbb{Z}^n$ in Lemma 4, we must first decompose $\hat{N}$ into its 'Sylow subgroups'. Briefly, the idea is as follows. For a prime p, put

$$\mathscr{S}_p = \{K \lhd N \,|\, N/K \text{ is a finite } p\text{-group}\},$$

and define the 'pro-p completion' of N to be

$$\hat{N}_p = \varprojlim_{K \in \mathscr{S}_p} N/K.$$

Most of the results discussed above about $\hat{G}$ have their analogues for $\hat{N}_p$, with the same proofs (in particular, since $\mathfrak{X}$-groups are always residually finite p groups, we have $N \leq \hat{N}_p$). Since N is nilpotent, each finite quotient of N is a direct product of p groups, for distinct primes p. So if $K \lhd_f N$ and $|N/K| = p_1^{e_1}\ldots p_r^{e_r}$, then

$$K = P_1 \cap \ldots \cap P_r$$

with $P_i \lhd N$ and $|N/P_i| = p_i^{e_i}$ for each i. The natural map

$$N/K \to N/P_1 \times \ldots \times N/P_r$$

will be an isomorphism (it is certainly injective, and the two groups have the same order). As K runs over all normal subgroups of finite index in N, the corresponding isomorphisms fit together to give an isomorphism

$$\hat{N} = \varprojlim_{K \lhd_f N} N/K \to \prod_p \varprojlim_{p \in \mathscr{S}_p} N/P = \prod_p \hat{N}_p.$$

The subgroups N^{p^e}, $e \in \mathbb{N}$, are cofinal with the family $\mathscr{S}_p$; an automorphism α of $\hat{N}$ induces automorphisms on the N/N^{p^e}, and hence induces an automorphism, α_p say, on $\hat{N}_p$. Thus the isomorphism above enables us to identify $\mathrm{Aut}\,\hat{N}$ with $\prod_p \mathrm{Aut}\,\hat{N}_p$, making $\alpha \in \mathrm{Aut}\,N$ correspond to (α_p).

To prove the proposition, we shall show that, making suitable identifications, we have

$$\text{Aut } \hat{N}_p = \mathscr{G}(\mathbb{Z}_p).$$

Put $G = N\beta_N \le \text{Tr}_1(n, \mathbb{Z})$. Thus $L = \log G$ is a lattice in the Lie algebra $\mathscr{L}(G) = \mathbb{Q} \log(N\beta_N)$, and $\mathscr{G} = \mathscr{A}ut\ N$ is the automorphism group of the Lie algebra $\mathscr{L}(G) \otimes \mathbb{C}$. In Chapter 6, we made $\mathscr{G}(\mathbb{Z})$ act on N by

$$x^\gamma \beta_N = \exp((\log x\beta_N)^\gamma), \ x \in N, \ \gamma \in \mathscr{G}(\mathbb{Z}).$$

We must extend this to an action of $\mathscr{G}(\mathbb{Z}_p)$ on $\hat{N}_p$. To do this, we need

(a) β_N extends to an isomorphism of $\hat{N}_p$ onto $\hat{G}_p$;
(b) $\hat{G}_p = \exp(L\mathbb{Z}_p)$;
(c) the mapping

$$\gamma \mapsto \log \circ \gamma \circ \exp = \dot{\gamma}, \text{ say,}$$

gives a bijection of $\mathscr{G}(\mathbb{Z}_p)$ onto $\text{Aut } \hat{G}_p$.

Of course, (a) is clear. For (b), note that the congruence topology on $\text{Tr}_1(n, \mathbb{Z}_p)$ is the same as the 'pro-p' topology (this is very easy, and does not depend on the results of Chapter 4). So just as in Proposition 5, we may identify $\hat{G}_p$ with the congruence closure of G in $\text{Tr}_1(n, \mathbb{Z}_p)$. Since exp and log are given by polynomials, it follows that $\log \hat{G}_p$ is just the congruence closure of $L = \log G$ in the module $(n-1)!^{-1}\text{Tr}_0(n, \mathbb{Z}_p)$; and this closure is exactly $L\mathbb{Z}_p$. This establishes (b).

To prove (c), note first that if $\gamma \in \mathscr{G}(\mathbb{Z}_p)$ then $\dot{\gamma}$ is a bijection of $\hat{G}_p$ onto itself, and $\dot{\gamma}$ is a group endomorphism because of the Baker–Campbell–Hausdorff formula (Theorem 1 of Chapter 6). Thus '·' maps $\mathscr{G}(\mathbb{Z}_p)$ into $\text{Aut } \hat{G}_p$. Conversely, let $\alpha \in \text{Aut } \hat{G}_p$. Then $\alpha^\dagger = \exp \circ \alpha \circ \log$ is a bijection of $L\mathbb{Z}_p$ onto itself, and in fact it is a homeomorphism w.r.t. the p-adic topology ($= $ the congruence topology), because α is a homeomorphism of $\hat{G}_p$ w.r.t. its own topology, and exp and log are homeomorphisms. Since $\alpha^\dagger$ certainly respects the operation of $\mathbb{Z}$, by scalar multiplication, on $L\mathbb{Z}_p$, and $\mathbb{Z}$ is dense in $\mathbb{Z}_p$, it follows that $\alpha^\dagger$ respects the action of $\mathbb{Z}_p$ by scalar multiplication on $L\mathbb{Z}_p$. One can now show, exactly as in the proof of Proposition 4 of Chapter 6, section C, that $\alpha^\dagger$ respects addition and the Lie bracket operation in $L\mathbb{Z}_p$ (i.e., that the extension of $\alpha^\dagger$ to a map of $L\mathbb{Q}_p$ onto $L\mathbb{Q}_p$ is a Lie algebra automorphism). Thus $\alpha^\dagger \in \mathscr{G}(\mathbb{Z}_p)$, and so $\alpha = \dot{\gamma}$ for some $\gamma \in \mathscr{G}(\mathbb{Z}_p)$. This establishes (c) and finishes the proof.

Just as Lemma 4 yielded Corollary 4, Proposition 6 gives

Corollary 5 If a polycyclic-by-finite group G acts on a lattice $\mathfrak{X}$-group N, then $C_{\hat{G}}(\hat{N}) = (C_G(N))\hat{\ }$.

Exercise 7   Show that this holds even if N is not a lattice group.

To conclude this section, here are some remarks which will be needed in section D. Using Proposition 6, we can define a Jordan decomposition for automorphisms of $\hat{N}$, when N is a lattice $\mathfrak{X}$-group, as we did in Chapter 7 for Aut N. If $x = (x(p)) \in GL_h^\infty$, define

$$x_s = (x(p)_s) \in \prod_p GL_h(\mathbb{Q}_p),$$

$$x_u = (x(p)_u) \in \prod_p GL_h(\mathbb{Q}_p).$$

If $x \in$ Aut $\hat{N}$ we identify Aut $\hat{N}$ with $\mathscr{G}^\infty$, where $\mathscr{G} = $ Aut N, and obtain

$$x_s \in \prod_p \mathscr{G}(\mathbb{Q}_p), \quad x_u \in \prod_p \mathscr{G}(\mathbb{Q}_p);$$

if it so happens that $x(p)_s \in \mathscr{G}(\mathbb{Z}_p)$ for every prime p, then $x_s \in \mathscr{G}^\infty$, $x_u = xx_s^{-1} \in \mathscr{G}^\infty$, and we may consider x_u and x_s as automorphisms of $\hat{N}$.

A little caution must be exercised in the use of this notation. The matrix group $\mathscr{G} = \mathscr{A}ut(N)$ is uniquely determined, up to conjugacy in $GL_h(\mathbb{Z})$, by the $\mathfrak{X}$-group N. This is clear from the construction of $\mathscr{G}$, and it shows that for $x \in$ Aut N, x_s and x_u are uniquely determined as automorphisms of $N^\mathbb{Q}$ (the radicable hull of N), because the Jordan decomposition of a matrix is preserved under conjugation. To show that the Jordan decomposition in Aut $\hat{N}$ is uniquely defined (when it exists), we similarly have to check that the subgroup $\mathscr{G}^\infty$ of GL_h^∞ is uniquely determined, up to conjugacy in GL_h^∞, by the group $\hat{N}$. To do this, suppose $\hat{N} = \hat{N}_1$ where N and N_1 are two lattice $\mathfrak{X}$-groups. Put $L = \log N\beta_N$ and $L_1 = \log N_1\beta_{N_1}$. Now we may identify $\hat{N}_p$ with $(\hat{N}_1)_p$, for each prime p (by identifying corresponding finite p-images of N and N_1); as in (c) in the proof of Proposition 6, we then obtain an isomorphism θ_p, say, of $L\mathbb{Z}_p$ onto $L_1\mathbb{Z}_p$, for each p. It follows that $L \cong L_1 \cong \mathbb{Z}^h$, say. Let $\alpha_p \in GL_h(\mathbb{Z}_p)$ be the matrix of θ_p w.r.t. the chosen bases of L and L_1. Then $\alpha = (\alpha_p) \in GL_h^\infty$ conjugates $\mathscr{G}^\infty$ to $\mathscr{G}_1^\infty$, where $\mathscr{G} = \mathscr{A}ut\ N$ and $\mathscr{G}_1 = \mathscr{A}ut\ N_1$. This is what we had to show, and it ensures that the notation x_s, x_u for $x \in$ Aut $\hat{N}$ is unambiguous.

C. Pickel's second theorem

If we want Theorem 1 to help us towards the main result, the obvious next step is to see how knowledge of $\mathscr{F}(G)$, for a polycyclic-by-finite group G, can be used to determine $\mathscr{F}(N)$ where $N = $ Fitt(G). Here the ideas of section B really come into their own; we shall prove

Theorem 2   Let G be polycyclic-by-finite with Fitting subgroup N. Then $\operatorname{Fitt}(\hat{G}) = \hat{N}$.

As usual, we here identify $\hat{N}$ with the closure of N in $\hat{G}$.

Suppose to begin with that G/N is finite. Then $\hat{G} = \hat{N}G$, $\hat{N} \lhd \hat{G}$ and $\hat{N}$ is nilpotent. So if $F = \operatorname{Fitt}(\hat{G})$ then

$$F = \hat{N}(F \cap G).$$

But $F \cap G$ is evidently contained in N, and we deduce

$$F = \hat{N}(F \cap G) \le \hat{N}N = \hat{N} \le F.$$

Thus Theorem 2 is established in this special case.

The general case is not so simple, and depends ultimately on the results of Chapter 4. We start with some reduction steps. Let K/N be a free abelian normal subgroup of finite index in G/N, and put $I = \tau_2(N)$, so that $I/N' = \tau(N/N')$.

Lemma 5 K/N acts faithfully as a unipotent-free group of automorphisms on N/I.

Proof   Suppose $x \in K$ acts nilpotently on N/I. Since I/N' is finite, x^m centralises I/N' for some $m \ne 0$, and then x^m acts nilpotently on N/N'. By Proposition 14 of Chapter 1, x^m acts nilpotently on N. Hence $N\langle x^m \rangle$ is a nilpotent group. But $N\langle x^m \rangle \lhd K$ since K/N is abelian, so

$$N\langle x^m \rangle \le \operatorname{Fitt}(K) \le \operatorname{Fitt}(G) = N.$$

Thus $x^m \in N$, and since K/N is free abelian we have $x \in N$. The lemma follows.

The heart of the proof lies in

Proposition 7   Let $A \lhd B$ be groups, with A and B/A free abelian of finite rank. If the action of B/A on A faithful and unipotent-free, then so is the action of $\hat{B}/\hat{A}$ on $\hat{A}$.

This will be proved later. Together with Lemma 5, it shows that

$$\text{if } x \in \hat{K} \text{ and } x \text{ acts nilpotently on } \hat{N}/\hat{I} \text{ then } x \in \hat{N}. \tag{8}$$

We can now quickly finish the

Proof of Theorem 2   Put $H = C_G(N/I)$. By Lemma 5, $H \cap K = N$, so H/N is finite. Since $H \lhd G$, $\operatorname{Fitt}(H) = N$, so by the first case we dealt with we have

$$\hat{N} = \operatorname{Fitt}(\hat{H}).$$

Now put $F = \text{Fitt}(\hat{G})$. Then $\hat{N} \lhd F$, and each element of F acts nilpotently on $\hat{N}/\hat{I}$ (by Fitting's Theorem, see Chapter 1). Hence by (8), $F \cap \hat{K} = \hat{N}$. Therefore $F/\hat{N}$ is finite. Since a group which acts faithfully and nilpotently on a torsion-free abelian group is necessarily torsion-free (e.g. by Corollary 6 of Chapter 1), it follows that F centralises $\hat{N}/\hat{I}$. Corollary 4 in section B shows that $C_{\hat{G}}(\hat{N}/\hat{I}) = \hat{H}$, so we have $F \leq \hat{H}$. Consequently

$$F = \text{Fitt}(\hat{H}) = \hat{N}$$

as claimed.

We still have to prove Proposition 7. This will be a consequence of a general result, to which we alluded in section D of Chapter 9, and which will also be needed later.

Proposition 8   Let X be a nilpotent subgroup of $GL_n(\mathbb{Z})$ such that X' is unipotent and $X_s \leq GL_n(\mathbb{Z})$. Write $^-$ to denote closure w.r.t. the congruence topology in GL_n^∞. Then

$$(\bar{X})_s = (X_s)^-, \quad (\bar{X})_u = (X_u)^-, \quad \text{and} \quad (\bar{X})_N = (X_N)^-.$$

As usual, x_s, x_u denote the Jordan components of a matrix x (see Chapter 7, section A). For $x = (x(p)) \in \prod_p GL_n(\mathbb{Q}_p)$, x_s is by definition just $(x(p)_s)$, and $x_u = (x(p)_u)$. For any set Y, $Y_s = \{y_s | y \in Y\}$ and $Y_u = \{y_u | y \in Y\}$. Also

$$Y_N = Y \cap Y_u = \{y \in Y | y \text{ is unipotent}\}.$$

Note that for $x \in GL_n^\infty$, it is not necessarily the case that $x_s, x_u \in GL_n^\infty$; for $x \in X$, this is a *consequence* of the proposition.

Proof   Define $\theta: X \to X_u$ and $\varphi: X \to X_s$ by $x\theta = x_u$, $x\varphi = x_s$. Since X is nilpotent, these maps are homomorphisms, and they extend to homomorphisms $\hat{\theta}: \hat{X} \to \hat{X}_u$ and $\hat{\varphi}: \hat{X} \to \hat{X}_s$. By Proposition 5 in section B, we may identify $\hat{X}$ with $\bar{X}$, $\hat{X}_N$ with $\overline{X_N}$, $\hat{X}_u$ with $\overline{X_u}$ and $\hat{X}_s$ with $\overline{X_s}$; thus we obtain $\bar{\theta}: \bar{X} \to \overline{X_u}$ and $\bar{\varphi}: \bar{X} \to \overline{X_s}$, extending θ and φ respectively. Since θ and φ are surjective, $\bar{\theta}$ and $\bar{\varphi}$ are also surjective, and since $\ker \varphi = X_N$ we have $\ker \bar{\varphi} = \overline{X_N}$ (see Proposition 3 and Corollary 1 in section B).

Now every element of $\overline{X_u}$ is unipotent: if $\hat{u} \in GL_n^\infty$ is the limit of a sequence of unipotent matrices $u(i) \in M_n(\mathbb{Z})$, then $(u(i) - 1)^n = 0$ for each i implies $(\hat{u} - 1)^n = 0$ in $\prod_p M_n(\mathbb{Z}_p)$. Also every element of X_s is diagonalizable; for since X' is unipotent, X_s is a finitely generated abelian group, and so X_s^α is diagonal for some $\alpha \in GL_n(k)$, k a suitable algebraic number field. If $\hat{s}$ is the limit of a sequence $(s(i))$ in X_s, then each $s(i)^\alpha$ is a diagonal matrix, and it follows that s^{α} is diagonal (that is, for each prime p the matrix s_p^{α} is diagonal, where $s = (s_p) \in \prod_p GL_n(\mathbb{Z}_p)$; to see this, observe that each off-diagonal entry in the matrix s_p^{α} is congruent to zero modulo p^m for all positive integers m).

Let $\hat{x} \in \bar{X}$ be the limit of a sequence $x(i)$ in X. For each i we have

$$x(i) = x(i)\theta \cdot x(i)\varphi = x(i)\varphi \cdot x(i)\theta,$$

consequently

$$\hat{x} = \hat{x}\bar{\theta} \cdot \hat{x}\bar{\varphi} = \hat{x}\bar{\varphi} \cdot \hat{x}\bar{\theta}. \tag{9}$$

Since $\hat{x}\bar{\theta} \in \overline{X_u}$ and $\hat{x}\bar{\varphi} \in \overline{X_s}$, the preceding paragraph shows that $\hat{x}\bar{\theta} = \hat{x}_u$ and $\hat{x}\bar{\varphi} = \hat{x}_s$ (consider each p-component separately). Thus

$$(\bar{X})_u \subseteq \bar{X}\bar{\theta} = \overline{X_u}, \quad (\bar{X})_s \subseteq \bar{X}\bar{\varphi} = \overline{X_s}.$$

For the reverse inclusion, take $\hat{u} \in \overline{X_u}$. Since $\bar{\theta}$ is surjective, $\hat{u} = \hat{x}\bar{\theta}$ for some $\hat{x} \in \bar{X}$. Putting $\hat{s} = \hat{x}\bar{\varphi}$ we have (from (9))

$$\hat{x} = \hat{u}\hat{s} = \hat{s}\hat{u},$$

whence (as before) $\hat{u} = \hat{x}_u$. Thus $\overline{X_u} \subseteq (\bar{X})_u$, and an identical argument shows that $\overline{X_s} \subseteq (\bar{X})_s$.

The first two equalities in Proposition 8 are now established. Now the argument of the last paragraph shows that for $\hat{x} \in \hat{X}$, $\hat{x}_s = \hat{x}\bar{\varphi}$. Therefore $(\bar{X})_N = \ker \bar{\varphi}$. But we saw above that $\ker \bar{\varphi} = \overline{X_N}$, so $(\bar{X})_N = \overline{X_N}$ and the proof is complete.

Proposition 7 now follows easily. We have $A \triangleleft B$ with A and B/A both free abelian of finite rank; B/A acts faithfully and unipotent-freely on A. Say $A \cong \mathbb{Z}^n$, and let Q be the image of B/A in $GL_n(\mathbb{Z})$, via the action on A. Corollary 4 in section B shows that $\hat{Q}$ acts faithfully on $\hat{A}$. Now let X be a subgroup of finite index in Q such that $X_u X_s \leq GL_n(\mathbb{Z})$. By hypothesis, $X_N = 1$. Hence, by Proposition 8, $(\bar{X})_N = 1$. Therefore $(\bar{Q})_N \cap \bar{X} = 1$. But $\bar{Q}/\bar{X} \cong \hat{Q}/\hat{X} \cong Q/X$, so $\bar{Q}/\bar{X}$ is finite and consequently $(\bar{Q})_N$ is finite. Since unipotent matrices in characteristic zero have infinite order, it follows that $(\bar{Q})_N = 1$. Thus no element $\neq 1$ of $\hat{Q}$ acts unipotently on $\hat{A}$, which is what we had to prove.

To conclude this section, we now prove the following lemma, which was postponed from section D of Chapter 9:

Lemma 6 Let A and B be abelian subgroups of $GL_n(\mathbb{Z})$, with A_u, A_s, B_u and B_s all contained in $GL_n(\mathbb{Z})$. Suppose $\hat{g} \in GL_n^{\infty}$ satisfies $A^{\theta}\pi_m = B\pi_m$ for all $m \in \mathbb{N}$. Then

$$A_u^{\theta}\pi_m = B_u\pi_m, \quad A_s^{\theta}\pi_m = B_s\pi_m, \quad A_N^{\theta}\pi_m = B_N\pi_m$$

for every $m \in \mathbb{N}$.

Proof   Let $a \in A$. Then for each m, $a^{\hat{g}}\pi_m \in B\pi_m$, so $a^{\hat{g}} \in \bar{B}$. Hence, by Proposition 8, $(a^{\hat{g}})_s \in (\bar{B})_s = \bar{B}_s$ and $(a^{\hat{g}})_u \in (\bar{B})_u = \bar{B}_u$. Considering each p-component separately, it is easy to see that $(a^{\hat{g}})_u = (a_u)^{\hat{g}}$ and $(a^{\hat{g}})_s = (a_s)^{\hat{g}}$. Putting these facts together we conclude that for each m,

$$A_u^{\hat{g}}\pi_m \subseteq \bar{B}_u\pi_m = B_u\pi_m, \quad A_s^{\hat{g}}\pi_m \subseteq \bar{B}_s\pi_m = B_s\pi_m.$$

Arguing similarly with $\hat{g}^{-1}$ in place of $\hat{g}$, we obtain the reverse inclusions, and this establishes the first two claims of the lemma.

Finally, if $a \in A_N$ then $a^{\hat{g}}$ is unipotent, so $a^{\hat{g}} \in (\bar{B})_N$; thus, by Proposition 8,

$$(A_N)^{\hat{g}} \subseteq (\bar{B})_N = \bar{B}_N,$$

and so $(A_N)^{\hat{g}}\pi_m \subseteq \bar{B}_N\pi_m = B_N\pi_m$ for each m. The reverse inclusion is obtained by arguing with $\hat{g}^{-1}$ in place of $\hat{g}$, and this completes the proof. (Note that for $X, Y \subseteq GL_n^{\infty}$, we have

$$X^{\hat{g}}\pi_m \supseteq Y\pi_m \Leftrightarrow X\pi_m^{g_m} \supseteq Y\pi_m \Leftrightarrow X\pi_m \supseteq Y\pi_m^{g_m^{-1}} \Leftrightarrow X\pi_m \supseteq Y^{\hat{g}^{-1}}\pi_m,$$

where $g_m = \hat{g}\pi_m \in GL_n(\mathbb{Z}/m\mathbb{Z})$; we applied this with $X = A_u$, A_s, A_N and $Y = B_u$, B_s, B_N in turn.)

D. Proof of the main theorem – a special case

Proposition 4 showed that for polycyclic-by-finite groups G and H, $\mathscr{F}(G) = \mathscr{F}(H)$ if and only if $\hat{G} \cong \hat{H}$. When these conditions hold G and H are said to belong to the same ^-*class* (pronounced 'hat-class'). The main result of the chapter is

Theorem 3   Every ^-class of polycyclic-by-finite groups is the union of finitely many isomorphism classes.

This includes Theorem 1 as a special case, and the proof is modelled on the argument of section A. Ultimately, the result depends on the main theorem of Chapter 9; to apply that theorem we must embed our groups in a canonical fashion into a suitable $GL_n(\mathbb{Z})$. Now we already have a canonical embedding, described in Chapter 7. It works best for groups which are *splittable* in the sense of that chapter, and in this section we prove the special case of Theorem 3 for a ^-class of splittable polycyclic groups.

The proof of this case rests on two key results. The first is an elaborate version of Theorem 4 of Chapter 3:

Proposition 9   Let G be a polycyclic group and N a $\mathfrak{X}$-group with $G' \leq N \leq G$. For each subgroup H of $\hat{G}$ put

$$V_H = H \cap \hat{N},$$

and denote the set of all maximal nilpotent supplements for V_H in H by $\mathscr{C}(H, V_H)$. Let

$$\mathscr{X}(G, N) = \{H \leq \hat{G} | H \text{ is polycyclic}, \hat{H} = \hat{G} \text{ and } \hat{V}_H = \hat{N}].$$

Then the set

$$\bigcup_{H \in \mathscr{X}(G,N)} \{\hat{C} | C \in \mathscr{C}(H, V_H)\}$$

breaks up into finitely many classes under conjugation in $\hat{G}$.

The other key result is an elaboration of Lemma 1:

Lemma 7 For $i = 1$, 2, let M_i be a $\mathfrak{X}$-group, T_i an abelian subgroup of Aut M_i, and G_i a subgroup of $M_i]T_i$. Suppose there exists an isomorphism $\theta : (M_1]T_1)\hat{} \to (M_2]T_2)\hat{}$ such that $\hat{M}_1\theta = \hat{M}_2$, $\hat{T}_1\theta = \hat{T}_2$ and $\hat{G}_1\theta = \hat{G}_2$. Then $n(M_1) = n(M_2) = n$, say, and $G_1\beta_{M_1} \approx G_2\beta_{M_2}$ in $GL_n(\mathbb{Z})$.

Here, $\beta_{M_i} : M_i]T_i \to GL_{n(M_i)}(\mathbb{Z})$ is the canonical embedding described in section B of Chapter 5, and '$\approx$' is the relation '$\approx_{GL_n}$' of Chapter 9. Before proving these results, let us see how they give the main theorem; to simplify the argument, we consider only splittable groups G such that Fitt(G) is a *lattice $\mathfrak{X}$-group*, or 'lattice-splittable' groups.

Proposition 10 Every $\hat{}$-class of lattice-splittable polycyclic groups is the union of finitely many isomorphism classes.

Proof Let $\mathscr{C}$ be a family of lattice-splittable polycyclic groups all belonging to the same $\hat{}$-class. We will show that $\mathscr{C}$ contains only finitely many non-isomorphic groups. Choosing $G \in \mathscr{C}$, we may clearly suppose that in fact $\hat{H} = \hat{G}$ for every group $H \in \mathscr{C}$. For $H \in \mathscr{C}$, put

$$N_H = \text{Fitt}(H),$$

and put $N = N_G$. Then by Theorem 2,

$$\hat{N}_H = \text{Fitt}(\hat{H}) = \text{Fitt}(\hat{G}) = \hat{N},$$

from which it follows that

$$H \cap \hat{N} = H \cap \hat{N}_H = N_H$$

for each $H \in \mathscr{C}$. Thus in the notation of Proposition 9, $N_H = V_H$ for each $H \in \mathscr{C}$, and so $\mathscr{C} \subseteq \mathscr{X}(G, N)$.

By hypothesis, each group $H \in \mathscr{C}$ is splittable. Thus there exists a maximal nilpotent supplement C_H for N_H in H, such that $(C_H^*)_s \leq \text{Aut } N_H$ where $* : H \to \text{Aut } N_H$ is the action of H by conjugation; see Lemma 10 of Chapter 7. By Proposition 9, the groups $\hat{C}_H$ for $H \in \mathscr{C}$ lie in finitely many conjugacy classes in $\hat{G}$; we may therefore, without loss of generality, assume that the groups $\hat{C}_H$ are in fact identical as H runs through $\mathscr{C}$ (thus $\mathscr{C}$ is replaced by one of finitely many subsets which partition $\mathscr{C}$, and each $H \in \mathscr{C}$ is replaced by a suitable conjugate in $\hat{G}$ – of course the original assumption that $G \in \mathscr{C}$ must henceforth be abandoned.)

Now for $H \in \mathscr{C}$ let

$$\bar{H} = M_H] T_H = H] T_H$$

be the semi-simple splitting of H corresponding to the subgroup C_H (see Chapter 7, section B). Thus

$$M_H = \mathrm{Fitt}\,(\bar{H}) \geq N_H;$$

T_H acts faithfully on N_H, M_H and H;

T_H centralizes C_H and acts like $(C_H^*)_s$ on N_H.

Now the action of T_H on H extends to an action of $\hat{T}_H$ on $\hat{H} = \hat{G}$, giving a homomorphism $\rho_H \colon \hat{T}_H \to \mathrm{Aut}\,\hat{G}$. Note that ρ_H is injective, by Corollary 5 in section B.

Let us work out that $\hat{T}_H \rho_H$ looks like. Put

$$X = \hat{C}_H \leq \hat{H} = \hat{G};$$

we have arranged that X is independent of $H \in \mathscr{C}$. Since $H = N_H C_H$ and $\widehat{N_H} = \hat{N}$ we have

$$\hat{G} = \hat{N} X.$$

Put $\mathscr{H} = \mathscr{A}ut\,N_H$ and $\mathscr{G} = \mathscr{A}ut\,N$; we make the identifications

$$\mathscr{H}^\infty = \mathrm{Aut}\,\hat{N}_H = \mathrm{Aut}\,\hat{N} = \mathscr{G}^\infty,$$

by Proposition 6. Let $^\dagger \colon \hat{G} \to \mathrm{Aut}\,\hat{N}$ denote the action by conjugation, so that † extends the action $* \colon H \to \mathrm{Aut}\,N_H$. Then

$$\hat{T}_H \rho_H |_{\hat{N}} = \hat{T}_H \rho_H |_{\hat{N}_H} = ((C_H^*)_s)^-,$$

where '$-$' denotes congruence closure in $\mathscr{H}^\infty = \mathscr{G}^\infty$, by Proposition 5. Proposition 8 shows that

$$((C_H^*)_s)^- = (\bar{C}_H^*)_s,$$

and this group is the same as

$$(\hat{C}_H^\dagger)_s = (X^\dagger)_s,$$

by Proposition 5 again (it is here that the mysterious remarks at the end of section B are relevant).

Thus $\hat{T}_H \rho_H$ acts like $(X^\dagger)_s$ on $\hat{N}$. Since T_H centralizes C_H, $\hat{T}_H \rho_H$ acts as the identity on $\hat{C}_H = X$. We conclude that *the group* $\hat{T}_H \rho_H \leq \mathrm{Aut}\,\hat{G}$ *is independent of* $H \in \mathscr{C}$.

Now let $H, K \in \mathscr{C}$. Since ρ_H and ρ_K are injective, there is an isomorphism $\theta: \hat{T}_H \to \hat{T}_K$ making the diagram

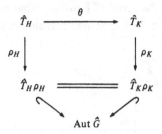

commute. This extends to an isomorphism

$$\theta: \hat{G}] \hat{T}_H \to \hat{G}] \hat{T}_K$$

which is the identity on $\hat{G}$. But

$$\hat{G}] \hat{T}_H = \hat{H}] \hat{T}_H = (H] T_H)\hat{\,} = (M_H] T_H)\hat{\,}$$

and similarly

$$\hat{G}] \hat{T}_K = (M_K] T_K)\hat{\,}.$$

Also, $\hat{M}_H = \mathrm{Fitt}\,(M_H] T_H)\hat{\,}$ and $\hat{M}_K = \mathrm{Fitt}\,(M_K] T_K)\hat{\,}$, by Theorem 2. So in fact we have

$$\theta:(M_H] T_H)\hat{\,} \overset{\sim}{\to} (M_K] T_K)\hat{\,}$$
$$\hat{T}_H \theta = \hat{T}_K, \quad \hat{M}_H \theta = \hat{M}_K, \quad \text{and} \quad \hat{H}\theta = \hat{G}\theta = \hat{K}.$$

Thus we may apply Lemma 7: this shows that $n(M_H) = n(M_K) = n$, say, and that $H\beta_{M_H} \approx K\beta_{M_K}$ in $GL_n(\mathbb{Z})$. Since H and K were arbitrary members of $\mathscr{C}$ it follows that there is a fixed natural number n such that the set $\{H\beta_{M_H}|H \in \mathscr{C}\}$ is contained in a single $\approx$-class in $GL_n(\mathbb{Z})$. Theorem 6 of Chapter 9 shows that the groups $H\beta_{M_H}$ for $H \in \mathscr{C}$ lie in finitely many conjugacy classes in $GL_n(\mathbb{Z})$. Since each β_{M_H} is injective, it follows that the groups in $\mathscr{C}$ lie in finitely many isomorphism classes. This completes the proof of Proposition 10.

Proof of Lemma 7 we have $\mathfrak{X}$-groups M_1, M_2, subgroups T_1 of Aut M_1 and T_2 of Aut M_2, and an isomorphism $\theta:(M_1] T_1)\hat{\,} \to (M_2] T_2)\hat{\,}$, sending $\hat{M}_1$ to $\hat{M}_2$ and $\hat{T}_1$ to $\hat{T}_2$. Put $\beta_i = \beta_{M_i}: M_i] T_i \to GL_{n_i}(\mathbb{Z})$ for the canonical embedding $(i = 1, 2)$. By Lemma 1, $n_1 = n_2 = n$, say, and we have isomorphisms

$$V_1 \overset{\lambda}{\to} \mathbb{Z}^n \overset{\mu}{\leftarrow} V_2,$$

where V_i is the module $V(M_i) = \mathbb{Z}M_i/I_{M_i}$ as in Lemma 1. Let m be a positive integer and let $\theta_m : M_1/M_1^k \to M_2/M_2^k$ be the isomorphism induced by $\theta|_{M_1}$, where k is a suitably large multiple of m. Then θ_m induces an isomorphism $\varphi_m : V_1/mV_1 \to V_2/mV_2$; we showed in the proof of Lemma 1 that if $x \in M_1$ and $y \in M_2$ satisfy $(M_1^k x)\theta_m = M_2^k y$, then the following diagram commutes:

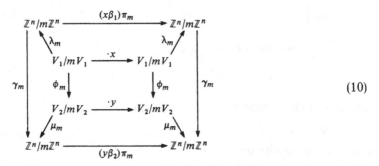

$$(10)$$

(here λ_m, μ_m are induced by λ, μ respectively; γ_m is defined by the diagram; and $\cdot x$, $\cdot y$ denote the actions of x, y coming from the right regular representations of M_i on $\mathbb{Z}M_i$).

The next step is to compare the action of T_1 on V_1/mV_1 with that of T_2 on V_2/mV_2. The action of T_i on $\mathbb{Z}M_i$ is given by linearly extending its action on M_i, and the action on V_i (which gives $\beta_i|_{T_i}$) is induced from this action on $\mathbb{Z}M_i$. Now given m, choose k as before; then choose l so that $T_i^l \le C_{T_i}(M_i/M_i^k)$ for $i = 1, 2$, and choose h so that $(M_i]T_i)^h \le M_i^k T_i^l$ for $i = 1,\ 2$. Then θ induces an isomorphism of $(M_1]T_1)\hat{\ }/((M_1]T_1)\hat{\ })^h$ onto $(M_2]T_2)\hat{\ }/((M_2]T_2)\hat{\ })^h$, hence also an isomorphism

$$\theta_m : (M_1/M_1^k)](T_1/T_1^l) \to (M_2/M_2^k)](T_2/T_2^l).$$

This isomorphism extends the θ_m of the preceding paragraph, and since $\hat{T}_1\theta = \hat{T}_2$ we have $(T_1/T_1^l)\theta_m = T_2/T_2^l$.

Let $s \in T_1$ and $t \in T_2$ satisfy $(T_1^l s)\theta_m = T_2^l t$. Then it is perfectly clear, though messy to justify in writing, that we again get a commutative diagram if in (10) we replace $x\beta_1$ by $s\beta_1$, $y\beta_2$ by $t\beta_2$, $\cdot x$ by the action of s on V_1/mV_1, and $\cdot y$ by the action of t on V_2/mV_2. It follows that whenever $x \in M_1$, $s \in T_1$, $y \in M_2$ and $t \in T_2$ satisfy

$$(M_1^k x)(T_1^l s)\theta_m = (M_2^k y)(T_2^l t), \tag{11}$$

then

$$\gamma_m^{-1} \cdot ((xs)\beta_1)\pi_m \cdot \gamma_m = ((yt)\beta_2)\pi_m. \tag{12}$$

Now recall that $G_1 \le M_1]T_1$ and $G_2 \le M_2]T_2$ satisfy $\hat{G}_1\theta = \hat{G}_2$. This implies that θ_m maps $G_1 \cdot M_1^k T_1^l/M_1^k T_1^l$ onto $G_2 \cdot M_2^k T_2^l/M_2^k T_2^l$. Thus if xs

and yt satisfy (11) then $xs \in G_1 \cdot \ker \beta_1 \pi_m$ if and only if $yt \in G_2 \cdot \ker \beta_2 \pi_m$, and so (12) shows that

$$\gamma_m^{-1} \cdot (G_1 \beta_1) \pi_m \cdot \gamma_m = (G_2 \beta_2) \pi_m.$$

As m was an arbitrary positive integer and $\gamma_m \in GL_n(\mathbb{Z}/m\mathbb{Z})$, we have established that $G_1 \beta_1 \approx G_2 \beta_2$ in $GL_n(\mathbb{Z})$, and the lemma is proved.

To prove Proposition 9, we shall need a substitute for Theorem 2 of Chapter 3 (which was the finiteness result lying behind Theorem 4 of that chapter). A couple of simple lemmas are required first.

Lemma 8 Let G be a polycyclic group and $M \cong \mathbb{Z}^n$ a G-module. If $C_M(G) = 0$ then $C_{\hat{M}}(\hat{G}) = C_{\hat{M}}(G) = 0$.

Proof Let $x_1, \ldots, x_r \in GL_n(\mathbb{Z})$ be matrices representing the action on M of a set generators for G. Suppose $C_{\hat{M}}(G) \neq 0$. Then there exists

$$\hat{v} = (v_p) \in \hat{M} \setminus \{0\} = \prod_p M\mathbb{Z}_p \setminus \{0\}$$

such that

$$v_p(x_i - 1) = 0, \quad i = 1, \ldots, r, \quad \forall p.$$

For at least one prime p we have $v_p \neq 0$, so the system of homogeneous linear equations over $\mathbb{Z}$:

$$\mathbf{Y}(x_i - 1) = 0, \quad i = 1, \ldots, r$$

(in unknowns $\mathbf{Y} = (Y_1, \ldots, Y_n)$) has a non-zero solution in $\mathbb{Z}_p^n$. The system therefore has a non-zero solution in $\mathbb{Z}^n$. This solution represents a non-zero element of $C_M(G)$, and the lemma follows.

Definition Let G be a group and M a G-module. Then

$$H^1(G, M) = \mathrm{Der}\,(G, M)/\mathrm{Ider}\,(G, M).$$

Lemma 9 Let G be a group, E a G-module and F a G-submodule of E. Let $\iota : F \to E$ and $\pi : E \to E/F$ denote the inclusion and residue-class maps. Then the mappings $\delta \mapsto \delta \circ \iota\,(\delta \in \mathrm{Der}\,(G, F))$ and $\delta \mapsto \delta \circ \pi\,(\delta \in \mathrm{Der}\,(G, E))$ induce homomorphisms $\iota^* : H^1(G, F) \to H^1(G, E)$ and $\pi^* : H^1(G, E) \to H^1(G, E/F)$; and the sequence

$$H^1(G, F) \xrightarrow{\iota^*} H^1(G, E) \xrightarrow{\pi^*} H^1(G, E/F)$$

is exact.

The proof is an easy exercise (if we had defined the higher cohomology groups we could extend the exact sequence indefinitely to the right, getting the 'exact sequence of cohomology'; but this will not be needed here). The result we have been leading up to is now

Proposition 11   Let X be a finitely generated abelian group and $M \cong \mathbb{Z}^n$ an X-module. If $C_M(X) = 0$ then $H^1(\hat{X}, \hat{M})$ is finite.

Proof   Assuming that $M \neq 0$ we have $C_X(M) < X$. Choose $x \in X \backslash C_X(M)$ and put $B = C_M(x)$.
Case 1   Where $B > 0$. Then B is an X-submodule of M and $M/B \cong \mathbb{Z}^m$ for some $m < n$; also $B \cong \mathbb{Z}^r$ for some $r < n$, by the choice of x. Corollary 1 of Chapter 3 shows that

$$C_B(X) = C_{M/B}(X) = 0.$$

Arguing by induction on n, we may therefore suppose that $H^1(\hat{X}, \hat{B})$ and $H^1(\hat{X}, \hat{M}/\hat{B})$ are finite (identifying $(M/B)\hat{\ }$ with $\hat{M}/\hat{B}$). Lemma 9 now shows that $H^1(\hat{X}, \hat{M})$ is finite.
Case 2   Where $B = 0$. Then $M/M(x - 1)$ is finite and it follows that $\hat{M}/\hat{M}(x - 1)$ is finite also (if $mM \leq M(x - 1)$, then $m\hat{M} \leq \hat{M}(x - 1)$, and $\hat{M}/m\hat{M} \cong M/mM$). Lemma 9 gives the exact sequence

$$H^1(\hat{X}, \hat{M}(x - 1)) \xrightarrow{\imath^*} H^1(\hat{X}, \hat{M}) \xrightarrow{\pi^*} H^1(\hat{X}, \hat{M}/\hat{M}(x - 1)),$$

so it will suffice now to prove

(a) $\imath^*$ maps $H^1(\hat{X}, \hat{M}(x - 1))$ to zero, and
(b) $H^1(\hat{X}, \hat{M}/\hat{M}(x - 1))$ is finite.

Proof of (a)   Let $\delta : \hat{X} \to \hat{M}(x - 1)$ be a derivation. Then $x\delta = a(x - 1)$ for some $a \in \hat{M}$. If $y \in \hat{X}$ then $xy = yx$ implies

$$(y\delta)(x - 1) = (x\delta)(y - 1) = a(x - 1)(y - 1) = a(y - 1)(x - 1).$$

But $C_{\hat{M}}(x) = 0$, by Lemma 8 (with $\langle x \rangle$ for G); so $y\delta = a(y - 1)$ for every $y \in \hat{X}$, and $\delta \circ \imath$ is an inner derivation. This gives (a).

Proof of (b)   We know that $\hat{M}/\hat{M}(x - 1) = E$, say, is finite. Say $mE = 0$. Put $Y = C_{\hat{X}}(E)$. If $\delta \in \mathrm{Der}(\hat{X}, E)$ then $\delta|_Y \in \mathrm{Hom}(Y, E)$, so $Y^m \delta = 0$. Therefore δ factors through $\hat{X} \to \hat{X}/Y^m$, and we have an injective mapping

$$\mathrm{Der}(\hat{X}, E) \to \mathrm{Der}(\hat{X}/Y^m, E).$$

Now $\hat{X}/Y$ is finite, so $Y^m \geq \hat{X}^t$ for some positive integer t. Since $\hat{X}/\hat{X}^t \cong X/X^t$, it follows that $\hat{X}/Y^m$ is a finite group. Therefore $\mathrm{Der}(\hat{X}/Y^m, E)$ is finite, and (b) follows.

We are now ready for the

Proof of Proposition 9 G is a polycyclic group and N is a $\mathfrak{X}$-group with $G' \le N \le G$.

$$\mathfrak{X}(G,N) = \{H \le \hat{G} | H \text{ polycyclic, } \hat{H} = \hat{G}, \text{ and } \hat{V}_H = \hat{N}\}$$

where $V_H = H \cap \hat{N}$. For each such H, $\mathscr{C}(H, V_H)$ denotes the set of all maximal nilpotent supplements for V_H in H, and what we have to show is that the groups $\hat{C}$ with $C \in \mathscr{C}(H, V_H)$ and $H \in \mathfrak{X}(G, N)$ lie in finitely many conjugacy classes in $\hat{G}$; to do this we argue by induction on the Hirsch number of N.

If G acts nilpotently on $N/\tau_2(N)$ then G is nilpotent, by Proposition 3 of Chapter 3; so $\hat{G}$ is nilpotent and there is nothing to prove (this holds in particular if $N = 1$, so the induction can begin). Assume then that G does not act nilpotently on $N/\tau_2(N)$.

For $H \in \mathfrak{X}(G, N)$, define $W_H/\tau_2(V_H)$ to be the biggest H-submodule of $V_H/\tau_2(V_H)$ on which H acts nilpotently, and put $W = W_G$. By hypothesis, $W < N$. Now I make two claims:

(a) $\hat{W}_H = \hat{W}$ and $H \cap \hat{W} = W_H$ for every $H \in \mathfrak{X}(G, N)$;

(b) $\hat{W} = \hat{N} \cap (W_H C)^{\hat{\,}}$ for every $H \in \mathfrak{X}(G, N)$ and $C \in \mathscr{C}(H, V_H)$.

Granting (a) and (b) for the moment, we can finish the proof. Let $H \in \mathfrak{X}(G, N)$ and $C \in \mathscr{C}(H, V_H)$. Then $H = V_H C$ so

$$\hat{G} = \hat{H} = \hat{V}_H \hat{C} = \hat{N} \hat{C} = \hat{N}(W_H C)^{\hat{\,}}.$$

Thus, by (b), $(W_H C)^{\hat{\,}}/\hat{W}$ is a complement for $\hat{N}/\hat{W}$ in $\hat{G}/\hat{W}$:

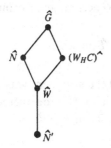

Now N/W is a free abelian group (see Exercise 4 of Chapter 3), and $C_{N/W}(G/N) = 1$ by the definition of W. So Proposition 11 ensures that

$$H^1(\hat{G}/\hat{N}, \hat{N}/\hat{W}) = H^1((G/N)^{\hat{\,}}, (N/W)^{\hat{\,}})$$

is finite. It follows that the complements for $\hat{N}/\hat{W}$ in $\hat{G}/\hat{W}$ lie in finitely many conjugacy classes (see Chapter 3, section A). Thus there exist $G_1, \ldots, G_n \in$

$\mathscr{X}(G,N)$ and $C_i \in \mathscr{C}(G_i, V_{G_i})(i = 1, \ldots, n)$, say, such that

$$H \in \mathscr{X}(G,N), \quad C \in \mathscr{C}(H, V_H) \quad \Rightarrow$$

$$((W_H C)\hat{\,})^\alpha = (W_i C_i)\hat{\,}, \text{ some } i \in \{1, \ldots, n\} \text{ and some } \alpha \in \hat{G} \tag{13}$$

(where $W_i = W_{G_i}$).

Fix $i \in \{1, \ldots, n\}$, put $\Gamma_i = W_i C_i$, and for $K \leq \hat{\Gamma}_i$ write $U_K = K \cap \hat{W}_i$. By (a),

$$(N/W_i)\hat{\,} \cong \hat{N}/\hat{W} \cong (N/W)\hat{\,}.$$

Since N/W is infinite, it follows that $h(W_i) < h(N)$. So by inductive hypothesis, the groups $\hat{C}$ with $C \in \mathscr{C}(K, U_K)$ and $K \in \mathscr{X}(\Gamma_i, W_i)$ lie in finitely many conjugacy classes in $\hat{\Gamma}_i$.

Now suppose $H \in \mathscr{X}(G,N)$ and $C \in \mathscr{C}(H, V_H)$. Let i and α be as in (13), and put $K = (W_H C)^\alpha$. Using (a), it is easy to check that

$$K \in \mathscr{X}(\Gamma_i, W_i), \quad C^\alpha \in \mathscr{C}(K, U_K).$$

Thus there are only finitely many possibilities for $\hat{C}^\alpha = (C^\alpha)\hat{\,}$ up to conjugacy in $\hat{\Gamma}_i$, and consequently only finitely many possibilities for $\hat{C}$ up to conjugacy in $\hat{G}$.

Let us prove (a). Put $T = \tau_2(N)$, and use $^-$ to denote closure in the group $\hat{N}$. The first point to note is that

$$\hat{T} = (\tau_2(\hat{N}))^-. \tag{14}$$

For $\hat{N}/\hat{T} \cong (N/T)\hat{\,}$ is torsion-free and abelian, so $\hat{T} \geq \tau_2(\hat{N})$; since $\hat{T}$ is closed in $\hat{N}$, we have $\hat{T} \geq (\tau_2(\hat{N}))^-$. On the other hand, $T = \tau_2(N) \leq \tau_2(\hat{N})$, so the reverse inclusion follows from $\hat{T} = \bar{T}$. This establishes (14). Now G acts nilpotently on W/T and $C_{N/W}(G) = 1$; therefore $\hat{G}$ acts nilpotently on $\hat{W}/\hat{T}$, and $C_{\hat{N}/\hat{W}}(\hat{G}) = 1$ by Lemma 8. Thus

$$\hat{W}/\hat{T} \text{ is the biggest } \hat{G}\text{-submodule of } \hat{N}/\hat{T} \text{ on which}$$
$$\hat{G} \text{ acts nilpotently.} \tag{15}$$

Now let $H \in \mathscr{X}(G,N)$. Then $\hat{H} = \hat{G}$, $\hat{V}_H = \hat{N}$, and the argument which gave (14) shows equally that $(\tau_2(V_H))\hat{\,} = (\tau_2(\hat{N}))^- = \hat{T}$. Thus (15) together with its analogue for H shows that $\hat{W}_H = \hat{W}$, as claimed. Then $H \cap \hat{W} = H \cap \hat{W}_H = W_H$, so (a) is established.

Finally, we prove (b). Take H and C as above. From (a),

$$\hat{N} \cap (W_H C)\hat{\,} = \hat{N} \cap \hat{W}_H \hat{C} = \hat{N} \cap \hat{W} \hat{C} = \hat{W}(\hat{N} \cap \hat{C}). \tag{16}$$

Now $\hat{G} = \hat{H} = \hat{N}\hat{C}$, $\hat{N}$ centralizes $\hat{N}/\hat{W}$, and $\hat{C}$ is nilpotent; consequently $\hat{G}$ acts nilpotently on the factor $\hat{W}(\hat{N} \cap \hat{C})/\hat{W}$. Hence, by (15), $\hat{W}(\hat{N} \cap \hat{C}) = \hat{W}$. With (16) this gives (b), and the proof is complete.

E. The general case

To reduce the main theorem to the special case of the preceding section, we need

Proposition 12   Let $\mathscr{C}$ be a family of polycyclic-by-finite groups belonging to a single ^-class. Then there exists a positive integer d such that every group in $\mathscr{C}$ has a subgroup of index at most d which is lattice-splittable.

Before embarking on this, let us finish the

Proof of Theorem 3   Let $\mathscr{C}$ be a family of polycyclic-by-finite groups belonging to the same ^-class. Let d be the integer provided by Proposition 12, and for each $G \in \mathscr{C}$ let G_1 be a lattice-splittable subgroup of index at most d in G. We may assume that $\hat{G} = \hat{H} = \Gamma$, say, for all G and $H \in \mathscr{C}$. Then for $G \in \mathscr{C}$ we have

$$\Gamma^{d!} = (G^{d!})^{\widehat{}} \leq \hat{G}_1 \leq \hat{G} = \Gamma.$$

Since $\Gamma/\Gamma^{d!} \cong G/G^{d!}$ is a finite group, there are only finitely many distinct subgroups of Γ which contain $\Gamma^{d!}$, and consequently only finitely many distinct groups $\hat{G}_1$ as G runs through $\mathscr{C}$. Hence, by Proposition 10, there are only finitely many non-isomorphic groups among the G_1s. But Theorem 6 of Chapter 8 shows that there are, up to isomorphism, only finitely many possibilities for a group G containing a given G_1 as a subgroup of index at most d. Hence $\mathscr{C}$ contains only finitely many non-isomorphic groups, and the theorem is proved.

We proceed now with the proof of Proposition 12. For a group G we write

$$N_G = \mathrm{Fitt}\,(G).$$

Henceforth, $\mathscr{C}$ will denote a family of polycyclic-by-finite groups lying in one ^-class.

Lemma 10   Let G, $H \in \mathscr{C}$ and let e, f be positive integers. Then

$$(N_G^e G^f)^{\widehat{}} \cong (N_H^e H^f)^{\widehat{}}$$

and

$$G/N_G^e G^f \cong H/N_H^e H^f.$$

Proof Theorem 2 and Lemma 3 show that

$$(N_G^e G^f)^\widehat{} = (N_G^e)^\widehat{}(G^f)^\widehat{} = \hat{N}_G^e \hat{G}^f = N_{\hat{G}}^e \hat{G}^f.$$

Any isomorphism $\hat{G} \overset{\sim}{\to} \hat{H}$ therefore maps $(N_G^e G^f)^\widehat{}$ onto $(N_H^e H^f)^\widehat{}$, and this gives the first claim. Since $G/N_G^e G^f \cong \hat{G}/(N_G^e G^f)^\widehat{}$, and analogously for H, the second claim also follows.

We prove Proposition 12 in a series of reduction steps.

Step 1 We may assume that N_G is a $\mathfrak{X}$-group for every G.

Proof Choose $G \in \mathscr{C}$, and let f be a positive integer such that G^f is torsion-free. Then $N_{G^f} = N_G \cap G^f$ is a $\mathfrak{X}$-group. If $H \in \mathscr{C}$ then

$$\hat{N}_{H^f} = N_{\hat{H^f}} = N_{\hat{H}^f} \cong N_{\hat{G}^f} = \hat{N}_{G^f},$$

so N_{H^f} is torsion-free. By Lemma 10, we may replace each $H \in \mathscr{C}$ by H^f to achieve the reduction. (*Exercise*: why is $\hat{N}_{G^f}$ torsion-free?)

Step 2 We may assume that the groups N_H for $H \in \mathscr{C}$ are all isomorphic.

Proof By Theorem 2, the groups $N_H, H \in \mathscr{C}$ lie in a single $\frown$-class. Hence by Theorem 1 they lie in finitely many isomorphism classes. Dividing $\mathscr{C}$ into finitely many subsets and dealing with each one separately, we may therefore suppose that the groups N_H are all in fact isomorphic.

Step 3 We may assume that for each $H \in \mathscr{C}$, N_H is a lattice $\mathfrak{X}$-group and H/N_H is free abelian.

Proof Choose $G \in \mathscr{C}$. By Theorem 5 of Chapter 6, N_G^e is a lattice $\mathfrak{X}$-group for some positive integer e. Let f be a positive integer such that $G^f \cap N_G \le N_G^e$ and $G^f N_G/N_G$ is free abelian. For each $H \in \mathscr{C}$, put

$$H_1 = N_H^e H^f.$$

By Lemma 10, we may replace each $H \in \mathscr{C}$ by the corresponding H_1; what has to be checked now is that the groups N_{H_1} for $H \in \mathscr{C}$ are isomorphic lattice $\mathfrak{X}$-groups, and that H_1/N_{H_1} is free abelian for each H. This will follow once we have proved that for each H,

$$N_{H_1} \cong N_G^e \quad \text{and} \quad (H_1/N_{H_1})^\widehat{} \cong (G_1/N_{G_1})^\widehat{}. \tag{17}$$

The proof of (17) is very similar to the proof of Lemma 10, and is left to the reader.

Henceforth, we keep fixed a lattice $\mathfrak{X}$-group N, and for each $H \in \mathscr{C}$ an isomorphism

$$\nu_H : N_H \overset{\sim}{\to} N.$$

Put $\mathscr{G} = \mathscr{A}ut\, N$ and for $H \in \mathscr{C}$ define

$$\psi_H : H \to \mathscr{G}(\mathbb{Z})$$

to be the composition of the maps

$$H \to \operatorname{Aut} N_H \xrightarrow{v_H^*} \operatorname{Aut} N = \mathscr{G}(\mathbb{Z}),$$

where v_H^* is the map induced by v_H, and the first map is conjugation.

The next step depends on

Lemma 11   Let G and H be in $\mathscr{C}$ and let $\alpha : \hat{G} \to \hat{H}$ be an isomorphism. Put $\alpha_1 = \alpha|_{\hat{N}_G} : \hat{N}_G \to \hat{N}_H$ and

$$\gamma = \hat{v}_G^{-1} \alpha_1 \hat{v}_H : \hat{N} \to \hat{N}.$$

(Thus $\gamma \in \operatorname{Aut} \hat{N} = \mathscr{G}^\infty$.) Then for each positive integer m,

$$(H\psi_H)\pi_m = (\gamma^{-1} \cdot G\psi_G \cdot \gamma)\pi_m.$$

(Here π_m is the usual map $\mathscr{G}^\infty \to GL_n(\mathbb{Z}/m\mathbb{Z})$.) Postponing the proof till later, let us carry on with

Step 4   We may assume that there is a subgroup $\Gamma \le \mathscr{G}(\mathbb{Z})$ such that $H\psi_H = \Gamma$ for all $H \in \mathscr{C}$.

Proof   Lemma 11 shows that the groups $H\psi_H$, $H \in \mathscr{C}$, lie in a single $\approx_{\mathscr{g}}$-class in $\mathscr{G}(\mathbb{Z})$. Hence, by Theorem 6 of Chapter 9, they lie in finitely many conjugacy classes in $\mathscr{G}(\mathbb{Z})$. Dividing $\mathscr{C}$ into finitely many subsets and restricting attention to one of them, we may therefore suppose that the groups $H\psi_H$ are all conjugate in $\mathscr{G}(\mathbb{Z})$. Now $\mathscr{G}(\mathbb{Z}) = \operatorname{Aut} N$; so if we change each of the isomorphisms v_H by composing it with a suitable automorphism of N, and adjust ψ_H accordingly, we can arrange it so that the subgroups $H\psi_H$ of $\mathscr{G}(\mathbb{Z})$ are all in fact identical. This achieves Step 4.

Let us pause here to take stock. For $H \in \mathscr{C}$, write

$$Z_H = \zeta_1(N_H).$$

Then

$$Z_H = C_H(N_H) = \ker \psi_H$$

(by Exercise 2 in Chapter 2). So for G and H in $\mathscr{C}$, we have isomorphisms

$$v_{GH} = v_G v_H^{-1} : N_G \to H_H$$

$$\varphi_{GH} = \varphi_G \varphi_H^{-1} : G/Z_G \to H/Z_H,$$

where $\varphi_H : H/Z_H \to H\psi_H = \Gamma$ and $\varphi_G : G/Z_G \to G\psi_G = \Gamma$ are the isomorphisms inducted by ψ_H and ψ_G respectively. Moreover, the maps v_{GH} and

φ_{GH} fit together in the best possible way:

$$(a v_{GH})^{(gZ_G)\varphi_{GH}} = (a^g) v_{GH} \quad \forall a \in N_G, \forall g \in G; \tag{18}$$

this is immediate from the definitions. Since each coset of Z_H in H is determined by how it acts, via conjugation, on N_H, (18) implies that

$$\varphi_{GH}|_{N_G/Z_G} \text{ is the map induced by } v_{GH} \text{ from } N_G/Z_G$$

$$\text{onto } N_H/Z_H. \tag{19}$$

Now choose $G \in \mathscr{C}$. By Theorem 3 of Chapter 3, G has a nilpotent subgroup D such that $|G : N_G D|$ is finite. Choose D to be maximal among such subgroups, and put

$$\Delta = D\psi_G \leq \Gamma.$$

The key step in the whole argument is now

Lemma 12   There exist a positive integer l, and for each $H \in \mathscr{C}$ a nilpotent subgroup D_H in H, such that

$$D_H \psi_H \leq \Delta \quad \text{and} |\Delta : D_H \psi_H| \leq l.$$

The proof of Lemma 12 depends on a quantitative version of Theorem 1 of Chapter 3. This will be proved at the end. It is

Proposition 13   Let X be a finitely generated nilpotent group and $A \cong \mathbb{Z}^n$ an X-module with $C_A(X) = 0$. Then there is a positive integer $l = l(A, X)$ such that every extension E of A by X contains a subgroup S with $A \cap S = 1$ and $|E : AS| \leq l$.

Proof of Lemma 12   Recall that we chose a group $G \in \mathscr{C}$ and a maximal nilpotent subgroup D of G such that $|G : N_G D|$ is finite. We defined $\Delta = D\psi_G \leq \Gamma$. Put

$$P = N_G \cap Z_G D = Z_G(N_G \cap D),$$

and let A be the $Z_G D/P$-module

$$A = P/(N_G \cap D).$$

Since D is a maximal nilpotent subgroup of G, it is easy to see that D can centralise no element $\neq 1$ in A. Now let

$$l = l(A, Z_G D/P)$$

be the number provided by Proposition 13.

Given $H \in \mathscr{C}$, we must find a nilpotent subgroup D_H of H such that $D_H \psi_H \leq \Delta$ and $|\Delta : D_H \psi_H| \leq l$.

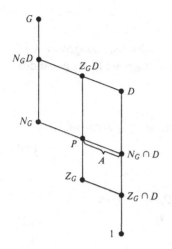
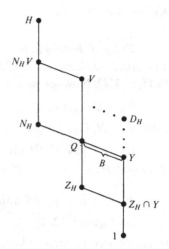

The picture on the left represents G. On the right we have a picture of H. Here

$$Y = (N_G \cap D)v_{GH}, \quad Q = Pv_{GH}$$

and

$$V/Z_H = (Z_G D/Z_G)\varphi_{GH}.$$

Using (19) one can easily verify that

$$Q = V \cap N_H = Z_H Y,$$

so the picture doesn't lie. Let B be the V/Q-module Q/Y. If we identify B with A via v_{GH}, then (18) shows that the isomorphism $Z_G D/P \to V/Q$ induced by φ_{GH} respects the module structure; thus we may consider V/Y as an extension of A by $Z_G D/P$, with B playing the role of A. Proposition 13 therefore ensures that V/Y has a subgroup D_H/Y, say, such that $D_H \cap Q = Y$ and $|V : QD_H| \le l$.

Let us verify that D_H has the required properties.

(a) D_H *is nilpotent.* $D_H/(Z_H \cap Y)$ is isomorphic to a subgroup of V/Z_H, and $V/Z_H \cong Z_G D/Z_G$. So $D_H/(Z_H \cap Y)$ is nilpotent. Since D is nilpotent, we also know that $Z_G D$ acts nilpotently on $Z_G \cap D$; hence, by (18), V acts nilpotently on $Z_H \cap Y$. As $D_H \le V$, it follows that D_H is nilpotent.

(b) $D_H \psi_H \le \Delta$ *and* $|\Delta : D_H \psi_H| \le l$. From the definition of V, we have

$$V\psi_H = (Z_G D)\psi_G = D\psi_G = \Delta.$$

So (b) follows from the choice of D_H, the fact that

$$QD_H = Z_H Y D_H = Z_H D_H,$$

and the fact that $Z_H = \ker \psi_H$. This completes the proof.

We can now give the

Proof of Proposition 12   We have a nilpotent subgroup Δ of $\mathscr{G}(\mathbb{Z}) = \operatorname{Aut} N$. By Lemma 7 of Chapter 7, Δ has a subgroup Π of finite index with $\Pi_s \leq \mathscr{G}(\mathbb{Z})$. Now for each $H \in \mathscr{C}$ put

$$C_H = D_H \cap \Pi \psi_H^{-1},$$

and let $H_1 = N_H C_H$. To prove the proposition, we will show (a) that

$$|H:H_1| \leq |G:N_G D| \cdot l \cdot |\Delta:\Pi| \tag{20}$$

and (b) that H_1 is a lattice-splittable group.

Proof of (a)   By definition, $C_H \psi_H = D_H \psi_H \cap \Pi$, so

$$|\Delta:C_H \psi_H| \leq |\Delta:D_H \psi_H| \cdot |\Delta:\Pi| \leq l \cdot |\Delta:\Pi|. \tag{21}$$

Also $N_H \psi_H = N_G \psi_G$ (this follows from (19)), so

$$(N_H \psi_H)\Delta = (N_G \psi_G)\Delta = (N_G D)\psi_G.$$

Since $G\psi_G = \Gamma$ and $\ker \psi_G \leq N_G$, we have

$$|\Gamma:(N_H \psi_H)\Delta| = |\Gamma:(N_G D)\psi_G| = |G:N_G D|, \tag{22}$$

Similarly,

$$|H:H_1| = |H:N_H C_H| = |H\psi_H:(N_H C_H)\psi_H|$$
$$\leq |\Gamma:(N_H \psi_H)\Delta| \cdot |\Delta:C_H \psi_H|.$$

Thus (20) follows from (21) and (22).

Proof of (b)   Since H/N_H is abelian, $H_1 \triangleleft H$ and so $\operatorname{Fitt}(H_1) = N_H$ is a lattice $\mathfrak{X}$-group. Moreover $H_1/N_H \leq H/N_H$ is free abelian, and C_H is a nilpotent supplement for N_H in H_1. Since $(C_H \psi_H)_s \leq \Pi_s \leq \mathscr{G}(\mathbb{Z}) = \operatorname{Aut} N$, it follows from the definition of ψ_H that $(C_H^*)_s \leq \operatorname{Aut} N_H$, where $*: H \to \operatorname{Aut} N_H$ is the action of H by conjugation on N_H. Thus H_1 is splittable, by Theorem 1 of Chapter 7, and (b) follows.

Proof of Lemma 11   This is similar to Lemma 7 in section D. We have an isomorphism $\alpha: \hat{G} \to \hat{H}$, which restricts to $\alpha_1: \hat{N}_G \to \hat{N}_H$, and we have isomorphisms

$$N_G \overset{v_G}{\to} N \overset{v_H}{\leftarrow} N_H$$

where N is a third lattice $\mathfrak{X}$-group. Thus

$$\gamma = \hat{v}_G^{-1} \alpha_1 \hat{v}_H$$

is an automorphism of $\hat{N}$; and we simply have to check that 'modulo congruences', γ translates the action on $\hat{G}$ on $\hat{N}$ (via $\hat{v}_G$) to the action of $\hat{H}$ on

$\hat{N}$ (via $\hat{v}_H$). This is just a matter of looking carefully at the definitions.

It is clear from the way that $\mathcal{G}^{\infty}$ is supposed to act on $\hat{N}$ (where $\mathcal{G} = \mathcal{A}ut\,N$) that for each positive integer m there exists a positive integer k such that, for $\xi, \eta \in \text{Aut}\,\hat{N}$,

$\xi\pi_m = \eta\pi_m$ *if* ξ *and* η *induce the same automorphism on* $\hat{N}/\hat{N}^k$.

So given $m \in \mathbb{N}$, take such a k; then let l be a positive integer such that

$$G^l \cap N_G \le N_G^k \text{ and } H^l \cap N_H \le N_H^k.$$

The isomorphism $\alpha \colon \hat{G} \to \hat{H}$ induces an isomorphism $\bar{\alpha} \colon G/G^l \to H/H^l$; and to prove the lemma, it will suffice to show that whenever $g \in G$ and $h \in H$ satisfy

$$(G^l g)\bar{\alpha} = H^l h,$$

then the middle square in the following diagram commutes:

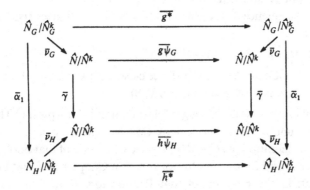

Here g^* and h^* represent the actions of g, h by conjugation on N_G, N_H respectively; $\widehat{N_G}/\hat{N}_G^k$ is being identified with N_G/N_G^k and $\hat{N}_H/\hat{N}_H^k$ with N_H/N_H^k; and '$^-$'s denote the maps induced on the exhibited quotient groups.

The four outer quadrilaterals all commute by definition of the various maps. Since all the maps are isomorphisms, the inner square will commute if and only if the outer square commutes. But the outer square does commute: for if $x \in \hat{N}_G$ then

$$((x^{g^*})\alpha_1 = (x^g)\alpha_1 = (x\alpha_1)^{g\alpha} \in \hat{N}_H;$$

by hypothesis we have $g\alpha \equiv h \bmod \hat{H}^l$, and by the choice of l, $\hat{H}^l$ centralizes $\hat{N}_H/\hat{N}_H^k$. Therefore

$$(x^{g^*})\alpha_1 \equiv (x\alpha_1)^{h^*} \bmod \hat{N}_H^k,$$

which is what we had to show. This finishes the proof of Lemma 11.

All that remains to be done is the proof of Proposition 13. This concerns a finitely generated nilpotent group X and an X-module $A \cong \mathbb{Z}^n$, with $C_A(X) = 0$. Theorem 1 of Chapter 3 shows that every extension of A by X 'almost splits', and what we have to do is to bound the finite index implied in the 'almost'. We shall call two extensions E and E_1 of A by X *isomorphic* (as extensions) if there is an isomorphism $\theta : E \to E_1$ with $A\theta = A$. It is clear then that if $S \leq E$ and $A \cap S = 1$, then the subgroup $S\theta$ of E_1 satisfies $A \cap S\theta = 1$, $|E_1 : A \cdot S\theta| = |E : AS|$; so Proposition 13 will follow if it can be shown that the extensions of A by X (inducing the given X-module structure on A) lie in finitely many isomorphism classes (as extensions).

If we are prepared to accept three elementary facts about the cohomology of groups, this is easy to establish. The facts are

(a) The second cohomology group $H^2(X, A)$ is a finitely generated abelian group (for X and A as in Proposition 13);

(b) The exact sequence $0 \to A \to M = \mathbb{Q} \otimes A \to M/A \to 0$ gives rise to an exact sequence

$$H^1(X, M/A) \to H^2(X, A) \to H^2(X, M);$$

(c) For any X-module B, the equivalence classes of extensions of B by X are in bijective correspondence with $H^2(X, B)$.

(*Equivalence* of extensions is a stronger relation than isomorphism.) These results can be found in Gruenberg's book *Cohomological Topics in Group Theory*. For the definition of H^2 and for (b), see Chapter 2 of that book; for (c), see Chapter 5, and for (a), see Proposition 6 in Chapter 3 of that book (and note that, in the notation of that Proposition, R/R' is a finitely generated G-module, since $G (= X)$ is a finitely presented group; since $A \cong \mathbb{Z}^n$ it then follows that $\text{Hom}_G(R/R', A)$ is finitely generated).

First Proof of Proposition 13   We show that the extensions of A by X lie in finitely many isomorphism classes. In view of (c), it will suffice to show that $H^2(X, A)$ is finite, and by (a) this will follow once we know that $H^2(X, A)$ is periodic. Now Theorem 1* of Chapter 3 says that every extension of $M = \mathbb{Q} \otimes A$ by X splits; all split extensions are equivalent, so (by (c) again) $H^2(X, M) = 0$. Hence, by (b), $H^2(X, A)$ is a homomorphic image of $H^1(X, M/A)$, and we are reduced to proving that this latter group is periodic. But $H^1(X, M/A) = \text{Der}(X, M/A)/\text{Ider}(X, M/A)$. As X is finitely generated, $\text{Der}(X, M/A)$ can be embedded in the direct sum of finitely many copies of $M/A \cong (\mathbb{Q}/\mathbb{Z})^n$. This is a periodic group, so the result follows.

In keeping with the spirit of Chapter 3, here is an alternative argument; less elegant, but self-contained.

Second proof of Proposition 13   Again, let $M = \mathbb{Q} \otimes A$, and consider A as a submodule of M. Let E be an extension of A by X, and embed E in $\bar{E} = ME$, as in section A of Chapter 3. By Theorem 1* of that chapter, $\bar{E} = M]C$ for some subgroup $C \cong X$. Identifying C with X, we thus embed E into $M]X = \Gamma$, say. Now $M \cap E = A$ and $ME = \Gamma$, so E/A is a complement for M/A in Γ/A. Let E_1 be another extension of A by X and embed E_1 likewise into Γ; then E_1/A is another complement for M/A in Γ/A. If E_1/A is conjugate to E/A in Γ/A, then E_1 and E are isomorphic as extensions: for if $\mu \in M$ and $x \in E$ give

$$(E_1/A)^{A \cdot \mu x} = E/A,$$

then $E_1^\mu = E^{x^{-1}} = E$ and $A^\mu = A$. So it will suffice now to show that the complements for M/A in Γ/A lie in finitely many conjugacy classes, and in view of Proposition 2 of Chapter 3 this comes down to showing that $H^1(X, M/A)$ is finite (of course the argument so far is just the group-theoretic version of statement (b) above).

To show that $H^1(X, M/A)$ is finite, we follow the proof of Theorem 2** in Chapter 3. There are two cases to consider.

Case 1   Suppose A has an X-submodule $B \neq 0$, with A/B infinite. Replacing B by its isolator, we may assume that A/B is again $\mathbb{Z}$-free. Put $C = A/B$, $N = \mathbb{Q} \otimes B$, and note that the inclusion $B \to A$ induces an exact sequence

$$0 \to N/B \to M/A \to (\mathbb{Q} \otimes C)/C \to 0.$$

By Corollary 1 of Chapter 3, $C_C(X) = C_B(X) = 0$. So arguing by induction on n, we may suppose that $H^1(X, N/B)$ and $H^1(X, (\mathbb{Q} \otimes C)/C)$ are both finite. The exact cohomology sequence (Lemma 9 in section D, above) then shows that $H^1(X, M/A)$ is finite.

Case 2   Suppose A is rationally irreducible as X-module. Put $K = C_X(M)$ and let $x \in X \backslash K$ be central in X modulo K. Then $C_A(x) = 0$, and Corollary 1 of Chapter 3 shows that $rA \subseteq A(x-1)$ for some positive integer r.

The first thing to show is that for every $\delta \in \mathrm{Der}(X, M/A)$,

$$r^c \cdot K\delta = 0, \tag{23}$$

where c is the nilpotency class of X. For this, note that $\delta|_K$ is a group homomorphism, so it induces a homomorphism $\delta_1 : K/K' \to M/A$, say. If $g \in X$ and $h \in K$, then $h^g \in K$ and so

$$h^g \delta = -(g\delta)g^{-1} \cdot hg + (h\delta)g + g\delta$$
$$= (h\delta)g.$$

Thus δ_1 is actually an X-module homomorphism. Since $[K, x, \ldots, x] = 1$ (with c entries x), it follows that $(x-1)^c$ annihilates the module K/K', and hence that

$$(K\delta)(x - 1)^c = (K/K')\delta_1 \cdot (x - 1)^c = 0.$$

Now $K\delta = L/A$ for some $\mathbb{Z}$-submodule L of M, and what we have established is that

$$L(x - 1)^c \leq A.$$

Therefore

$$r^c L(x - 1)^c \leq r^c A \leq A(x - 1)^c,$$

and since $C_M(x) = 0$ this implies that $r^c L \leq A$, which is the same as (23).

The next step is to show that for $\delta \in \mathrm{Der}\,(X, M/A)$, $r^{c+1}\delta \in \mathrm{Ider}\,(X, M/A)$. Now Lemma 1 of Chapter 3 shows that $M = M(x - 1)$; it follows that $M/A = (M/A)(x - 1)$, so

$$x\delta = \mu(x - 1)$$

for some $\mu \in M/A$. If $g \in X$, $xg = gxy$ for some $y \in K$, and

$$(x\delta)g + g\delta = (g\delta)xy + (x\delta)y + y\delta.$$

Using the fact that $y \in C_G(M)$, we re-arrange this to give

$$\begin{aligned}(g\delta)(x - 1) &= (x\delta)(g - 1) - y\delta \\ &= \mu(x - 1)(g - 1) - y\delta \\ &= \mu(g - 1)(x - 1) - y\delta.\end{aligned}$$

Say $g\delta = p + A$, $\mu = q + A$ with $p, q \in M$. Then

$$(p - q(g - 1))(x - 1) + A = y\delta.$$

Since $y \in K$, (23) gives

$$r^c(p - q(g - 1))(x - 1) \in A$$

whence

$$r^{c+1}(p - q(g - 1))(x - 1) \in rA \leq A(x - 1).$$

As before, we deduce that $r^{c+1}(p - q(g - 1)) \in A$, whence $r^{c+1}g\delta = r^{c+1}\mu(g - 1)$. As $g \in X$ was arbitrary, this shows that $r^{c+1}\delta$ is the inner derivation on $r^{c+1}\mu$.

Thus $H^1(X, M/A)$ has finite exponent. Now $H^1(X, M/A)$ is a factor group of $\mathrm{Der}\,(X, M/A)$, which itself is isomorphic to a subgroup of $(\mathbb{Q}/\mathbb{Z})^{nd}$ if X is a d-generator group. It follows that $H^1(X, M/A)$ is finite as required (see Exercise 8 below). This completes the proof.

Exercise 8   Let $V = \mathbb{Q}^s$. If $U \leq V$ and $e \in \mathbb{N}$, show that $|U/eU| \leq e^s$.

(*Hint*: Suppose $u_1, \ldots, u_k$ lie in distinct cosets of eU in U. Show that

$U_1 = \langle u_1, \ldots, u_k \rangle$ is a free abelian group of rank at most s, and that $u_1, \ldots, u_k$ lie in distinct cosets of eU_1 in U_1.)

F. Notes

The main theorem is from Grunewald, Pickel & Segal (1980). I have managed to simplify and (I hope) clarify the proof to some extent; but there still remains a certain lack of elegance in the way the whole argument of section D seems to be repeated, with variations, in section E. My feeling is that one should be able to deduce the theorem 'in one go' from the result of Chapter 9; perhaps the methods of Donkin (1982) will turn out to provide the right approach, but this is for the future to show.

The question 'does $\mathscr{F}(G)$ determine G' for polycyclic groups G was raised by Hirsch some years ago, and soon gave rise to counterexamples (see Chapter 11). The present approach, leading to a finiteness theorem, originated with Borel, who pointed out that his Galois cohomology result (Theorem 2 in Chapter 9 above) implies a finiteness theorem for commensurability classes of $\mathscr{X}$-groups (via Exercise 4 in Chapter 9). This inspired Fred Pickel, who proved Theorem 1 in Pickel (1971b). The result was pushed further in Pickel (1973) (but this paper was more complicated than it had to be, because Theorem 6 of Chapter 8, above, had not been noticed at that time). Another special case of the main theorem is in Grunewald & Segal (1978b). For an outline of the whole argument, see the announcement Grunewald, Pickel & Segal (1979).

A good outline of Pickel's approach to Theorem 1, which is in some ways more elegant than the one I have presented here, is given in his announcement Pickel (1971a). Pickel (1974) gives example of infinitely many non-isomorphic finitely generated metabelian groups all having isomorphic finite quotients.

'Pickel's second theorem', about Fitting subgroups, is from Pickel (1976). The proof there is based on a neat trick with unipotent matrices, which I have perhaps rather obscured by putting it in the form of Proposition 8 in section C; as we have seen, it also plays an important role in the argument of Chapter 9.

Much of the material on profinite completions, in section B, is probably 'folklore'. A different approach to Proposition 5 will be found in Pickel's 1971 papers. Yet another way to get this result would be to use the natural isomorphism

$$\mathbb{Z}_p G / (\mathbb{Z}_p \mathfrak{g})^{c+1} \cong \mathbb{Z}_p \hat{G}_p / (\mathbb{Z}_p \hat{\mathfrak{g}}_p)^{c+1},$$

where G is a $\mathfrak{X}$-group of class c, and $\hat{\mathfrak{g}}_p$ is the augmentation ideal of $\hat{G}_p$. The reader might like to prove this as an exercise.

One should remark that $\hat{G}$ is the completion of G with respect to the profinite topology, a point which I failed to emphasize in the text. A discussion of inverse limits, including a self-contained proof of Proposition 1, will be found in Kegel & Wehrfritz (1973), section 1–K.

11
Examples

It is easy to construct polycyclic groups, and in this final chapter I shall describe various ways of doing so which illustrate points discussed so far. In a way, this chapter may be seen as a continuation of Chapter 4, section B: to construct groups with specified properties, we must in each case elucidate the arithmetical meaning of those properties, and then have to solve the resulting problem of number theory.

Section A constructs non-isomorphic polycyclic groups with isomorphic finite quotients; the simplest examples of these are directly related to algebraic number fields with class number greater than 1. In section B we consider groups which do not split over their Fitting subgroup, hence cannot be isomorphic to arithmetic groups.

The rest of the chapter concentrates on torsion-free nilpotent groups of class 2. These groups have a particularly simple description in 'arithmetical' terms, which is explained in section C. Those of Hirsch number at most 6 can be completely classified, in terms of equivalence classes of binary integral quadratic forms. This classification is given in section D; it illustrates a theme which I have tried to bring out throughout the book, by isomorphically embedding a small segment of the theory of polycyclic groups into an interesting and classical branch of number theory.

Finally, section E discusses unitriangular groups of matrices over rings of algebraic integers. Using the results of the previous section, we shall see that 'most' $\mathfrak{T}$-groups of class 2 and Hirsch number 6 arise as subgroups of finite index in $\mathrm{Tr}_1(2, \mathfrak{o})$, where $\mathfrak{o}$ is the ring of integers in a quadratic number field.

A. Groups with isomorphic finite quotients

The finiteness theorems of Chapters 9 and 10 were derived from certain finiteness properties of arithmetic groups. These properties are generalisations of two classical results about rings of algebraic integers: the finite generation of the group of units, and the finiteness of the number of ideal classes (i.e. isomorphism classes of ideals). The existence of number fields with class number greater than 1 implies that none of those finiteness theorems can be replaced with a 'uniqueness' theorem, in general.

Proposition 1   In $GL_2(\mathbb{Z})$ there exist $\approx$ -classes of infinite cyclic subgroups containing arbitrarily many (though $< \infty$) non-conjugate subgroups.

Proof   Here, $A \approx B$ means that $A\pi_m$ and $B\pi_m$ are conjugate in $GL_2(\mathbb{Z}/m\mathbb{Z})$ for every positive integer m. Now let $\mathfrak{o}$ be the ring of integers in a real quadratic number field $\mathbb{Q}(\sqrt{d})$, d a square-free positive integer. Provided that d is divisible by sufficiently many primes, the class number of this field can be arbitrarily large (see Borevich & Shafarevich: *Number Theory*, Chapter 3, section 8). Let $\mathfrak{a}_1 = \mathfrak{o}$, $\mathfrak{a}_2, \ldots, \mathfrak{a}_h$ be ideals of $\mathfrak{o}$ lying in distinct ideal classes, and let u be an element of infinite order in $\mathfrak{o}^*$ (such a u exists by the Units Theorem). The subring $\mathbb{Z}[u, u^{-1}] = S$, say, has finite index in $\mathfrak{o}$, and I claim that for $i \neq j$, $\mathfrak{a}_i$ and $\mathfrak{a}_j$ are non-isomorphic as S-modules. For suppose $\theta : \mathfrak{a}_i \to \mathfrak{a}_j$ is an S-module isomorphism; if $r \in \mathfrak{o}$ then $tr \in S$ for some $t \in \mathbb{N}$, and so for each $a \in \mathfrak{a}_i$ we have

$$t \cdot (ar)\theta = (a \cdot tr)\theta = a\theta \cdot tr = t \cdot (a\theta)r,$$

which implies that $(ar)\theta = (a\theta)r$. Thus θ is an $\mathfrak{o}$-module isomorphism, whence i must be equal to j. On the other hand, for each $m \in \mathbb{N}$ and for each i there is an $\mathfrak{o}/m\mathfrak{o}$-module isomorphism

$$\mathfrak{a}_i/m\mathfrak{a}_i \cong \mathfrak{o}/m\mathfrak{o},$$

because $\mathfrak{a}_i/m\mathfrak{a}_i$ is a 1-generator $\mathfrak{o}/m\mathfrak{o}$ -module and $|\mathfrak{a}_i/m\mathfrak{a}_i| = m^2 = |\mathfrak{o}/m\mathfrak{o}|$.

Now for each i choose a $\mathbb{Z}$-basis for $\mathfrak{a}_i$, and let x_i be the matrix of 'multiplication by u' w.r.t. this basis. The results of the last paragraph imply that

(i) x_i is not conjugate to x_j in $GL_2(\mathbb{Z})$ for $i \neq j$;
(ii) for each $m \in \mathbb{N}$ and each i, $x_i\pi_m$ is conjugate to $x_1\pi_m$ in $GL_2(\mathbb{Z}/m\mathbb{Z})$.

It follows that the subgroups $\langle x_1 \rangle, \ldots, \langle x_h \rangle$ of $GL_2(\mathbb{Z})$ lie in a single $\approx$ -class, but at least $[h/2]$ of them are pairwise non-conjugate (for if $\langle x_i \rangle$ is conjugate to $\langle x_j \rangle$ then x_i is conjugate to $x_j^{\pm 1}$). This establishes the proposition.

Proposition 2   For each positive integer n, there exists a $\frown$-class of torsion-free polycyclic groups, of Hirsch number 3, containing at least n non-isomorphic groups.

Proof   Let $x_1, \ldots, x_n \in GL_2(\mathbb{Z})$ be matrices of infinite order, generating pairwise non-conjugate subgroups in $GL_2(\mathbb{Z})$, such that $x_i\pi_m$ is conjugate to $x_1\pi_m$ in $GL_2(\mathbb{Z}/m\mathbb{Z})$ for every positive integer m and each i; the existence of suitable matrices was established in the proof of Proposition 1. For each i, let M_i be $\mathbb{Z}^2$ made into an $\langle x_i \rangle$-module in the natural way, and put $G_i = M_i]\langle x_i \rangle$.

Fix i and $j \le n$ and let m be a positive integer. By hypothesis, there exists $\gamma \in GL_2(\mathbb{Z}/m\mathbb{Z})$ such that $x_i \pi_m^\gamma = x_j \pi_m$. The map

$$G_i(m) = (M_i/mM_i)]\langle x_i \rangle \to (M_j/mM_j)]\langle x_j \rangle = G_j(m)$$

$$\mu \cdot x_i^e \mapsto \mu\gamma \cdot x_j^e$$

is easily seen to be an isomorphism. Since for each l,

$$G_l/G_l^m \cong G_l(m)/G_l(m)^m,$$

it follows that $G_i/G_i^m \cong G_j/G_j^m$. As i, j and m were arbitrary, this shows that the groups $G_1, \ldots, G_n$ all belong to the same $\wedge$-class.

Now suppose there is an isomorphism $\theta : G_i \to G_j$. If θ maps M_i onto M_j, we can represent $\theta|_{M_i}$ by a matrix $\gamma \in GL_2(\mathbb{Z})$, w.r.t. the chosen bases of M_i and M_j, and it is easy to see that then

$$\langle x_j \rangle = \gamma^{-1} \langle x_i \rangle \gamma.$$

By hypothesis, this can only happen if $i = j$. Thus to ensure that the groups $G_1, \ldots, G_n$ be pairwise non-isomorphic, it will suffice to arrange things so that M_i is the Fitting subgroup of G_i for each i. But if the matrices x_i are those given in Proposition 1, this will already be the case; for it is clear that no power of such a matrix can be unipotent. This finishes the proof.

Exercise 1 Take $\mathfrak{o}$, $\mathfrak{a}_i$ and $u \in \mathfrak{o}^*$ as in Proposition 1. Put

$$H_i = \begin{bmatrix} 1 & \mathfrak{a}_i \\ 0 & \langle u \rangle \end{bmatrix} \le \mathrm{Tr}_2(\mathfrak{o}).$$

Show that H_i is isomorphic to the group G_i of Proposition 2. Deduce that the groups G_i of Proposition 2 are isomorphic to commensurable arithmetic groups.

(*Hint*: See Proposition 3 in section A of Chapter 9.)

B. Non-arithmetic groups

At several points in the book, notably in Chapters 8 and 9, we have seen that arithmetic groups and polycyclic groups have rather similar properties; and in Chapter 6 we saw that every finitely generated nilpotent group is isomorphic to an arithmetic group. However, this is not true for polycyclic groups in general. To demonstrate this fact, we have to pick out some abstract structural feature which is universal to arithmetic groups, and then find a polycyclic group which does not have it.

Exercise 2 Let G be a soluble subgroup of $GL_n(\mathbb{Z})$ which has finite index in its Zariski closure in $GL_n(\mathbb{Z})$. Show that G has a subgroup H of finite index of the form

$$H = N]T,$$

where N is a $\mathfrak{X}$-group and T is abelian.
(*Hint*: Let $G_1 \leq_f G$ be triangularizable, and put $N = u(G_1)$; see Chapter 8, section B, Exercise 9. Let C be a nilpotent almost-supplement for N in G_1. Find $D \leq_f C$ such that $D_u D_s \leq NC$, and put $T = D_s$; see Chapter 7, section A.)

With Lemma 7 of Chapter 8, the exercise shows that if a polycyclic group G is to be isomorphic to an arithmetic group, then G must have a subgroup of finite index of the form $N]T$, with $N \in \mathfrak{X}$ and T abelian. Our task therefore is to construct polycyclic groups having no such subgroup of finite index.

Lemma 1 Let F be a group and $M \cong \mathbb{Z}^n$ an F-module, with $n \geq 2$. Assume that $C_F(M) \leq \zeta_1(F)$ and that M is rationally irreducible as an F_1-module for every subgroup F_1 of finite index in F. Let G be the semi-direct product $M]F$. If G has a subgroup H of finite index with $H = N]T$, $N \in \mathfrak{X}$ and T abelian, then F is abelian-by-finite.

Proof Put $Z = \zeta_1(N)$. Since $H \cap F \leq_f F$ and $Z \cap M$ is H-invariant, either $Z \cap M = 1$ or $|M : Z \cap M|$ is finite.
Case 1 where $Z \cap M = 1$. Then $N \cap M = 1$, since N is nilpotent, and therefore $H' \cap M = 1$. Consequently

$$[M \cap H, H] \leq H' \cap M = 1,$$

so H centralizes $M \cap H$. Since $M \cong \mathbb{Z}^n$ and $M \cap H$ has finite index in M, it follows that H centralizes M. But this is impossible: for $M \cong \mathbb{Z}^n$ with $n \geq 2$ and M is supposed to be rationally irreducible for $H \cap F$. So this case cannot occur, and we are left with
Case 2 where $Z \cap M \leq_f M$. Since N centralizes $Z \cap M$, N must now centralize M. Hence MN centralizes M. Put $K = C_F(M)$, and note that $C_G(M) = MK$. Now

$$K(MT \cap F) = KMT \cap F = C_G(M)T \cap F \geq MNT \cap F \geq H \cap F,$$

so $|F : K(MT \cap F)|$ is finite. But $MT \cap F$ is abelian, since T is abelian and $M \cap F = 1$; since, by hypothesis, $K \leq \zeta_1(F)$, it follows that $K(MT \cap F)$ is abelian. This now gives the result.

Let us set about finding a suitable F and M. It is easiest to construct modules M for which $M/C_F(M)$ is abelian, but we don't want F to be abelian-by-finite. The simplest candidate for F is then the free nilpotent group of class 2 on two generators:

$$F = \langle x, y; [x, y] \text{ central} \rangle$$
$$\cong \text{Tr}_1(3, \mathbb{Z}).$$

Exercise 3 By establishing this isomorphism, or otherwise, prove that F is not abelian-by-finite, that $F' = \zeta_1(F)$, and that F/F' is free abelian of rank 2.

Now put $A = F/F'$. To make our A-module, we proceed as in Chapters 2 and 4 by embedding A in the group of units $\mathfrak{o}^*$ of some ring of algebraic integers $\mathfrak{o}$. The condition on M stipulated in Lemma 1 then has a simple arithmetical interpretation:

Lemma 2 Let $\mathfrak{o}$ be the ring of integers in an algebraic number field k, and let B be a subgroup of $\mathfrak{o}^*$. Then $\mathfrak{o}$ is rationally irreducible as a B-module if and only if $k = \mathbb{Q}(B)$.

Proof Note that $k = \mathbb{Q}(B)$ if and only if B generates k as a $\mathbb{Q}$-algebra, since a finite-dimensional $\mathbb{Q}$-algebra which is an integral domain is necessary a field. Suppose $k = \mathbb{Q}(B)$. If L is a non-zero B-submodule of $\mathfrak{o}$, then

$$\mathbb{Q} \cdot L = L \cdot \mathbb{Q}B = L \cdot k = k = \mathbb{Q} \cdot \mathfrak{o};$$

thus $\mathfrak{o}$ is r. i. as a B-module. Conversely, suppose $k > \mathbb{Q}(B) = k_1$, say. Then $k_1 \cap \mathfrak{o}$ is a non-zero B-submodule of $\mathfrak{o}$, of smaller $\mathbb{Z}$-rank, and so $\mathfrak{o}$ is rationally reducible as B-module.

It is now a simple matter to find examples of the required kind. For the smallest possible one, let k be a *totally real cubic number field*: i.e. $(k:\mathbb{Q}) = 3$, and not only k but also each conjugate of k over $\mathbb{Q}$ is contained in $\mathbb{R}$. Let $\mathfrak{o}$ be the ring of integers of k, so $\mathfrak{o} \cong \mathbb{Z}^3$. By the Units Theorem, $\mathfrak{o}^* = T \times U$, say, with T finite and U free abelian of rank $3 + 0 - 1 = 2$. Now if $V \leq_f U$ then $\mathbb{Q}(V)$ must be equal to k, since there are no fields strictly intermediate between k and $\mathbb{Q}$; so Lemma 2 shows that $\mathfrak{o}$ is r.i. as a V-module for every such V. Putting it all together, what we have established is

Proposition 3 Let k be a totally real cubic number field with ring of integers $\mathfrak{o}$. Let F be the free nilpotent group of class 2 on 2 generators, let U be a free abelian subgroup of rank 2 in $\mathfrak{o}^*$, and make $\mathfrak{o}$ into an F-module M via

$$F \to F/F' \cong U,$$

where U acts on $\mathfrak{o}$ by multiplication. The polycyclic group

$$G = M \,]\, F$$

is then isomorphic to no arithmetic group.

Exercise 4   Find some totally real cubic number fields and construct corresponding explicit matrix representations of F in $GL_3(\mathbb{Z})$.

Exercise 5   Does there exist a polycyclic group of Hirsch number ≤ 5 which is isomorphic to no arithmetic group? (I don't know the answer to this.)

Exercise 6   Show that the group G given in Proposition 3 is splittable, in the sense of Chapter 7, section B; construct its semi-simple splitting. What is the degree of its canonical embedding (i.e. the number n_G of Chapter 7, secion C) going to be? (See Chapter 5, section B, Proposition 2 for help.)

C. $\mathfrak{T}$-groups of class 2

In Chapter 4, I called groups of the kind discussed in section A, above, 'group-theoretically trivial', and contrasted them with 'arithmetically trivial' nilpotent groups. Of course, this slur on nilpotent groups is not to be taken seriously; the point is simply that the interesting structure of a nilpotent group is not to be found by looking at the module-theoretic structure of its upper central factors, say. Instead, one has to study the nature of the *extensions* occurring within the group, of one factor by the next one one up. This is easy to do only when the group has class 2, in which case it is essentially a matter of linear algebra.

Let G be a $\mathfrak{T}$-group of class 2 and put $T = \tau_2(G)$ (the isolator of G'). Then $G/T \cong \mathbb{Z}^n$ and $T \cong \mathbb{Z}^m$ for certain positive integers n and m. The class of all such groups G will be denoted

$$\mathfrak{T}(n, m).$$

Since G' has finite index in T, $T \cong G'$ as abelian groups. Let $x_1, \ldots, x_n \in G$ be elements whose images modulo T form a basis for G/T. Since $T \leq \zeta_1(G)$ (see Chapter 1), the commutators $[x_i, x_j]$ with $1 \leq i < j \leq n$ generate G'. Therefore G' has rank at most $\frac{1}{2}n(n-1)$ as abelian group, and consequently

$$m \leq \tfrac{1}{2}n(n-1).$$

Now let F be the free nilpotent group of class 2 on n generators; that is,

$F = X/\gamma_3(X)$ where X is an (ordinary) free group on n generators. Let M be the free abelian group of rank m. There is an isomorphism

$$\pi_1 : M \to T.$$

and there is an epimorphism

$$\pi_2 : F \to \langle x_1, \ldots, x_n \rangle = H, \quad \text{say}.$$

Since $TH = G$ and T is central in G, these maps fit together to give an epimorphism

$$\pi : F \times M \to G.$$

Put

$$K = \ker \pi.$$

Let us describe K. Put $A = G/T$, and define a map

$$\psi : A \times A \to T$$

$$\psi(Tx, Ty) = [x, y], \quad x, y \in G.$$

One verifies immediately that ψ is

(i) *bilinear*, i.e. $\psi(a + b, c) = \psi(a, c) + \psi(b, c)$, $\psi(a, b + c) = \psi(a, b) + \psi(a, c)$;

(ii) *alternating*, i.e. $\psi(a, b) = -\psi(b, a)$;

(iii) *full*, i.e. $\langle \operatorname{Im} \psi \rangle$ has finite index in T.

Now since F/F' and $G/T = HT/T \cong H/(T \cap H)$ are both free abelian of rank n, the epimorphism π_2 induces an isomorphism

$$\pi_3 : \bar{F} = F/F' \to G/T = A.$$

We obtain now a map

$$\varphi = (\pi_3, \pi_3) \circ \psi \circ \pi_1^{-1} : \bar{F} \times \bar{F} \to M \tag{1}$$

and this map is also bilinear, alternating and full.

Lemma 3   Write $^- : F \to \bar{F} = F/F'$ for the natural map. Then

$$K = K_\varphi = \langle [x, y]^{-1} \cdot \varphi(\bar{x}, \bar{y}) \mid x, y \in F \rangle \leq F \times M. \tag{2}$$

Proof   Let K_φ be defined by (2). One computes directly that K_φ is mapped to the identity in G, so $K_\varphi \leq K$. On the other hand, it is easy to see that $K_\varphi M = F'M$. Since π_3 is an isomorphism, $K \leq F'M$. It follows that $K = K_\varphi(K \cap M)$. But $K \cap M = 1$ since π_1 is also an isomorphism, so $K = K_\varphi$ as claimed.

The upshot of the discussion is that every group in $\mathfrak{T}(n, m)$ can be realised, up to isomorphism, as one of the groups

$$G_\varphi = (F \times M)/K_\varphi \tag{3}$$

for some full alternating bilinear map $\varphi:\bar{F} \times \bar{F} \to M$, where K_φ is defined by (2) above. Conversely, every such G_φ does belong to $\mathfrak{X}(n,m)$:

Exercise 7   Let $\varphi:\bar{F} \times \bar{F} \to M$ be alternating, bilinear and full. Put $G = (F \times M)/K$ where K is given by (2). Show that $K \cap M = 1$. Deduce that $\tau_2(G) = KM/K \cong M$ and that $G/\tau_2(G) \cong F/F'$.
(*Hint*:   We need the fact that if F is freely generated by $y_1, \ldots, y_n$, then F' is the free abelian group with basis $\{[y_i, y_j] \mid 1 \leq i < j \leq n\}$; see Hall's *Nilpotent Groups*, Chapter 4. Now using the alternating and bilinear properties of the commutator map $\bar{F} \times \bar{F} \to F'$, show that for $u_1, \ldots, u_r, v_1, \ldots, v_r \in F$, we have

$$\prod_s [u_s, v_s] = 1 \Leftrightarrow \sum_s \varphi(\bar{u}_s, \bar{v}_s) = 0.$$

This shows that $K \cap M = 1$. For the rest, use the hypothesis that φ is full.)

In order to classify the groups in $\mathfrak{X}(n,m)$, we must elucidate the relation between bilinear maps which corresponds to the groups being isomorphic. The formula (1) gives the clue.

Definition   Let E, M be $\mathbb{Z}$-modules and $\varphi, \varphi':E \times E \to M$ bilinear maps. Then φ and φ' are *equivalent*, $\varphi \sim \varphi'$, if there exist automorphisms η of E and μ of M such that

$$\varphi' = (\eta, \eta) \circ \varphi \circ \mu.$$

This is clearly an equivalence relation; if φ is alternating (respectively full) and $\varphi \sim \varphi'$, then φ' is likewise alternating (respectively full). Now to define the map φ in (1), starting from a given group G, we chose certain maps $\pi_1: M \to \tau_2(G) = T$ and $\pi_2:F \to H \leq G$; it is easy to see that a different choice of these maps would have the effect of replacing π_1 by $\mu\pi_1$, for some $\mu \in \text{Aut } M$, and π_3 by $\eta\pi_3$, for some $\eta \in \text{Aut } \bar{F}$. Hence φ is uniquely determined up to equivalence by G. Similarly, if $G_1 \cong G$ and $\varphi_1:\bar{F} \times \bar{F} \to M$ is the map corresponding to G_1, we find that $\varphi_1 \sim \varphi$; for we can map $M \to \tau_2(G) \rightarrowtail \tau_2(G_1)$ and $F \to G \rightarrowtail G_1$ to get a suitable 'π_1' and 'π_2' for G_1. Thus *isomorphic groups give equivalent bilinear maps.*

Conversely, suppose $\varphi, \varphi':\bar{F} \times \bar{F} \to M$ are equivalent full alternating bilinear maps. I claim that the groups G_φ and $G_{\varphi'}$ are then isomorphic. This follows from

Lemma 4   Let $\varphi, \varphi':\bar{F} \times \bar{F} \to M$ be full alternating bilinear maps. Let $\eta \in \text{Aut } \bar{F}$ and $\mu \in \text{Aut } M$ be such that

$$\varphi' = (\eta, \eta) \circ \varphi \circ \mu.$$

Then there is an automorphism γ of $F \times M$ such that

$$K_\varphi^\gamma = K_{\varphi'}.$$

Proof Let $\{y_1, \ldots, y_n\}$ be a set of free generators for F, and for each i choose $w_i \in F$ so that $\bar{y}_i \eta^{-1} = F' w_i$. Since F is free nilpotent, there is a homomorphism $\sigma : F \to F$ with $y_i \sigma = w_i$ for $i = 1, \ldots, n$. Since η^{-1} maps $\bar{F}$ onto $\bar{F}$, we have

$$F = F'\langle w_1, \ldots, w_n \rangle = F' \cdot F\sigma;$$

as F is nilpotent it follows that $F\sigma = F$. Consideration of the Hirsch number shows that $\ker \sigma$ is then finite, so since F is torsion-free we conclude that $\sigma \in \operatorname{Aut} F$. By construction, σ induces the automorphism η^{-1} on $\bar{F} = F/F'$. We now put

$$\gamma = (\sigma, \mu) : F \times M \to F \times M,$$

and have to verify that $K_\varphi^\gamma = K_{\varphi'}$. This is simply a matter of checking what γ does to the given generators

$$[x, y]^{-1} \cdot \varphi(\bar{x}, \bar{y})$$

of K_φ, and is left to the reader.

Let us sum up the results:

Proposition 4 Let F be the free nilpotent group of class 2 on $n \geq 2$ generators, and let M be the free abelian group on m generators, where n and $m \leq \frac{1}{2}n(n-1)$ are positive integers. The assignment $\varphi \mapsto G_\varphi$ defined at (3) and (2) induces a bijective correspondence between the set of all equivalence classes of full alternating bilinear maps $\bar{F} \times \bar{F} \to M$ and the set of all isomorphism classes of groups in $\mathfrak{X}(n, m)$, where $\bar{F} = F/F'$.

This reduces the theory of $\mathfrak{X}$-groups of class 2 to linear algebra over $\mathbb{Z}$. In fact the correspondence we have set up can be carried further. Suppose $\varphi, \varphi' : E \times E \to M$ are bilinear maps; say $\varphi \sim_\mathbb{Q} \varphi'$ if there exist $\eta \in \operatorname{Aut}(\mathbb{Q} \otimes E)$ and $\mu \in \operatorname{Aut}(\mathbb{Q} \otimes M)$ such that $\bar{\varphi}' = (\eta, \eta) \circ \bar{\varphi} \circ \mu$, where $\bar{\varphi}, \bar{\varphi}' : (\mathbb{Q} \otimes E) \times (\mathbb{Q} \otimes E) \to (\mathbb{Q} \otimes M)$ are the $\mathbb{Q}$-bilinear extensions of φ and φ'. Analogously, for each prime p one defines a relation $\varphi \sim_{\mathbb{Z}_p} \varphi'$. It can then be shown, by arguments similar to the above, that for full alternating bilinear maps $\varphi, \varphi' : \bar{F} \times \bar{F} \to M$, we have

$$G_\varphi^\mathbb{Q} \cong G_{\varphi'}^\mathbb{Q} \Leftrightarrow \varphi \sim_\mathbb{Q} \varphi' \tag{4}$$

$$(\hat{G}_\varphi)_p \cong (\hat{G}_{\varphi'})_p \Leftrightarrow \varphi \sim_{\mathbb{Z}_p} \varphi' \tag{5}$$

$$\hat{G}_\varphi \cong \hat{G}_{\varphi'} \Leftrightarrow \varphi \sim_{\mathbb{Z}_p} \varphi' \quad \text{for all primes } p. \tag{6}$$

For the details, see the reference in section F. (Here $G^\mathbb{Q}$ denotes the radicable hull, $\hat{G}_p$ the pro-p completion, and $\hat{G}$ the profinite completion of a $\mathfrak{X}$-group G; see Chapter 10, section B.)

D. 𝔗-groups of small Hirsch number

If we want to get specific classification results for 𝔗-groups of class 2, we must set about classifying bilinear maps from $\mathbb{Z}^n \times \mathbb{Z}^n \to \mathbb{Z}^m$. Such a map is represented by an m-tuple of $n \times n$ matrices over $\mathbb{Z}$, skew-symmetric ones if the map is alternating:

$$(\mathbf{X}, \mathbf{Y}) \in \mathbb{Z}^n \times \mathbb{Z}^n \mapsto (\mathbf{X}^t A^1 \mathbf{Y}, \dots, \mathbf{X}^t A^m \mathbf{Y}) \in \mathbb{Z}^m$$

where $\mathbf{X}^t$ denotes the transpose of the column vector $\mathbf{X}$, and $A^1, \dots, A^m$ are $n \times n$ matrices. It is convenient to consider this m-tuple of matrices as a single $n \times n$ matrix with entries in $\mathbb{Z}^m$, so we write

$$(A^1, \dots, A^m) = \mathbf{A}.$$

The bilinear maps represented by $\mathbf{A}$ and by $\mathbf{B}$ are equivalent (in the sense of section C) if there exist $P \in GL_n(\mathbb{Z})$ and $Q \in GL_m(\mathbb{Z})$ such that

$$P^t A^i P = \sum_{j=1}^m B^j Q_{ji}, \quad i = 1, \dots, m \tag{7}$$

or in abbreviated notation

$$P^t \mathbf{A} P = \mathbf{B} \cdot Q.$$

The classification of alternating bilinear maps in general seems to be a rather difficult problem; in fact it is completely open, as far as I know, for $m > 2$. The case $m = 1$, however, is straightforward and has long been known:

Lemma 5   Let A be a skew-symmetric $n \times n$ matrix over $\mathbb{Z}$. Then there exists $P \in GL_n(\mathbb{Z})$ such that

$$P^t A P = \begin{cases} \mathrm{diag}(D_1, \dots, D_r) & n = 2r \\ \mathrm{diag}(D_1, \dots, D_r, 0) & n = 2r + 1 \end{cases} \tag{8}$$

where $D_i = \begin{pmatrix} 0 & d_i \\ -d_i & 0 \end{pmatrix}$ for each i, and $d_1, \dots, d_r$ are non-negative integers with $d_1 | d_2 | \dots | d_r$. Moreover, the numbers $d_1, \dots, d_r$ are uniquely determined by A.

For the proof, see M. Newman, *Integral Matrices*, Theorem IV.2. The equivalance relation which we are considering allows also the operation of $GL_1(\mathbb{Z})$ on the entries of the given matrix. But this is merely multiplication by ± 1; since

$$\begin{pmatrix} -1 & 0 \\ 0 & 1 \end{pmatrix} \begin{pmatrix} 0 & d \\ -d & 0 \end{pmatrix} \begin{pmatrix} -1 & 0 \\ 0 & 1 \end{pmatrix} = \begin{pmatrix} 0 & -d \\ d & 0 \end{pmatrix},$$

this operation can be 'swallowed up' by the action of $GL_n(\mathbb{Z})$ 'on the other side'. Thus in view of Proposition 4, the lemma provides a complete classification of groups in $\mathfrak{T}(n, 1)$, for every n:

Proposition 5   Let n be an integer ≥ 2. Then every group in $\mathfrak{X}(n, 1)$ is isomorphic to exactly one of the following groups:

(a) $n = 2r$.

$$G(d_1, \ldots, d_r) = \langle x_1, \ldots, x_r, y_1, \ldots, y_r, z; [x_i, y_j] = [x_i, x_j] = [y_i, y_j]$$
$$= 1 \text{ for } i \neq j, [x_i, y_i] = z^{d_i}, \text{ each } i, z \text{ central} \rangle$$

where $d_1 | d_2 | \ldots | d_r$, each d_i is a non-negative integer, and $d_1 \neq 0$.

(b) $n = 2r + 1$.

$$G(d_1, \ldots, d_r) \times C_\infty$$

with $d_1, \ldots, d_r$ as above.

Exercise 8   Prove Proposition 5 in detail.

Corollary 1   If G and H are groups in $\mathfrak{X}(n, 1)$ and $\hat{G} \cong \hat{H}$ then $G \cong H$.

Proof   This depends on (6) in section C. What it comes down to is this: let

$$A = \text{diag}(D_1, \ldots, D_r, (0))$$
$$B = \text{diag}(D_1', \ldots, D_r', (0))$$

be as in (8); suppose that for each prime p there exist $P_p \in GL_n(\mathbb{Z}_p)$ and $u_p \in \mathbb{Z}_p^*$ such that

$$P_p^t A P_p = u_p B;$$

then we must deduce that $A = B$. Now the invariant factors of A as a matrix over the P.I.D. $\mathbb{Z}_p$ (i.e. the diagonal entries when A is put into 'Smith normal form') are easily seen to be $(d_1, d_1, d_2, d_2, \ldots, d_r, d_r, (0))$, while those of B are the corresponding numbers d_i'. Consequently, for each $i \leq r$ and for each prime p there exists a unit $v_p \in \mathbb{Z}_p^*$ such that $d_i v_p = d_i'$. Since

$$\mathbb{Q} \cap \bigcap_p \mathbb{Z}_p^* = \{\pm 1\}$$

and the numbers d_i, d_i' are non-negative, it follows that $d_i = d_i'$ for each i. Thus $A = B$ as required.

Using *ad hoc* arguments, one can similarly classify $\mathfrak{X}(3, 2)$ and $\mathfrak{X}(3, 3)$; see section F for references (or do it as an exercise). In these cases also, one finds that $\hat{G} \cong \hat{H}$ implies $G \cong H$. Things get more interesting when we come to consider $\mathfrak{X}(4, 2)$.

Let s_{ij}^1, s_{ij}^2 ($1 \leq i < j \leq 4$) be 12 independent indeterminates, let R be the polynomial ring $\mathbb{Z}[s_{11}^1, \ldots, s_{34}^1, s_{11}^2, \ldots, s_{34}^2]$, and let $\mathbf{S} = (S^1, S^2)$ be the skew-symmetric 4×4 matrix over R^2 whose (i,j)-entry is the vector (s_{ij}^1, s_{ij}^2). Take two new indeterminates X, Y and consider the matrix $S^1 X + S^2 Y$; this has entries in the polynomial ring $R[X, Y]$. If F is the field of fractions of $R[X, Y]$, there exists $W \in GL_4(F)$ such that

$$W^t(S^1 X + S^2 Y)W = \operatorname{diag}\left(\begin{pmatrix} 0 & 1 \\ -1 & 0 \end{pmatrix}, \begin{pmatrix} 0 & 1 \\ -1 & 0 \end{pmatrix}\right).$$

Then

$$\det(S^1 X + S^2 Y) = (\det W)^{-2}$$

is a square in the field F. It follows (by Gauss's Lemma) that $\det(S^1 X + S^2 Y)$ is equal to $\Phi(X, Y)^2$ for some $\Phi \in R[X, Y]$. As $\det(S^1 X + S^2 Y)$ is homogeneous of degree 4 in X and Y, we see that Φ must be homogeneous of degree 2, i.e. a quadratic form in X and Y (we can fix the sign of Φ, but this doesn't really matter here). Now if $A = (A^1, A^2)$ is a skew-symmetric 4×4 matrix over $\mathbb{Z}^2$, we define

$$\operatorname{Pf}_A(X, Y)$$

to be the quadratic form, over $\mathbb{Z}$, obtained on substituting the entries a_{ij}^1, a_{ij}^2 of A for the indeterminates s_{ij}^1, s_{ij}^2 in the coefficients of Φ. This is the *Pfaffian* of the matrix $A^1 X + A^2 Y$; see Lang's *Algebra*, Chapter XIV section 10.

Suppose B is another skew-symmetric 4×4 matrix over $\mathbb{Z}^2$ and that B is equivalent to A in the sense of (7), above. Then

$$P^t(A^1 X + A^2 Y)P = (B^1 Q_{11} + B^2 Q_{21})X + (B^1 Q_{12} + B^2 Q_{22})Y$$
$$= B^1(Q_{11} X + Q_{12} Y) + B^2(Q_{21} X + Q_{22} Y).$$

Here $P \in GL_4(\mathbb{Z})$ and $Q \in GL_2(\mathbb{Z})$; so $\det P = \pm 1$, and consequently

$$\operatorname{Pf}_A(X, Y) = \pm \operatorname{Pf}_B(X', Y')$$

where

$$\begin{pmatrix} X' \\ Y' \end{pmatrix} = Q \begin{pmatrix} X \\ Y \end{pmatrix}.$$

In the usual terminology, the integral quadratic form Pf_A is *equivalent* to $\pm \operatorname{Pf}_B$. If we call this relation between quadratic forms 'weak equivalence', we see that to each equivalence class of alternating bilinear maps $\mathbb{Z}^4 \times \mathbb{Z}^4 \to \mathbb{Z}^2$ there is associated a unique weak equivalence class of binary quadratic forms.

In fact all of this works as well for alternating bilinear maps $\mathbb{Z}^n \times \mathbb{Z}^n \to \mathbb{Z}^m$: if n is even one obtains a class of forms of degree $n/2$ in m variables, while if n is odd one gets the form which is identically zero. The remarkable feature of the case $n = 4, m = 2$ is that here, conversely, the Pfaffian actually determines the bilinear map up to equivalence. The key to this result is a 'canonical form' theorem:

Lemma 6   Let A^1 and A^2 be skew-symmetric 4×4 matrices over $\mathbb{Z}$, and assume that the vectors $\mathbf{a}_{ij} = (a_{ij}^1, a_{ij}^2)$, $1 \le i < j \le 4$, span $\mathbb{Z}^2$. Then there exists a matrix $P \in GL_4(\mathbb{Z})$ such that

$$P^t A P = \begin{bmatrix} 0 & 0 & (0\ 1) & (1\ 0) \\ 0 & 0 & (\alpha\ \beta) & (0\ \gamma) \\ -(0\ 1) & -(\alpha\ \beta) & 0 & 0 \\ -(1\ 0) & -(0\ \gamma) & 0 & 0 \end{bmatrix}. \tag{9}$$

I shall give the proof later. Let us denote the matrix on the right-hand side of (9) by $\mathbf{C}(\alpha, \beta, \gamma) = (C^1, C^2)$, say. Then

$$\mathrm{Pf}_{\mathbf{A}}(X, Y) = \pm \mathrm{Pf}_{\mathbf{C}(\alpha, \beta, \gamma)}(X, Y) = \pm \sqrt{\det(C^1 X + C^2 Y)}$$

$$= \pm \det\begin{pmatrix} Y & X \\ \alpha X + \beta Y & \gamma Y \end{pmatrix}$$

$$= \pm (\alpha X^2 + \beta XY - \gamma Y^2);$$

so the 'canonical form' (9) is uniquely determined by $\mathbf{A}$ up to a sign.

The classification we are seeking now follows easily. For simplicity, I shall give the result only for bilinear maps $\varphi : \mathbb{Z}^4 \times \mathbb{Z}^4 \to \mathbb{Z}^2$ for which $\langle \mathrm{Im}\, \varphi \rangle = \mathbb{Z}^2$; these correspond via Proposition 4 to groups $G \in \mathfrak{T}(4, 2)$ with $\tau_2(G) = G'$, and the matrix $\mathbf{A}_\varphi$ representing such a map φ satisfies the condition in the statement of Lemma 6. The general case is a little more involved; see the reference in section F.

Proposition 6   The assignment $\varphi \mapsto \mathrm{Pf}_{\mathbf{A}_\varphi}$ induces a bijective correspondence between the set of all equivalence classes of alternating bilinear maps $\varphi : \mathbb{Z}^4 \times \mathbb{Z}^4 \to \mathbb{Z}^2$ for which $\langle \mathrm{Im}\, \varphi \rangle = \mathbb{Z}^2$ and the set of all weak equivalence classes of binary integral quadratic forms.

Proof   Here $\mathbf{A}_\varphi$ is the matrix representing φ, and to save space I will write Pf_φ for $\mathrm{Pf}_{\mathbf{A}_\varphi}$. We have already seen that if $\varphi \sim \varphi'$ then $\mathrm{Pf}_\varphi \sim \pm \mathrm{Pf}_{\varphi'}$, so the assignment $\varphi \mapsto \mathrm{Pf}_\varphi$ induces a well defined map of equivalence classes. It is easy to see that this map is surjective: given a quadratic form $\Phi(X, Y) = \alpha X^2 + \beta XY - \gamma Y^2$, take φ to be the map represented by the matrix $\mathbf{C}(\alpha, \beta, \gamma)$; then $\mathrm{Pf}_\varphi = \pm \Phi$, and since $(0\ 1)$ and $(1\ 0)$ belong to $\mathrm{Im}\, \varphi$ we also have $\langle \mathrm{Im}\, \varphi \rangle = \mathbb{Z}^2$.

To show that our correspondence is $1:1$ on equivalence classes, let φ, φ' be alternating bilinear maps with $\langle \mathrm{Im}\, \varphi \rangle = \langle \mathrm{Im}\, \varphi' \rangle = \mathbb{Z}^2$, and suppose that $Pf_\varphi \sim \pm Pf_{\varphi'}$. Let $\mathbf{A}$, $\mathbf{B}$ be the matrices representing φ and φ' respectively. Then by hypothesis, there exists $Q \in GL_2(\mathbb{Z})$ such that

$$Pf_\mathbf{B}(\mathbf{X}) = \pm Pf_\mathbf{A}(Q\mathbf{X}),$$

where $\mathbf{X}$ is the column vector (X, Y). By Lemma 6, there exist $P, R \in GL_4(\mathbb{Z})$ such that

$$P^t(\mathbf{A} \cdot Q)P = \mathbf{C}(\alpha, \beta, \gamma) = \mathbf{C}, \text{ say}$$
$$R^t\mathbf{B}R = \mathbf{C}(\alpha', \beta', \gamma') = \mathbf{C}', \text{ say}.$$

Comparing the Pfaffians, we find that

$$(\alpha, \beta, \gamma) = \varepsilon(\alpha', \beta', \gamma')$$

with $\varepsilon = \pm 1$. If $\varepsilon = 1$ we have $\mathbf{C}' = \mathbf{C}$, so

$$\mathbf{B} = R^{-t}\mathbf{C}'R^{-1} = R^{-t}\mathbf{C}R^{-1} = S^t(\mathbf{A} \cdot Q)S$$

where $S = PR^{-1}$. If $\varepsilon = -1$, put $D = \mathrm{diag}\,(1\ -1\ 1\ 1)$ and observe that

$$D^t\mathbf{C}D = \mathbf{C}'.$$

Thus we get the same result on taking $S = PDR^{-1}$. In each case, the conclusion is that $\varphi \sim \varphi'$, as required.

With Proposition 4, this now gives

Theorem 1   The assignemnt $G_\varphi \mapsto Pf_\varphi$ induces a bijective correspondence between the set of all isomorphism classes of groups $G \in \mathfrak{T}(4, 2)$ with G/G' free abelian and the set of all weak equivalence classes of binary integral quadratic forms (as φ runs over all alternating bilinear maps $\mathbb{Z}^4 \times \mathbb{Z}^4 \to \mathbb{Z}^2$ such that $\langle \mathrm{Im}\,\varphi \rangle = \mathbb{Z}^2$, and G_φ is as defined in section C).

Exercise 9   Give a presentation by generators and relations for the group corresponding to the quadratic form $\alpha X^2 + \beta XY - \gamma Y^2$.

There are analogues of Lemma 6 with $\mathbb{Q}$ and with $\mathbb{Z}_p$ in place of $\mathbb{Z}$, with essentially the same proof. The argument of Proposition 6 then shows that for alternating bilinear maps φ, $\varphi' : \mathbb{Z}^4 \times \mathbb{Z}^4 \to \mathbb{Z}^2$ with $\langle \mathrm{Im}\,\varphi \rangle = \langle \mathrm{Im}\,\varphi' \rangle = \mathbb{Z}^2$, one has $\varphi \sim_\mathbb{Q} \varphi'$ if and only if Pf_φ is rationally equivalent to $u \cdot Pf_{\varphi'}$ for some $u \in \mathbb{Q}^*$, and $\varphi \sim_{\mathbb{Z}_p} \varphi'$ if and only if Pf_φ is p-adically equivalent to $u_p \cdot Pf_{\varphi'}$ for some p-adic unit u_p. Now the classification of binary quadratic forms under these equivalence relations is very simple. Define the *discriminant* of $f = aX^2 + bXY + cY^2$ to be

$$D(f) = b^2 - 4ac,$$

and the *content* to be

$$C(f) = \mathrm{h.c.f.}(a, b, c).$$

Then the following hold:

(i) $f \sim_\mathbb{Q} u \cdot g$, some $u \in \mathbb{Q}^* \Leftrightarrow D(f)/D(g) \in \mathbb{Q}^{*^2}$;

(ii) $f \sim_{\mathbb{Z}_p} u_p \cdot g$, some $u_p \in \mathbb{Z}_p^* \Leftrightarrow C(f)/C(g) \in \mathbb{Z}_p^*$ and $D(f)/D(g) \in \mathbb{Z}_p^{*^2}$

(here, $0/0$ is to be interpreted as 1). to establish (i) is an elementary exercise; (ii) follows from the classification of binary forms over $\mathbb{Z}_p$: see Cassells, *Rational Quadratic Forms*, Chapter 8.

If a group G is isomorphic to G_φ, define

$$C(G) = C(\mathrm{Pf}_\varphi), \quad D(G) = D(\mathrm{Pf}_\varphi).$$

This is a well defined invariant for groups in $\mathfrak{X}(4, 2)$; for it is easy to see that both C and D are invariants of binary quadratic forms with respect to weak equivalence. If we accept the (unproved) statements (4) and (6) at the end of section C, we can now infer

Theorem 2   Let G and H be groups in $\mathfrak{X}(4, 2)$ with G/G' and H/H' free abelian. Then

$$G^\mathbb{Q} \cong H^\mathbb{Q} \Leftrightarrow D(G)/D(H) \in \mathbb{Q}^{*^2} ;$$
$$\hat{G} \cong \hat{H} \Leftrightarrow C(G) = C(H) \text{ and } D(G) = D(H).$$

(For the second claim, note that

$$\mathbb{Q} \cap \bigcap_p \mathbb{Z}_p^{*^2} = \{1\}.)$$

Corollary 2   (i) Let $G, H \in \mathfrak{X}(4, 2)$ have free abelian derived factor groups. Then

$$G \cong H \Rightarrow \hat{G} \cong \hat{H} \Rightarrow G^\mathbb{Q} \cong H^\mathbb{Q}.$$

(ii)   There exist ^-classes of groups in $\mathfrak{X}(4, 2)$ containing arbitrarily many non-isomorphic groups.

Proof   (i) is immediate from Theorem 2. (ii) follows from the fact that for each n one can find integers D such that there are at least $2n$ inequivalent binary quadratic forms of discriminant D (hence at least n such which are not weakly equivalent); see Cassells, *op. cit.* page 359, or Borevich & Shafarevich, *Number Theory*, page 244. (it was the same fact which lay behind the construction in section A).

To conclude this section, here is some old-fashioned matrix algebra:

Proof of Lemma 6   Given $\mathbf{A} = (A^1, A^2)$, a skew-symmetric matrix with entries in $\mathbb{Z}^2$, such that

$$\sum_{i,j} \mathbb{Z}\mathbf{a}_{ij} = \mathbb{Z}^2, \tag{10}$$

we must find $P \in GL_4(\mathbb{Z})$ such that

$$P^t A P = \mathbf{C}(\alpha, \beta, \gamma).$$

Here $\mathbf{a}_{ij} = (a_{ij}^1, a_{ij}^2)$ denotes the (i, j)-entry of $\mathbf{A}$. I shall use the notation

$$\mathbf{A} = [\mathbf{a}_{12}, \mathbf{a}_{13}, \mathbf{a}_{14}, \mathbf{a}_{23}, \mathbf{a}_{24}, \mathbf{a}_{34}]$$

to save space (it specifies $\mathbf{A}$ completely because $\mathbf{A}$ is skew-symmetric). In this notation,

$$\mathbf{C}(\alpha, \beta, \gamma) = [0, (0\ 1), (1\ 0), (\alpha\ \beta), (0\ \gamma), 0]$$

(but the reader is advised to write out the matrices in the usual form when checking the following computations!).

Step 1 Put A^1 into canonical form (see Lemma 5). Thus we find $P_1 \in GL_4(\mathbb{Z})$ such that

$$P_1^t A^1 P_1 = \operatorname{diag}\left[\begin{pmatrix} 0 & r \\ -r & 0 \end{pmatrix}, \begin{pmatrix} 0 & s \\ -s & 0 \end{pmatrix} \right]$$

with $r|s$ and $r \geq 0$. Then (10) ensures that $r = 1$.

Let $M \in M_2(\mathbb{Z})$ be the 'top right-hand corner' of $P_1^t A^2 P_1$, i.e.

$$P_1^t A^2 P_1 = \begin{bmatrix} ? & M \\ -M^t & ? \end{bmatrix}.$$

Step 2 Put M into Smith normal form. That is, we find $S, T \in SL_2(\mathbb{Z})$ such that $SMT = \operatorname{diag}(a, b)$ with $a|b$. Put

$$P_2 = P_1 \cdot \begin{pmatrix} S^t & 0 \\ 0 & T \end{pmatrix}.$$

We then compute:

$$P_2^t A P_2 = [(1\ \lambda), (0\ a), (0\ 0), (0\ 0), (0\ b), (s\ \mu)]$$

for some $\lambda, \mu \in \mathbb{Z}$.

Step 3 Subtract row 3 from row 1 and add $s \times$ (row 2) to row 4, and do the analogous operation on columns. If Q is the elementary matrix representing this operation and $P_3 = P_2 Q$, we obtain

$$P_3^t A P_3 = [(1\ \lambda), (0\ a), (0\ s\lambda - \mu), (0\ 0), (0\ b), (s\ \mu)].$$

Step 4 As in Step 2, find $S', T' \in SL_2(\mathbb{Z})$ such that

$$S'\begin{pmatrix} a & s\lambda - \mu \\ 0 & b \end{pmatrix} T' = \begin{pmatrix} a_1 & 0 \\ 0 & b_1 \end{pmatrix}$$

with $a_1 \geq 0$ and $a_1|b_1$. Note that then $a_1 = \text{h.c.f.}(a, b, s\lambda - \mu)$. Now put

$$P_4 = P_3 \cdot \begin{pmatrix} S'' & 0 \\ 0 & T' \end{pmatrix}$$

to get

$$P_4^t A P_4 = [(1\ \lambda), (0\ a_1), \mathbf{0},\ \mathbf{0}, (0\ b_1), (s\ \mu)].$$

Since $a_1|s\lambda - \mu$, we have $(s\ \mu) = x(0\ a_1) + s(1\ \lambda)$ for some $x \in \mathbb{Z}$; therefore, from (10), $\det\begin{pmatrix} 0 & a_1 \\ 1 & \lambda \end{pmatrix} = \pm 1$, and so we have $a_1 = 1$.

Step 5 Interchange row 4 with row 2 and column 4 with column 2. If P_5 is P_4 times the corresponding permutation matrix, we then have

$$P_5^t A P_5 = [\mathbf{0}, (0\ 1), (1\ \lambda),\ -(s\ \mu),\ -(0\ b_1),\ \mathbf{0}].$$

Step 6 Subtract $\lambda \times$ (column 3) from column 4, subtract $\lambda s \times$ (row 1) from row 2, and do the analogous row (respectively column) operations. If P is P_5 times the corresponding elementary matrix, we find that

$$P^t A P = [\mathbf{0}, (0\ 1), (1\ 0),\ -(s\ \mu + \lambda s), (0\ \lambda\mu - b_1), \mathbf{0}]$$
$$= \mathbf{C}(\alpha, \beta, \gamma)$$

where $\alpha = -s$, $\beta = -(\mu + \lambda s)$, $\gamma = \lambda\mu - b_1$. This completes the reduction of A to canonical form.

E. Some matrix groups

An obvious supply of polycyclic groups is provided by the subgroups of $\text{Tr}_n(\mathfrak{o})$ where $\mathfrak{o}$ is the ring of integers in an algebraic number field. Here we shall analyse some of these in detail. Let R be a commutative ring with 1, I an ideal of R, and M an additive subgroup of R containing I. Then

$$G(R, I, M) = \begin{pmatrix} 1 & R & M \\ 0 & 1 & I \\ 0 & 0 & 1 \end{pmatrix} \leq \text{Tr}_1(3, R)$$

is a subgroup of $\text{Tr}_1(3, R)$, so it is nilpotent of class at most 2. I shall use the notation

$$(x, y, z) = \begin{pmatrix} 1 & x & z \\ 0 & 1 & y \\ 0 & 0 & 1 \end{pmatrix}$$

$$(X, Y, Z) = \{(x, y, z,)|x \in X, y \in Y, z \in Z\}.$$

Matrix calculation shows that

$$[(x, y, z), (u, v, w)] = (0, 0, xv - yu); \tag{11}$$

with this it is easy to verify

Lemma 7 Let $G = G(R, I, M)$. Then $G' = (0, 0, I)$, $\zeta_1(G) =$ $(0, 0, M)$, and $G/G' \cong R \oplus I \oplus M/I$; if R is additively torsion-free and M/I is periodic, then also $\tau_2(G) = (0, 0, M) \cong M$, $G/\tau_2(G) \cong R \oplus I$ and $\tau_2(G)/G' \cong M/I$.

Exercise 10 (i) Show that if $I \cong J$ as R-modules then $G(R, I, I) \cong$ $G(R, J, J)$. (ii) Suppose $I/mI \cong J/mJ$ as R/mR-modules for every $m \in \mathbb{N}$; show that then $\mathscr{F}(G(R, I, I)) = \mathscr{F}(G(R, J, J))$.
(*Hint*: for (ii), write $G = G(R, I, I)$ and $H = G(R, J, J)$; show that $(mR, mI, mI) \leq G^m$, $(mR, mJ, mJ) \leq H^m$, and that $G/(mR, mI, mI) \cong H/(mJ, mJ, mJ)$.)

If $R \cong \mathbb{Z}^a$ and $I \cong M \cong \mathbb{Z}^b$, we see from Lemma 7 that $G(R, I, M) \in$ $\mathfrak{X}(a + b, b)$. Henceforth we specialise to the case $a = b = 2$, and concentrate mainly on what happens for $M = I$.

Consider, then, a ring R with additive group $\mathbb{Z} \oplus \mathbb{Z}$, and an ideal I of finite index in R. The group $G = G(R, I, I)$ belongs to $\mathfrak{X}(4, 2)$ and $\tau_2(G) = G'$. So let us see where G fits in the classification of Theorem 1, in the previous section. Now $G \cong G_\varphi$ where $\varphi : \mathbb{Z}^4 \times \mathbb{Z}^4 \to \mathbb{Z}^2$ is the bilinear map induced by the commuatator $G/G' \times G/G' \to G'$, and we have to compute the quadratic form Pf_φ. First, the structure of the ring R must be clarified:

Exercise 11 Let $R \cong \mathbb{Z} \oplus \mathbb{Z}$ be a commutative ring with 1. (i) Show that $R = \mathbb{Z} \oplus \omega\mathbb{Z}$ where ω satisfies an equation

$$F(\omega) := \omega^2 - \alpha\omega - \beta = 0 \tag{12}$$

with $\beta \in \mathbb{Z}$ and $\alpha = 1$ or 0. (ii) Let I be an additive subgroup of finite index in R. Show that $I = a\mathbb{Z} + (b - c\omega)\mathbb{Z}$ for some $a, b, c \in \mathbb{Z}$ with $ac \neq 0$. (iii) Show that $I \lhd R$ if and only if

$$c|a, \quad c|b, \quad \text{and} \quad a/c \,|\, F(b/c). \tag{13}$$

We keep the notation of Exercise 11, and assume that (12) and (13) hold. In view of Exercise 10 (i), no generality is lost if we replace I by the ideal $c^{-1}I$; so we may as well take $c = 1$. Then the calculation which gives (13) also gives

Lemma 8 Let $G = G(R, I, I)$. Then $G \cong G_\varphi$ where $\varphi : \mathbb{Z}^4 \times \mathbb{Z}^4 \to$ $\mathbb{Z}^2$ is represented by the matrix

$$\begin{bmatrix} 0 & 0 & (0\ 1) & (1\ 0) \\ 0 & 0 & (-a\ b) & (\alpha - b\,F(b)/a) \\ (0\ -1) & (a\ -b) & 0 & 0 \\ (-1\ 0) & (b - \alpha\ -F(b)/a) & 0 & 0 \end{bmatrix}$$

Also

$$\pm \operatorname{Pf}_\varphi(X, Y) = aX^2 + (\alpha - 2b)XY + (F(b)/a)Y^2 \tag{14}$$
$$= F_{R,I}(X, Y) \text{ say},$$
$$D(G) = D(\operatorname{Pf}_\varphi) = \alpha^2 + 4\beta. \tag{15}$$

Proof This is just a matter of applying the formula (11) to suitable basis elements in G/G'. As representatives for a basis of G/G', take $(1\ 0\ 0)$, $(\omega\ 0\ 0)$, $(0\ a\ 0)$, and $(0\ b - \omega\ 0)$; as basis for G', take $\{(0\ 0\ b - \omega), (0\ 0\ a)\}$.

One interesting conclusion may be drawn straight away. Since $D(G)$ is an invariant of the group G, (15) shows that $\alpha^2 + 4\beta$ is determined by the isomorphism type of G. As α is either 0 or 1, $\alpha^2 + 4\beta$ actually determines both α and β; thus

> *the isomorphism type of $G(R, I, I)$ determines the ring R up to isomorphism.* (16)

To get a better understanding of the quadratic form (14), a slight digression is necessary. There is an automorphism $: R \to R$ sending ω to its conjugate $\alpha - \omega$. For an ideal $J = \lambda\mathbb{Z} \oplus \mu\mathbb{Z}$ of finite index in R, define the *norm form* of J to be

$$N_J(X, Y) = |R:J|^{-1} \cdot (\lambda X - \mu Y)(\lambda' X - \mu' Y)). \tag{17}$$

Exercise 12 (i) N_J is uniquely determined up to equivalence (in the sense of integral quadratic forms) by the ideal J. (ii) N_J has coefficients in $\mathbb{Z}$, (iii) $N_J = N_{J'}$ (iv) If $0 \neq c \in \mathbb{Z}$ then $N_{cJ} = N_J$.
(*Hint*: Part (i) is easy. For (ii), we may in view of (i) assume that $\lambda = a$ and $\mu = b - c\omega$ as in Exercise 11; then $|R:J| = |ac|$, and (ii) follows from (13) by a simple calculation.)

Having done the calculation for Exercise 12, the reader will have noticed a remarkable fact: in the notation of Lemma 8,

$$N_I = F_{R,I}. \tag{18}$$

Now $F_{R,I} = \pm Pf_\varphi$ is determined up to weak equivalence by the isomorphism type of G_φ (see section D); so the picture is (almost) completed now by

Lemma 9 Assume that R as above is an integral domain. Let I and J be ideals of finite index in R. Then $N_I \sim \pm N_J$ if and only if there exist $r, s \in R \backslash \{0\}$ such that $rI = sJ$ or $rI = s'J'$.
(One says that 'I and J belong to the same or conjugate ideal classes in R'.)

Proof '*If*': Suppose $rI = sJ$ with $r, s \neq 0$. If (λ, μ) is a basis for I and (σ, τ) is one for J, then

$$(r\lambda \quad -r\mu) = (s\sigma \quad -s\tau)Q$$

for some matrix $Q \in GL_2(\mathbb{Z})$. Writing

$$\begin{pmatrix} X_1 \\ Y_1 \end{pmatrix} = Q \begin{pmatrix} X \\ Y \end{pmatrix}$$

we have

$$rr'(\lambda X - \mu Y)(\lambda' X - \mu' Y) = ss'(\sigma X_1 - \tau Y_1)(\sigma' X_1 - \tau' Y_1).$$

So to show that $N_I \sim \pm N_J$ it will suffice to establish that

$$rr'|R:I| = \pm ss'|R:J|.$$

Now one calculates directly that rr' is equal to the determinant of the endomorphism $x \mapsto xr$ of R; hence it is also the determinant of the restriction of this endomorphism to I, and so

$$rr' = \pm |I:Ir|.$$

Similarly, $ss' = \pm |J:Js|$, and since $Ir = Js$ this gives

$$rr'|R:I| = \pm |I:Ir||R:I| = \pm |R:Ir| = \pm |R:Js| = \pm ss'|R:J|$$

as required.

If instead we assume that $rI = s'J'$, we get the same conclusion by Exercise 12 (iii).

'*Only if*': In view of Exercise 12 (i), (iv), we may take a basis of the form $(a, b - \omega)$ for I. By choosing a suitable basis (λ, μ) for J, we may assume that in fact $N_I = \pm N_J$. Putting $d = \pm |R:J|$, we then have

$$d(aX - (b - \omega)Y)(aX - (b - \omega')Y) = a(\lambda X - \mu Y)(\lambda' X - \mu' Y).$$

Equate coefficients to get

$$d(b - \omega)(b - \omega') = a\mu\mu'$$
$$ad((b - \omega) + (b - \omega')) = a(\lambda\mu' + \mu\lambda')$$
$$da^2 = a\lambda\lambda'.$$

The element $\xi = d(b - \omega)$ then satisfies

$$0 = \xi^2 - (\xi + \xi')\xi + \xi\xi' = (\xi - \lambda\mu')(\xi - \lambda'\mu),$$

so either $\xi = \lambda\mu'$ or $\xi = \lambda'\mu$. In the first case we get

$$\lambda J' = \lambda\lambda'\mathbb{Z} \oplus \lambda\mu'\mathbb{Z} = da\mathbb{Z} \oplus d(b - \omega)\mathbb{Z} = dI;$$

in the second case, $\lambda'J = dI$. Either way, this gives the result.

Most of the following theorem is now established:

Theorem 3 For $i = 1, 2$, let $R_i \cong \mathbb{Z} \oplus \mathbb{Z}$ be a commutative integral domain and I_i an ideal of finite index in R_i. Put $G_i = G(R_i, I_i, I_i)$. Then each $G_i \in \mathfrak{T}(4, 2)$, and $G_1 \cong G_2$ if and only if $R_1 \cong R_2$ by an isomorphism θ such that $I_1 \theta$ and I_2 lie in the same or in conjugate ideal classes of R_2. Moreover, every group $G \in \mathfrak{T}(4, 2)$ such that $\tau_2(G) = G'$ and $D(G)$ is not a perfect square arises in this way; in fact $G \cong G(R, I, I)$ where R is an order in the field $\mathbb{Q}(\sqrt{D})$ and I is an ideal of finite index in R.

Proof The first two claims follows from the preceding discussion and Theorem 1. For the last part, suppose $G \cong G_\varphi$ and put

$$D = D(\mathrm{Pf}_\varphi) = f^2 d$$

with $f \in \mathbb{Z}$ and d square-free. Recall that $D = q^2 - 4pr$ where $\mathrm{Pf}_\varphi(X, Y) = pX^2 + qXY + rY^2$. There are two possibilities to consider.

Case 1 $d \equiv 1 \bmod 4$. Take $R = \mathbb{Z} \oplus \omega\mathbb{Z}$ where $\omega = \frac{1}{2}(1 + \sqrt{d})f$, and put

$$I = p\mathbb{Z} \oplus (\tfrac{1}{2}(f - q) - \omega)\mathbb{Z}.$$

(The ring of integers of $\mathbb{Q}(\sqrt{D}) = \mathbb{Q}(\sqrt{d})$ is $\mathbb{Z} + \frac{1}{2}(1 + \sqrt{d})\mathbb{Z}$, and R is the unique subring of index f in this ring.)

Case 2 $d \not\equiv 1 \bmod 4$. Then f must be even; for if f is odd then $f^2 \equiv 1 \bmod 4$, which forces $d \equiv D \equiv 0$ or $1 \bmod 4$, while d being square-free is $\not\equiv 0 \bmod 4$. Say $f = 2f_1$. Take $R = \mathbb{Z} \oplus \omega\mathbb{Z}$ where $\omega = f_1\sqrt{d}$, and put

$$I = p\mathbb{Z} \oplus (\tfrac{1}{2}q + \omega)\mathbb{Z}.$$

(The ring of integers in $\mathbb{Q}(\sqrt{d})$ is now $\mathbb{Z} + \sqrt{d}\mathbb{Z}$, and R is the unique subring of index f_1.)

In either case, one can check via (13) that I is an ideal in R, and direct calculation shows that $F_{R,I} = pX^2 + qXY + rY^2$; Lemma 8 and Theorem 1 then show that $G_\varphi \cong G(R, I, I)$.

Theorem 3 does not tell us about all groups in $\mathfrak{T}(4, 2)$; the next few exercises fill in the gap. Recall that $C(G)$ denotes the h.c.f. of the coefficients of Pf_φ where $G \cong G_\varphi$, so '$C(G) = 0$' is a short way of saying '$G \cong G_\varphi$ where $\mathrm{Pf}_\varphi = 0$'.

Exercise 13 Let $G \in \mathfrak{T}(4, 2)$ and suppose that $\tau_2(G) = G'$. Show that $C(G) = 0$ if and only if $\zeta_1(G) \cong \mathbb{Z}^3$. Deduce that if $C(G) = 0$ then G cannot be embedded in $\mathrm{Tr}_1(3, R)$ for any commutative ring $R \cong \mathbb{Z} \oplus \mathbb{Z}$. (*Hint:* Look at Lemma 6. Note that there is only one such group G, up to isomorphism!)

Exercise 14   Let $G \in \mathfrak{T}(4, 2)$ satisfy $\tau_2(G) = G'$. (i) Suppose $D(G) = f^2 \neq 0$. Show that $G \cong G(R, I, I)$ where $R = \mathbb{Z} \oplus \omega\mathbb{Z}$ with $\omega^2 = f\omega$, and I is a suitable ideal of finite index in R. (ii) Suppose $D(G) = 0$ but $C(G) \neq 0$. Show that $G \cong G(R, I, I)$ where $R = \mathbb{Z} \oplus \omega\mathbb{Z}$ with $\omega^2 = 0$, and $I = a\mathbb{Z} + \omega\mathbb{Z}$ for a suitable integer $a \neq 0$.
Note that in case (i), R is the unique subring of index f in the ring $\mathbb{Z} \times \mathbb{Z}$; in case (ii), $R \cong \mathbb{Z} \times \mathbb{Z}^0$ where $\mathbb{Z}^0$ is the 'zero ring' on the additive group $\mathbb{Z}$.

Exercise 15   Let $G \in \mathfrak{T}(4, 2)$ satisfy $C(G) \neq 0$. Show that $G \cong G(R, I, M)$ for some commutative ring (with identity) $R \cong \mathbb{Z} \oplus \mathbb{Z}$, ideal I of finite index in R, and additive subgroup M of R containing I.
(*Hint*:   Say $G \cong G_\varphi$. If $\langle \mathrm{Im}\ \varphi \rangle = E \leq_f \mathbb{Z}^2$, let $Q \in M_2(\mathbb{Z})$ be a matrix such that
$$\mathbb{Z}^2 \cdot Q = E.$$
The map
$$\bar{\varphi} = \varphi \circ Q^{-1} : \mathbb{Z}^4 \times \mathbb{Z}^4 \to \mathbb{Z}^2$$
then corresponds to a group $G_{\bar{\varphi}}$ satisfying $\tau_2(G_{\bar{\varphi}}) = G'_{\bar{\varphi}}$. Hence $G_{\bar{\varphi}} \cong G(R, J, J)$ for suitable R and J. Now take $m \in \mathbb{N}$ so that $mQ^{-1} \in M_2(\mathbb{Z})$, put $I = mJ$ and $M = IQ^{-1} \subseteq R$.)

The last few exercises together now give

Corollary 3   The groups in $\mathfrak{T}(4, 2)$ with centre of rank 2 are up to isomorphism just the subgroups of finite index in $\mathrm{Tr}_1(3, R)$, as R runs over all commutative rings (with identity) additively isomorphic to $\mathbb{Z} \oplus \mathbb{Z}$.

Let us conclude with some further specific examples of $^\wedge$-classes. Take R in Theorem 3 to be the ring of integers in a quadratic number field. If R has exactly $\tilde{h}$ non-conjugate ideal classes, the theorem shows that there are exactly $\tilde{h}$ non-isomorphic groups among the $G(R, I, I)$ as I runs over all ideals of finite index in R(i.e., all non-zero ideals). Now if $G \cong G(R, I, I)$ then $D(G)$ determines R, and the $^\wedge$-class of G determines $D(G)$ (see Theorem 2, and the discussion leading up to (16), above). On the other hand, all the groups $G(R, I, I)$ for a given ring R lie in the same $^\wedge$-class; this follows from Exercise 10 (ii) and the fact that $I/mI \cong R/mR$ for $0 \neq I \lhd R$ and $m \in \mathbb{N}$ (see the proof of Proposition 1, in section A). So we infer the final

Corollary 4   Let R be the ring of integers in a quadratic number field, and suppose that R has exactly $\hat{h}$ non-conjugate ideal classes. Then the groups $G(R, I, I)$ with $0 \neq I \lhd R$ lie in exactly $\hat{h}$ isomorphism classes, and they form (a set of representatives for) exactly one $^\wedge$-class of $\mathfrak{T}$-groups.

As well as providing examples of non-isomorphic groups in a^-class, we have proved anew a very special case of 'Pickel's first theorem' (Chapter 10, section A). The argument has used none of the deep finiteness theorems about algebraic groups that we had to rely on in Chapters 9 and 10; it illustrates again the way in which those theorems can be seen as generalising classical number-theoretic results.

F. Notes

Groups with isomorphic finite quotients There are many examples in the literature. See Dyer (1969), Brigham (1971), Remeslennikov (1967), Remeslennikov (1971). The examples in section A are from Grunewald & Segal (1978*b*). The groups discussed in section E are the subject of Grunewald & Scharlau (1979); using a different approach, they prove the final corollary of this chapter, more generally, when R is the ring of integers in any algebraic number field (thereby producing ^-classes of groups in $\mathfrak{X}(2d,d)$ for values of d greater than 2).

Non-arithmetic polycyclic groups The fact that these exist was pointed out to me by Fred Pickel, and specific examples were constructed by Brian Hartley and myself. The argument of section B owes much to a suggestion of Peter Linnell. The result of Exercise 2, on the structure of soluble arithmetic groups, may also be derived from the structure theory of soluble algebraic groups: a connected such group is the semi-direct product of its unipotent radical by a torus. For this, see Borel (1969*b*), Chapter III Section 10, or Humphreys (1975), Chapter VII, section 19.3.

$\mathfrak{X}$-groups of class 2 Sections C, D and E are based on Grunewald, Segal & Sterling (1982). Using related methods, Grunewald and I can classify groups in $\mathfrak{X}(n,2)$ up to commensurability, for every n; but the whole area is still in an embryonic state, and it is quite unclear, for example, how one should tackle the classification of $\mathfrak{X}(n,m)$ for $m > 2$.

More groups of the form $G(R,I,I)$ are discussed in Grunewald & Scharlau (1979). An unpublished paper of Pickel, 'Finite quotients and multiplicities in nilpotent groups', studies these groups in connection with the representation theory of Lie groups and raises interesting questions.

Other examples Of course, many other kinds of examples have been studied. We discussed some in section B of Chapter 4. Bowers (1960) constructs torsion-free polycyclic groups which are not poly-C_∞, and gives examples of non-isomorphic 'composition series' in such groups. Humphreys (1969) classifies polycyclic groups in which every subgroup can be generated by two elements. Grunewald & Segal (1975) construct polycyclic groups which are not residually nilpotent even though all their two-generator subgroups are. Wilson (1982) constructs interesting examples related to quadratic extensions of number fields.

Appendix: further topics

A. Number of generators

Linnell & Warhurst (1981) prove the following beautiful

Theorem If G is a polycyclic group and every finite quotient group of G can be generated by n elements, then G can be generated by $n + 1$ elements.

This result is best possible (one really needs $n + 1$ sometimes). The proof uses the theory of *lattices over orders* (for which see Swan (1970)); this theory is undoubtedly important for the further study of polycyclic groups, though few such applications have emerged as yet.

B. Abelian subgroups of polycyclic groups

Wilson (1982) proves the following for a polycyclic group G: *If all subnormal abelian subgroups of G (or, equivalently, all abelian subgroups of $Fitt(G)$) have Hirsch number at most n, then (i) the same holds for all abelian subgroups of G; (ii) $G/Fitt(G)$ has Hirsch number at most $n-1$; (iii) $h(G) \leq \frac{1}{2}(n^2 + 3n - 2)$*. This is a powerful refinement of Mal'cev's theorem (Theorem 2 of Chapter 2), and depends on the quantitative Dirichlet Units Theorem. This interesting work is very much in the 'arithmetical' spirit of Chapters 2 and 4.

C. Polycyclic groups, Lie algebras and algebraic groups

In his recent paper of this title, S. Donkin associates to each polycyclic group a Lie algebra and an affine algebraic group over $\mathbb{Q}$, in a functorial manner. The constructions generalise the usual ones for $\mathfrak{T}$-groups (see Chapter 6). I think that this is an exciting development for the theory of polycyclic groups, but it is early yet to say where it will lead.

D. Group rings

There is now a large literature on the subject of group rings of polycyclic groups. It began with Hall (1954); a recent highlight is Roseblade (1978).

E. Polycyclic groups in topology

This is also a large subject, about which, however, I can say nothing. One might read Auslander (1973); Raghunathan (1972), Chapters II, III, IV; a recent deep contribution is Farrell & Hsiang (1981).

F. Automorphism towers

Hulse (1970) proves that the 'automorphism group tower' of a polycyclic group with trivial centre is necessarily countable. This is a rather long and intricate paper, combining Mal'cev's methods (see Chapter 2 above) with group-theoretical arguments about subnormal and ascendant subgroups.

G. Soluble groups of finite rank

Many theorems about polycyclic groups can be generalised to wider classes of soluble groups. The first major contribution here was Mal'cev (1951); for subsequent developments, see Robinson (1972). In his Freiburg dissertation (1979), D. Kilsch has extended Pickel's theorem (Theorem 1 of Chapter 10 above) to torsion-free nilpotent groups of finite rank which satisfy certain conditions, and has given examples to show that the result fails without such conditions. It is not known whether the main theorem of Chapter 10 can be similarly extended to soluble groups of finite rank which are not polycyclic.

References

Amayo, R. & Stewart, I. (1974). *Infinite-Dimensional Lie Algebras*. Noordhoff, Leyden.

Atiyah, M.F. & Macdonald, I.G. (1969). *Introduction to Commutative Algebra*. Addison-Wesley, Reading, Mass.

Auslander, L. (1967). On a problem of Philip Hall. *Annals of Math*. **86**, 112–16.

Auslander, L. (1969). The automorphism group of a polycyclic group. *Annals. of Math*. **89**. 314–22.

Auslander L. (1973). An exposition of the structure of solvmanifolds I. *Bull. Amer. Math. Soc*. **79**, 227–61.

Auslander L. & Baumslag, G. (1967). Automorphism groups of finitely generated nilpotent groups. *Bull. Amer. Math. Soc*. **73**, 716–17.

Baer, R. (1957). Überauflösbare Gruppen. *Abh. Math. Sem. Univ. Hamburg* **23**, 11–28.

Baer, R. (1974). Einbettungseigenschaften von Normalteilern: der Schluss vom Endlichen aufs Unendliche. In *Proc. 2nd Int. Conf. Group Theory, Canberra*, Springer Lecture Notes 372, 13–62.

Baumslag, G. (1969). Automorphism groups of nilpotent groups. *Amer. J. Math*. **91**, 1003–11.

Baumslag. G. (1971). *Lectures on Nilpotent Groups*. C.B.M.S. Regional Conference Series 2, A.M.S., Providence, R.I.

Blackburn, N. (1965). Conjugacy in nilpotent groups. *Proc. Amer. Math. Soc*. **16**, 143–8.

Borel, A. (1962). Arithmetic properties of linear algebraic groups. *Proc. Int. Congress Math. Stockholm*, 10–22.

Borel, A. (1963). Some finiteness properties of adele groups over number fields. *I.H.E.S. Pub. Math*. **16**, 101–26.

Borel, A. (1966). Density and maximality of arithmetic subgroups. *J. reine angew. Math*. **224**, 78–89.

Borel, A. (1969a). *Introduction aux Groupes Arithmétiques*. Hermann, Paris.

Borel, A. (1969b). *Linear Algebraic Groups*. Benjamin, New York.

Borel, A. & Harish-Chandra (1962). Arithmetic subgroups of algebraic groups. *Annals of Math*. **75**, 485–535.

Borel, A. & Serre, J.-P. (1964). Théorèmes de finitude en cohomologie galoisienne. *Comment. Math. Helv*. **39**, 111–64.

Borevich, Z. & Shafarevich, I.R. (1966). *Number Theory*. Academic Press, New York.

Bowers, J.F. (1960). On composition series of polycyclic groups. *J. London Math. Soc*. **35**, 433–44.

Brigham, R.C. (1971). On the isomorphism problem for just-infinite groups. *Comm. Pure Appl. Math*. **24**, 789–96.

Cassells, J.W.S. (1978). *Rational Quadratic Forms*. Academic Press, London.

Chevalley, C. (1951). Deux théorèmes d'arithmétique. *J. Math. Soc. Japan* **3**, 36–44.

Cohn, H. (1978). *A Classical Invitation to Algebraic Numbers and Class Fields.* Springer-Verlag, New York – Heidelberg – Berlin.

Cohn, P.M. (1977). *Algebra*, vol. 2. Wiley, London.

Donkin, S. (1982). Polycyclic groups, Lie algebras and algebraic groups. *J. reine angew. Math.* **326**, 104–23.

Dyer, J.L. (1969). On the isomorphism problem for polycyclic groups. *Math. Zeit.* **112**, 145–53.

Farrell, F. T. & Hsiang, W.C. (1981). Whitehead groups of poly-(cyclic or finite) groups. *J. London Math, Soc.* (2) **24**, 308–24.

Fogarty, J. (1969). *Invariant Theory.* Benjamin, New York.

Formanek, E. (1976). Conjugate separability of polycyclic groups. *J. Algebra* **42**, 1–10.

Gruenberg, K.W. (1957). Residual properties of infinite soluble groups. *Proc. London Math. Soc.* (3) **7**, 29–62.

Gruenberg, K.W. (1970). *Cohomological Topics in Group Theory.* Springer Lecture Notes 143.

Grunewald, F.J., Pickel, P.F. & Segal, D. (1979). Finiteness theorems for polycyclic groups. *Bull. Amer. Math. Soc. (N.S.)* **1**, 575–8.

    Grunewald, F.J., Pickel, P.F. & Segal, D. (1980). Polycyclic groups with isomorphic finite quotients. *Annals of Math.* **111**, 155–95.

Grunewald, F.J. & Scharlau, R. (1979). A note on torsion-free finitely generated nilpotent groups of class 2. *J. Algebra* **58**, 162–75.

Grunewald, F.J. & Segal, D. (1975). Residual nilpotence in polycyclic groups. *Math. Zeit.* **142**, 229–41.

    Grunewald, F.J. & Segal, D. (1978a). Conjugacy in polycyclic groups. *Comm. in Algebra* **6**, 775–98.

    Grunewald, F.J. & Segal, D. (1978b). On polycyclic groups with isomorphic finite quotients. *Math. Proc. Cambridge Phil. Soc.* **84**, 235–46.

    Grunewald, F.J. & Segal, D. (1978c). A note on arithmetic groups. *Bull. London Math. Soc.* **10**, 297–302.

    Grunewald, F.J. & Segal, D. (1979a). On congruence topologies in number fields. *J. reine angew. Math.* **311**, 389–96.

    Grunewald, F.J. & Segal, D. (1979b). The solubility of certain decision problems in algebra and arithmetic. *Bull. Amer. Math. Soc. (N.S.)* **1**, 915–18.

    Grunewald, F.J. & Segal, D. (1980). Some general algorithms. I: Arithmetic groups. *Annals of Math.* **112**, 531–83. II: Nilpotent groups. *Ibid.* **112**, 585–617.

    Grunewald, F.J. & Segal, D. (1982). Conjugacy of subgroups in arithmetic groups. *Proc. London Math. Soc.* (3) **44**, 47–70.

Grunewald, F.J., Segal, D. & Sterling, L.S. (1982). Nilpotent groups of Hirsch length six. *Math. Zeit.* **179**, 219–35.

Hall, P. (1954). Finiteness conditions for soluble groups. *Proc. London Math. Soc.* (3) **4**, 419–36.

    Hall, P. (1969). *Nilpotent Groups.* Queen Mary College Maths. Notes, London.

Hirsch, K.A. (1938–1954). On infinite soluble groups.

    I. *Proc. London Math. Soc.* (2) **44** (1938), 53–60.

    II. *Proc. London Math. Soc.* (2) **44** (1938), 336–44.

    III. *Proc. London Math. Soc.* (2) **49** (1946), 184–94.

    IV. *J. London Math. Soc.* **27** (1952), 81–5.

    V. *J. London Math. Soc.* **29** (1954), 250–1.

Hulse, J.A. (1970). Automorphism towers of polycyclic groups. *J. Algebra* **16**, 347–98.

Humphreys, J.E. (1975). *Linear Algebraic Groups.* Springer-Verlag, New York.

Humphreys, J.E. (1980). *Arithmetic Groups.* Springer Lecture Notes 789.

Humphreys, J.F. (1969). Two generator conditions for polycyclic groups. *J. London Math. Soc.* **1**, 21–9.

Huppert, B. (1967). *Endliche Gruppen I.* Springer-Verlag, Berlin – Heidelberg – New York.

Jacobson, N. (1962). *Lie Algebras.* Wiley, New York.

Janusz, G.J. (1973). *Algebraic Number Fields.* Academic Press, New York.

Jennings, S.A. (1955). The group ring of a class of infinite nilpotent groups. *Canadian J. Math.* **7**, 169–87.

Kargapolov, M.I. & Merzljakov, Ju.I. (1979). *Fundamentals of the Theory of Groups.* Springer, New York.

Kegel, O.H. (1966). Über den Normalisator von subnormalen und erreichbaren Untergruppen. *Math. Annalen.* **163**, 248–58.

Kegel, O.H. & Wehrfritz, B.A.F. (1973). *Locally Finite Groups.* North-Holland, Amsterdam – London.

Kelley, J.L. (1955). *General Topology.* Van Nostrand, Princeton.

Lang, S. (1965). *Algebra.* Addison-Wesley, Reading, Mass.

Lang, S. (1970). *Algebraic Number Theory.* Addison-Wesley, Reading, Mass.

Lennox, J.C. & Wilson, J.S. (1977). A note on permutable subgroups. *Archiv der Math.* **28**, 113–16.

Linnell, P.A. & Warhurst, D. (1981). Bounding the number of generators of a polycyclic group. *Archiv der Math.* **37**, 7–17.

Lyndon, R.C. & Schupp, P.E. (1977). *Combinatorial Group Theory.* Springer-Verlag, Berlin – Heidelberg – New York.

Magnus, W., Karrass, A. & Solitar, D. (1966). *Combinatorial Group Theory.* Wiley, New York.

Mal'cev, A.I. (1949). On a class of homogeneous spaces. *Izvestia Akad. Nauk SSSR Ser. Mat.* **13**, 9–32. (Russian) = *A.M.S. Translations* (1) **9** (1962) 276–307.

Mal'cev, A.I. (1951). On certain classes of infinite soluble groups. *Mat. Sbornik* **28**, 567–88. (Russian) = *A.M.S. Translations* (2) **2** (1956), 1–21.

Mal'cev, A.I. (1958). Homomorphisms onto finite groups. *Ivanov Gos. Ped. Inst. Učen. Zap.* **18**, 49–60. (Russian).

Merzlyakov, Yu. I. (1970). Integral representations of the holomorph of a polycyclic group. *Algebra i Logika* **9**, 539–58. (Russian) = *Algebra and Logic* **9**, 326–37.

Moore, C.C. (1965). Decomposition of unitary representations defined by discrete subgroups of nilpotent groups. *Annals of Math.* **82**, 146–82.

Newell, M.L. (1973). Homomorphs and formats in polycyclic groups. *J. London Math. Soc.* (2) **7**, 317–27.

Newell, M.L. (1975). Nilpotent projectors in S_1-groups. *Proc. Royal Irish Academy* **75**, 107–14.

Newman, M. (1972). *Integral Matrices.* Academic Press, New York.

Northcott, D.G. (1980). *Affine Sets and Affine Groups.* L.M.S. Lecture Notes 39, Cambridge.

Passi, I.B.S. (1979) *Group Rings and their Augmentation Ideals.* Springer Lecture Notes 715.

Passman, D.S. (1977). *The Algebraic Structure of Group Rings.* Wiley, New York.

Pickel, P.F. (1971a) Finitely generated nilpotent groups with isomorphic finite quotients. *Bull. Amer. Math. Soc.* **77**. 216–19.

Pickel, P.F. (1971b) Finitely generated nilpotent groups with isomorphic finite quotients. *Trans. Amer. Math. Soc.* **160**, 327–41.

Pickel, P.F. (1973). Nilpotent-by-finite groups with isomorphic finite quotients. *Trans. Amer. Math. Soc.* **183**, 313–25.

Pickel, P.F. (1974). Metabelian groups with the same finite quotients. *Bull. Austral. Math. Soc.* **11**, 115–20.

Pickel, P.F. (1976). Fitting subgroups and profinite completions of polycyclic groups. *J. Algebra* **42**, 41–5.

Raghunathan, M.S. (1972). *Discrete Subgroups of Lie Groups.* Springer-Verlag, Berlin – Heidelberg – New York.

Remeslennikov, V.N. (1967). Conjugacy of subgroups in nilpotent groups. *Algebra i Logika* **6**, 61–76. (Russian).

Remeslennikov, V.N. (1969). Conjugacy in polycyclic groups. *Algebra i Logika* **8**, 712–25. (Russian) = *Algebra and Logic* **8** (1969), 404–11.

Remeslennikov, V.N. (1971). Groups that are residually finite w.r.t. conjugacy. *Sibirski Mat. Ž.* **12**, 1085–99. (Russian) = *Siberian Math. J.* **12** (1971), 783–92.

Rips, E. (1972). On the fourth integer dimension subgroup. *Israel J. Math.* **12**, 342–6.

Robinson, D.J.S. (1968). A property of the lower central series of a group. *Math. Zeit.* **107**, 225–31.

Robinson, D.J.S. (1972). *Finiteness Conditions and Generalised Soluble Groups.* (2 Vols.) Springer-Verlag, Berlin – Heidelberg – New York.

Robinson, D.J.S. (1975). On the cohomology of soluble groups of finite rank. *J. Pure and Applied Algebra* **6**, 155–64.

Robinson, D.J.S. (1976a) The vanishing of certain homology and cohomology groups. *J. Pure and Applied Algebra* **7**, 145–67.

Robinson, D.J.S. (1976b) Splitting theorems for infinite groups. *Symposia Math.* **17**, 441–470.

Roseblade, J.E. (1973). The Frattini subgroup in infinite soluble groups. In *Three Lectures on Polycyclic Groups.* Queen Mary College Math. Notes, London.

Roseblade, J.E. (1978). Prime ideals in group rings of polycyclic groups. *Proc. London Math. Soc.* (3) **36**, 385–447.

Schmidt, F.K. (1930). Zur Klassenkörpertheorie im Kleinen. *J. reine angewandte Math.* **162**, 155–66.

Segal, D. (1975). Groups whose finite quotients are supersoluble. *J. Algebra* **35**, 56–71.

Segal, D. (1978). Two theorems on polycyclic groups. *Math. Zeit.* **164**, 185–7.

Segal, D. (1979). Congruence topologies in commutative rings. *Bull. London Math. Soc.* **11**, 186–90.

Serre, J.-P. (1964). *Cohomologie Galoisienne.* Springer Lecture Notes 5.

Serre, J.-P. (1973) *A Course in Arithmetic.* Springer-Verlag, New York.

Serre, J.-P. (1979). *Local Fields.* Springer-Verlag, New York.

Shafarevich, I.R. (1974). *Basic Algebraic Geometry.* Springer-Verlag, Berlin.

Šmel'kin, A.L. (1968). Polycyclic groups. *Sibirski Mat. Ž.* **9**, 234–5 (Russian) = *Siberian Math. J.* **9**, 178.

Stewart, I. & Tall, D. (1979). *Algebraic Number Theory.* Chapman & Hall, London.

Swan, R.G. (1967). Representations of polycyclic groups. *Proc. Amer. Math. Soc.* **18**, 573–4.

Swan, R.G. (1970). *K-Theory of Finite Groups and Orders.* Springer Lecture Notes 149.

Tolimieri, R. (1971). Structure of solvable Lie groups. *J. Algebra* **16**, 597–625.

Wang, H.-C. (1956). Discrete subgroups of solvable Lie groups I. *Annals of Math.* **64**, 1–19.

Warfield, R.B., Jr. (1976) *Nilpotent Groups.* Springer Lecture Notes 513.

Waterhouse, W.C. (1979). *Introduction to Affine Group Schemes.* Springer-Verlag, New York – Heidelberg – Berlin.

Wehrfritz, B.A.F. (1970). Groups of automorphisms of soluble groups. *Proc. London Math. Soc.* (3) **20**, 101–22.

Wehrfritz, B.A.F. (1971*a*). Supersoluble and locally supersoluble linear groups. *J. Algebra* **17**, 41–58.

Wehrfritz, B.A.F. (1971*b*). Remarks on centrality and cyclicity in linear groups. *J. Algebra* **18**, 229–36.

Wehrfritz, B.A.F. (1973*a*). Two examples of soluble groups that are not conjugacy separable. *J. London Math. Soc.* **7**, 312–16.

Wehrfritz, B.A.F. (1973*b*). *Infinite Linear groups.* Springer-Verlag, Berlin – Heidelberg – New York.

Wehrfritz, B.A.F. (1973*c*). The holomorph of a polycyclic group. In *Three Lectures on Polycyclic Groups.* Queen Mary College Math. Notes, London.

Wehrfritz, B.A.F. (1974). On the holomorphs of soluble groups of finite rank. *J. Pure Applied Algebra* **4**, 55–69.

Wehrfritz, B.A.F. (1976). Finitely generated groups of module automorphisms and finitely generated metabelian groups. *Symposia Math.* **17**, 262–75.

Wehrfritz, B.A.F. (1978). Nilpotence in groups of semi-linear maps I. *Proc. London Math. Soc.* (3) **36**, 448–79.

Wehrfritz, B.A.F. (1980). Modules over polycyclic groups. *Quarterly J. Math. (Oxford)* (2) **31**, 109–27.

Wilson, J.S. (1982). Abelian subgroups of polycyclic groups. *J. reine angew. Math.* **331**, 162–80.

Zassenhaus, H. (1938). Beweis eines Satzes über diskrete Gruppen. *Abh. Math. Sem. Univ. Hamburg* **12**, 289–312.

Index

Index of Symbols